P9-DCN-034

LIVING DOWNSTREAM

LIVING DOWNSTREAM

an ecologist looks at cancer and the environment

SANDRA STEINGRABER

A MERLOYD LAWRENCE BOOK

ADDISON-WESLEY PUBLISHING COMPANY, INC.
Reading, Massachusetts Menlo Park, California New York
Don Mills, Ontario Harlow, England Amsterdam Bonn
Sydney Singapore Tokyo Madrid San Juan
Paris Seoul Milan Mexico City Taipei

Library of Congress Cataloging-in-Publication Data

Steingraber, Sandra.
Living downstream : an ecologist looks at cancer and the environment / Sandra Steingraber.
p. cm.
"A Merloyd Lawrence book."
Includes bibliographical references and index.
ISBN 0-201-48303-3
1. Cancer—Environmental aspects. 2. Environmental toxicology.
I. Title.
RC268.25.S74 1997
616.99'4071—dc21 97-8164
CIP

Jacket design by Suzanne Heiser
Text design by Karen Savary
Production coordinated by Beth Burleigh Fuller
Set in 11-point Janson by NK Graphics

1 2 3 4 5 6 7 8 9-MA-0100999897
First printing, May 1997

for Jeannie Marshall

and for my mother,

whose original plan was to build

a laboratory in the north bedroom

CONTENTS

ACKNOWLEDGMENTS

During the years of researching and writing this book, I have enjoyed the encouragement, support, and direct assistance of considerable numbers of individuals and institutions. Foremost among these are the Bunting Institute of Radcliffe College, the Center for Research on Women and Gender at the University of Illinois, and the Women's Studies Program at Northeastern University in Boston. All variously provided me residencies, fellowship money, and communities of inspiring colleagues. For these necessities, I thank their respective directors, Drs. Florence Ladd, Alice Dan, and Christine Gailey. I am grateful also to the Radcliffe Research Partnership Program for sponsoring a coterie of Harvard University students to help me with the formidable task of library research: Rebecca Braun, Christine Chung, Theresa Esquerra, Palmira Gómez, Julie Nelson, Kathryn Patton, and Amy Stevens. I thank each of them for their enthusiastic and dogged assistance.

I owe a considerable debt to many librarians. I am especially grateful to those at the Harvard University Medical School and Widener Libraries, the National Library of Medicine in Bethesda, the Beinecke Library at Yale University, the Snell Library at Northeastern University, and the public libraries of Boston, Somerville, Peoria, and Pekin.

Several right-to-know experts aided in the search for my ecological roots. They are John Chelen at the Unison Institute, Lisa Damon at the Illinois Hazardous Waste Research and Information Center, Joe Goodner at the Illinois Environmental Protection Agency, Kathy Grandfield of Seattle, Ed Hopkins of Citizen Action, and Paul Orum at the Working Group on Community Right-to-Know. I am also, once again, indebted to Brian Burt for his redoubtable computer savvy and wholehearted resourcefulness.

Many colleagues in science, medicine, and policy-making contributed their expert knowledge to this project by reading parts of the manuscript. For their invaluable commentary and criticisms, I thank Ruth Allen, Ph.D., M.P.H., of the National Cancer Institute and U.S. Environmental Protection Agency; Dorothy Anderson, M.D., of Mason City, Illinois; Ann Aschengrau, Sc.D., of Boston University; Pierre Béland, Ph.D., of the St. Lawrence National Institute of Ecotoxicology in Québec; Judith Brady of San Francisco; Julia Brody, Ph.D., of Silent Spring Institute; Leslie Byster of the Silicon Valley Toxics Coalition; Kenneth Cantor, Ph.D., of the National Cancer Institute; Jackie Christensen of the Institute for Agriculture and Trade Policy; Richard Clapp, Sc.D., M.P.H., of Boston University; Brian Cohen of the Environmental Working Group; Penelope Fenner-Crisp, Ph.D., of the U.S. Environmental Protection Agency; Joan D'Argo of the National Coalition for Health and Environmental Justice; Devra Lee Davis, Ph.D., M.P.H., of the World Resources Institute; Samuel Epstein, M.D., of the University of Illinois, Chicago; James Davis, Ph.D., of St. Louis University; Thomas Downham, M.D., of the Henry Ford Medical Center; Jay Feldman of the National Coalition Against the Misuse of Pesticides; Vincent Garry, M.D., of the University of Minnesota; John W. Gephart, Ph.D., of Cornell University; Benjamin Goldman, Ph.D., of Boston; Joe Goodner of the Illinois Environmental Protection Agency; Ross Hall, Ph.D., professor emeritus of McMaster University; John Harshbarger, Ph.D., at the Registry of Tumors in Lower Animals; Monica Hargraves, Ph.D., of Ithaca, New York; Robert Hargraves, Ph.D., professor emeritus of Princeton University; Peter Infante, Ph.D., of the Occupational Safety and Health Administration; Frieda Knobloch, Ph.D., of St. Olaf's College; Nancy Krieger, Ph.D., of Harvard University; Philip Landrigan, M.D., of Mt. Sinai School of

Medicine; Linda Lear, Ph.D., of George Washington University and the Smithsonian Institution; Ronnie Levin of the U.S. Environmental Protection Agency; June Fessenden MacDonald, Ph.D., of Cornell University's Program on Breast Cancer and Environmental Risk Factors in New York State; Donald Malins, Ph.D., of the Pacific Northwest Research Foundation; Robert Millikan, Ph.D., of the University of North Carolina; Monica Moore of the Pesticide Action Network North America Regional Center; Mary O'Brien, Ph.D., of Eugene, Oregon; Maria Pellerano and Peter Montague, Ph.D., of the Environmental Research Foundation; Frederica Perera, Dr.P.H., of Columbia University; David Pimentel, Ph.D., of Cornell University; Mike Rahe of the Illinois Department of Agriculture; Edmund Russell III, Ph.D., of the University of Virginia; Arnold Schecter, M.D., M.P.H., of the State University of New York, Binghampton; Paul Schulte, Ph.D., of the National Institute of Occupational Safety and Health; Carl Shy, M.D., Dr.P.H., of the University of North Carolina; Carlos Sonnenschein, M.D., and Ana Soto, M.D., of Tufts University; William H. Smith, Ph.D., of Yale University; A. G. Taylor, C.P.S.S., of the Illinois Environmental Protection Agency; Paul Tessene and Susan Post of the Illinois Natural History Survey; Susan Teitelbaum, M.P.H., of Columbia University; Rebecca Van Beneden, Ph.D., of the University of Maine; Louis Verner, Ph.D., of the Antioch New England Graduate School; Tom Webster, Ph.D., of Boston University; Gail Williamson, M.D., of Brookfield, Illinois; Mary Wolff, Ph.D., of Mt. Sinai School of Medicine; and Sheila Hoar Zahm, Sc.D., of the National Cancer Institute. Of course, all responsibility for the accuracy and validity of the text rests with me alone.

Friends in arts and letters also offered their insights as readers and commentators. They are Karol Bennett, Anthony Brandt, Robert Currie, Joellen Masters, Kim McCarthy, Marnie McInnes, John McDonald, and Ann Patchett.

Many other scholars and researchers, too numerous to name here, cheerfully offered advice, shared data, fielded questions, and alerted me to important areas of study. I owe them all a great deal—as I do numerous public servants in an assemblage of government agencies. These include the U.S. and Illinois Environmental Protection Agencies; the National Cancer Institute; the National Institute of Occupational Safety and Health; the National Institute of Environmental Health Sciences; the Agency for Toxic Substances and Disease Registry; the National Program of Cancer Registries; the National Center for Health Statistics; the Illinois Geological, Water, and Natural History Surveys; the Illinois Department of Conservation; the Illinois and Massachusetts Departments of Public Health; and the Mason County Health Department. Once again, responsibility for the text is solely mine. The conclusions and recommendations expressed in this book may not be shared by those who assisted me in the research process nor by the agencies they represent.

All of us are indebted to those working, on local as well as national levels, for cancer prevention and environmental protection. I owe specific thanks to members of the Women's Community Cancer Project in Cambridge, Massachusetts; Nancy Evans of Breast Cancer Action; Cathie Ragovin, M.D., of the Massachusetts Breast Cancer Coalition; Barbara Balaban, Geri Barish, and Joan Swirsky of Long Island; Sandra Marquardt of Mothers and Others for a Liveable Planet; the staffs of Pesticide Action Network and the Northwest Coalition for Alternatives to Pesticides; and the citizens and scientists who serve with me in Washington on the National Action Plan on Breast Cancer.

Closer to home, I received assistance from Kevin Caveny, district supervisor of my hometown water utility; Elaine Hopkins of the *Peoria Journal Star*; state representative Ricca Slone; and Dr. Earl and Marge Melchers of Pekin, Illinois. My parents, Wilbur and Kathryn Steingraber, and sister, Julie Skocaj, read drafts, checked facts, and freely offered the authority of their memories and experience.

No writer could ask for a more steadfast pair of advocates than I have found in my literary agent, Charlotte Sheedy, and my editor, Merloyd Lawrence. Both have been unflagging in their enthusiasm and support—even in the face of tribulations and unforeseen delays on my part. Dropping a finished chapter through Merloyd's Beacon Hill mail slot was the most satisfying of gestures. For her patience, caring, and sharp editorial judgment, I am forever grateful.

Finally, I express my abiding appreciation to Bernice Bammann, to the extraordinary writers Valerie Cornell and Karen Lee Osborne, and to Jeff de Castro, whose intellect informs many of these chapters and continues to bless my life.

PROLOGUE

In the part of the country I come from, the names of high schools, retirement homes, and lumberyards sometimes contain the word *prairie*. It's a romantic sentiment. Almost none of us who live here has ever seen prairie—nor have our parents or grandparents—and we can scarcely imagine the familiar surfaces of this place covered in eight-foot grasses, rushes, gentians, blazing stars, and a hundred other flowering herbs. To be sure, Illinois was once overlain by twenty-three different prairie types. Their varying colors and textures would have thrown into sharp relief the topographic dips and swells that seem so imperceptible now. But Illinois inhabitants do not make these distinctions. *Prairie* is a word we use generically to refer to open tracts of land between towns, away from rivers, and devoid of trees—the aspect of the landscape that spawns tornadoes.

In October 1996, I brought my new husband, a Boston sculptor, home for a visit. Piloting the rental car through miles of what some would call prairie, I resisted the urge to offer botanical commentary. The soybean harvest was in full swing, and I watched him watch the combines, each the size of a small house, advance across the countryside. I watched him watch coal trains pass by in endless procession and then dissolve into black specks. And I watched him gaze across the blank, still Illinois River from the deck of a riverboat casino.

When we finally spoke, he said that Illinois made him look at space from a different part of his eye. The sky, he pointed out, provides the only vertical reference—mostly in the form of weather—while sense of scale on land constantly changes by virtue of large objects forever approaching and receding. Houses seem vulnerable, trees inconsequential. It was humbling. Even more than the specifics of these observations, his deference and regard touched me. He saw, from his perspective as an artist, what I have always felt: the extraordinary beauty of this profoundly altered place where I grew up.

But what I want to stress here is the unexceptional nature of central Illinois.

Federal right-to-know laws now make information about the ongoing contamination of our environment publicly available. This knowledge allows us—in ways not possible even a few years ago—to explore the extent to which toxic chemicals, including cancer-causing agents, have trespassed into our air, food, water, and soil. Our newly acknowledged right to know offers a chance to ask what connections might exist between these encroachments and the health problems of our families and communities. At the same time, cancer registries—also a recent phenomenon—provide us a view of cancer's trajectory through time and a map of its distribution across space. I attempt here to bring these two categories of information together—data on environmental contamination and data on cancer incidence—to see what patterns might exist, to identify questions for further inquiry, and to urge precautionary action, even in the face of incomplete answers.

Various published studies, gathered from far-flung corners of the biological literature, offer other glimpses of the connection between cancer and the environment. Woven into my discussion throughout this book, these range from reports on pesticides, river sediments, and trash incinerators to surveys of farmers, sport anglers, and nursing mothers. They include investigations of animals (wildlife, pets, livestock, and laboratory rats), as well as examinations of human tissues and cellular machinery (breasts, blood, hormone receptors, liver enzymes, and carcinogen-metabolizing genes). Few long-term, comprehensive studies on the environmental links to human cancers have been conducted—and I leave it to readers to judge the reasons for this neglect. However, the many small-scale, underfunded, and sometimes preliminary investigations that *do* exist create a startling picture when viewed together.

In the chapters "Time" and "Space," I start with the evidence now emerging of cancer's rising incidence rates and its tendency to concentrate in certain geographic regions. I then take a close look at our ecological surroundings, element by element. In "Earth," I look at our relationship to food and agriculture. In "Air," I consider not only the contaminants we inspire into our bodies through breathing but also the role the wind plays in conducting cancer-causing substances from industrial sites and farm fields to, for example, a river basin thousands of miles away. In "Water," I trace the pathways of such rivers, as well as the hidden journeys of the groundwater that feeds them and fills the wells from which we drink. In "Fire," I examine the misbegotten creation of one very potent but elusive carcinogen now believed to inhabit the tissues of every living person: dioxin.

What has emerged for me personally from all this study—and this book does include the deeply personal—is an appreciation for just how ordinary my Illinois home is. As in many other communities, the dramatic transformations of its industrial and agricultural practices that followed World War II had many unintended environmental consequences. In this, the story of central Illinois is utterly unexceptional. It receives my scientific attention not because its history is so unusual but because it is so typical. It receives my devotional attention because central Illinois is the source of my ecological roots, and my search for these roots is part of the story.

As I am writing these words, newspapers and science magazines are abuzz with the announcement that a direct link has been found between tobacco smoke and lung cancer. All of the warnings about cigarettes and cancer issued in the decades before this discovery were based on animal experiments and statistical associations. These demonstrated a consistent pattern but did not constitute absolute proof. Now, however, cell biologists have documented that a single component of cigarette smoke—called benzo[a]pyrene—causes genetic mutations in lung cells that are identical to those seen in the tumors of smokers with lung cancer. These mutations occur not only in the exact same gene—called p53—but in the exact same location within this particular gene. Those who seek to deny the connection between smoking and lung cancer are thus left with very little ground from which to wave the "no proof" flag.

I was four years old when the U.S. surgeon general first declared cigarettes a cancer hazard. My parents, never smokers themselves, struggled mightily to keep cigarette smoke away from me and my younger sister. I was seventeen when we first began segregating smokers and nonsmokers in the restaurants where I waited tables. I am now thirty-seven and can barely remember when I last breathed cigarette fumes in an airplane or hospital waiting room. For much of my life, I have been protected from a now-proven danger by those who had the courage to act on partial evidence.

There are individuals who claim, as a form of dismissal, that links between cancer and environmental contamination are unproven and unprovable. There are others who believe that placing people in harm's way is wrong—whether the exact mechanisms by which this harm is inflicted can be precisely deciphered or not. At the very least, they argue, we are obliged to investigate, however imperfect our scientific tools: with the right to know comes the duty to inquire.

Happily, the latter perspective is gaining esteem as many leading cancer researchers acknowledge the need for an "upstream" focus. As explained at a recent international conference, this image comes from a fable about a village along a river. The residents who live here, according to parable, began noticing increasing numbers of drowning people caught in the river's swift current and so went to work inventing ever more elaborate technologies to resuscitate them. So preoccupied were these heroic villagers with rescue and treatment that they never thought to look upstream to see who was pushing the victims in.

This book is a walk up that river.

Sandra Steingraber
November 1996

ONE

trace amounts

On a clear night after the harvest, central Illinois becomes a vast and splendid planetarium. This transformation amazed me as a child. In one of my earliest memories, I wake up in the back seat of the car on just such a night. When I look out the window, the black sky is so inseparable from the plowed, black earth—which dots are stars and which are farmhouse lights?—that it seems I am floating in a great, dark, glittering bowl.

Rural central Illinois still amazes me. Buried under the initial appearance of ordinariness are great mysteries. At least, I attempt to convince newcomers of that.

Were you to visit this countryside for the first time, its apparent flatness is probably what would impress you first—and indeed, for almost half the year, the landscape seems to consist of a simple plain of bare earth overlain by sky. But Illinois is not flat at all, I would insist, as I unfold geological survey maps that make visible the surprisingly contoured lay of the land. Parallel arcs of scalloped moraines slant across the state, each ridge representing the retreating edge of a glacier as it melted back into Lake Michigan and surrendered the tons of granulated rock and sand it had churned into itself.

Better than maps is a ground fog on a summer night when I drive you across these moraines and basins. Now you see how the shrouded bottomlands are distinguished from the uplands, the floodplains from the ridges, how the daytime perception of flatness belies a great depth. Out of the car and walking, I encourage you to feel, as we traverse land that appears to be utterly level, the slight tautness in the thighs that comes with ascending a long grade versus the looseness in our feet that indicates descent.

Then there is the issue of water. Consider your own body, how the blood does not pulse through your tissues in great tidal surges—as was presumed before the English physician William Harvey discovered circulation in 1628—but instead flows within a diffuse net of permeable vessels. So too in Illinois, a capillary bed of creeks, streams, forks, and tributaries lies over the land. Your newly found skill of walking downhill will help you locate it.

And this is only the water that is visible. Under your feet lie pools of groundwater held in shallow aquifers—interbedded lenses of sand and gravel—and in the bedrock valleys of ancient rivers that lie below. One of these is the Mahomet, part of a river system that once ran west across Ohio, Indiana, and Illinois. Thousands of tons of debris, let loose by melting glaciers, completely buried the Mahomet River at the end of the last ice age. It now flows underground. In Mason County you can stand over a place where the Mahomet once joined the Illinois River. Here, in an area called the Havana Lowlands, the groundwater lies just below the earth's surface. In times of heavy rain, lakes brim up from under the earth and reclaim whole fields and neighborhoods.

In the eastern half of my county, Tazewell, the ancestral Mississippi River cut a valley three miles wide and 450 feet deep before glaciers exiled it to the western border of Illinois, its current channel. Buried by soil, clay, silt, and stones, the old Mississippi River valley is still down there, connected to the same ancient tributaries, its fractures and pores full of water. Islands still rise from the bedrock channel. If you could see through dirt, imagine the dramatic view you would have.

Of course, what you do see are corn and soybean fields. About 89 percent of Illinois is cropland, meaning that if you fell to earth in Illinois, nine times out of ten you would land in a farm field. Illinois grows more soybeans than any other state in North America, and it produces more corn than any state but Iowa. Read any supermarket label. Corn syrup, corn gluten, cornstarch, dextrose, soy oil, and soy proteins are found in almost every processed food from soft drinks to sliced bread to salad dressing. These are also the ingredients of the food we feed to the animals we eat. Thus, you could say that we are standing at the beginning of a human food chain. The molecules of water, earth, and air that rearrange themselves to form these beans and kernels are the molecules that eventually become the tissues of our own bodies. You have eaten food that was grown here. You *are* the food that is grown here. You are walking on familiar ground.

Illinois is called the Prairie State, but you must really know where to look to find prairie. Most of it vanished after John Deere invented the self-scouring steel plow in 1836. To be exact, 99.99 percent went under the plow. The .01 percent that escaped occupies odd and neglected places: along railroad tracks, encircling gravestones in old pioneer cemeteries, on hillsides too awkward to plow. Of the original 281,900 acres of tallgrass prairie in my home county, an official 4.7 fragmented acres remain (equals .0017 percent). I have never found them. Illinois conceals not only its topography but its ecological past as well, and even though I went on to become a plant ecologist, I have no real relationship to the native plants of my native state.

Truthfully, the closest I have felt to the prairie is when looking at plain, unadorned dirt. There are plenty of opportunities to do this in central Illinois—although the fields look less naked between October and April than they did when I was a child, thanks to the switch to low-till and no-till farming. These practices have largely replaced the habit of turning the field completely over after the harvest. The newer techniques leave on the surface a certain fraction of stalks, leaves, and stems to serve as a thin blanket against the wind. It is a tricky business: Too much residue leaves the soil compressed, without air, and unable to warm up in time for spring planting; water puddles on the surface. Too little residue, and the soil refuses to clump up at all, is prone to blow away or run with meltwater into the nearest creek bed.

Thus, each September at the Farm Progress Show, farm equipment representatives demonstrate all the latest technology for striking the perfect balance between these two states. Popular among farmers in recent years has been the disc and chisel plow combination: parallel rows of slicing silver plates, like large pizza cutters, alternating with rows of beveled metal claws. These grids of discs and chisels are pulled, one by one, through an exhibition field as an announcer extols the virtues of each particular model. Observers, including me and my uncle, stand on either side of the tractor as it cuts a wide swath through corn stubble. We then step into the black wake and bend down to take a look. To assess depth of penetration, we are encouraged to poke yardsticks into the chiseled furrows. We heft clumps of dirt in our hands to check diameter and ease of crumbliness. We then walk ten yards over and form two lines on either side of the next tractor in the queue of tractors to cut a path through this field of stubble. We step in, bend down, heft clumps, stand up, walk over. And so on. It is a peculiar kind of country line dance. Each plowed strip is subtly different from the others.

There is no reason I should participate in this ritual except that my mother's family still farms the Illinois prairie and watching the earth being tilled offers me a connection to the past. Even though I live in New England now, it is important to me to maintain a relationship with both Illinoises—the present and familiar one as well as the Illinois that has vanished and is barely discernible. What remains of the twenty-two million acres of tallgrass prairie that once covered this state is the deep black dirt that those grasses produced from layers of sterile rock, clay, and silt dumped here by wind and glaciers. The molecules of earth contained in each plowed clod are the same molecules that once formed the roots and runners of countless species unfamiliar to me now. They died and became soil. This most obvious of realizations occurs to me every September as though for the first time. When I am touching Illinois soil, I am touching prairie grass.

Illinois soil holds darker secrets as well. To the 89 percent of Illinois that is farmland, an estimated 54 million pounds of synthetic pesticides are applied each year. Introduced into Illinois at the end of World War II, these chemical poisons quietly familiarized themselves with the landscape. In 1950, less than 10 percent of cornfields were sprayed with pesticides. In 1993, 99 percent were chemically treated.

Pesticides do not always stay on the fields where they are sprayed. They evaporate and drift in the jetstream. They dissolve in water and flow downhill into streams and creeks. They bind to soil particles and rise into the air as dust. They migrate into glacial aquifers and buried river valleys and thereby enter groundwater. They fall in the rain. They are detectable in fog. Little is known about how much goes where. In 1993, 91 percent of Illinois's rivers and streams showed pesticide contamination. These chemicals travel in pulses: pesticide levels in surface water during the months of spring planting—April through June—are sevenfold those during winter, although detections never fall to zero. Even less is known about pesticides in groundwater. A recent pilot study found that one-quarter of private wells tested in central Illinois contained agricultural chemicals. Those sampled in the Havana Lowlands region of Mason County showed some of the most severe contamination.

Some of the pesticides inscribed into the Illinois landscape promote cancer in laboratory animals. Some, including one of the most commonly used pesticides, atrazine, are suspected of causing breast and ovarian cancer in humans. Other probable carcinogens, such as DDT and chlordane, were banned for use years ago, but like the islands in preglacial river valleys, their presence endures.

A lot goes on in the 11 percent of Illinois that is not farmland. Approximately fifteen hundred hazardous waste sites are in need of remediation—a list that does not include several thousand pits, ponds, and lagoons containing liquid industrial waste. And each year Illinois injects some 250 million gallons of industrial waste—which, until recently, included pesticides—through five deep wells that penetrate into bedrock caverns. These geological formations are overlain by aquifers and farmland. Illinois exports hazardous waste but also imports it—almost 400,000 tons in 1992—from every state except Hawaii and Nevada. In this same year, Illinois industries legally released more than 100 million pounds of toxic chemicals into the environment.

Like pesticides, industrial chemicals have filtered into the groundwater and surface waters of streams and rivers. Metal degreasers and dry-cleaning fluids are among the most common contaminants of glacial aquifers. Both have been linked to cancer in humans. A recent assessment of the Illinois environment concluded that chemical contamination "has become increasingly dispersed and dilute (and thus less visible)," leaving residues that are "increasingly chemically exotic and whose health effects are not yet clearly understood."

I was born in 1959 and so share a birthdate with atrazine, which was first registered for market that year. In the same year DDT—dichloro diphenyl trichloroethane—reached its peak usage in the United States. The 1950s were also banner years for the manufacture of PCBs—polychlorinated biphenyls—the oily fluids used in electrical transformers, pesticides, carbonless copy paper, and small electronic parts. DDT was outlawed the year I turned thirteen and PCBs a few years later. Both have been linked to cancer.

I am compelled to learn what I can about the chemicals that presided over the industrial and agricultural transformations into which I was born. Certainly, all of these substances have an ongoing biological presence in my life. Atrazine remains among the most common contaminants of midwestern drinking water, and all of us in the United States carry detectable levels of DDT and PCBs in our tissues. PCBs lace the sediments of the river I grew up next to as well as the flesh of the fish that inhabit it. DDT can remain in soil for several decades.

I honestly have no memories of DDT. Instead, my images come from archival photographs and old film clips. In one shot, children splash in a swimming pool while DDT is sprayed above the water. In another, a picnicking family eats sandwiches, their heads engulfed in clouds of DDT fog. Old magazine ads are even more surreal: an aproned housewife in stiletto heels and a pith helmet aims a spray gun at two giant cockroaches standing on her kitchen counter. They raise their front legs in surrender. The caption reads, "Super Ammunition for the Continued Battle on the Home Front." DDT is a ruthless assassin. In another ad, the aproned woman appears in a chorus line of dancing farm animals who sing, "DDT is good for me!" DDT is a harmless pal.

During the 1940s and '50s, this chemical of multiple personalities found its way into all kinds of civic campaigns and household products. One Illinois town not far from where I grew up conducted aerial fumigations of DDT in an attempt to control polio, mistakenly thought to be spread by flies. Meanwhile, a paint company advertised a formulation that could be brushed onto porches, window screens, and baseboards. When dry, DDT crystals would rise to the surface, forming "a lethal film." Perfect for summer cottages and trailers. Perhaps I spent childhood vacations in some of them. And perhaps, while there, I slept soundly between pesticide-impregnated blankets. In 1952, researchers proudly announced that woolens could now be mothproofed by adding DDT to the dry-cleaning process.

Fellow baby boomers just a few years older do not rely on old magazine ads to recall DDT. From memory, they can describe the fogging trucks that rolled through their suburban neighborhoods as part of mosquito, Dutch elm disease, or gypsy moth control programs. Some can even describe childhood games that involved chasing these trucks. "Whoever could stay in the fog the longest was the winner," remembers one friend. "You had to drop back when you got too dizzy. I was good at it. I was almost always the winner." Says another, "When the pesticide trucks used to come through our neighborhood, the guys would haul their hoses into our backyard and spray our apple trees. Mostly we kids would throw the apples at each other. Sometimes we would eat them."

Hazards that are universally common or repetitive assume "the harmless aspect of the familiar," observed the wildlife biologist Rachel Carson in her book *Silent Spring*, published when I was three years old. "It is not my contention that chemical insecticides never be used," Carson emphasized. "I do contend we have put poisonous and biologically potent chemicals indiscriminately into the hands of persons wholly ignorant of their potentials for harm. We have subjected enormous numbers of people to contact with these poisons, without their consent and often without their knowledge." She went on to predict that future generations would not condone this lack of prudent concern.

Reading *Silent Spring* as a member of this generation, across a distance of more than three decades, I gain another view of DDT. What impresses me most is just how much was known about the harmful aspects of this familiar and seemingly harmless substance. As Carson made clear, the scientific case against DDT—even by the late 1950s—was damning. It was not objective science, nor was it blissful ignorance, that created the impression that DDT was somehow both our most lethal weapon against undesirable life forms ("killer of killers," "the atomic bomb of the insect world") and a completely benign helpmate. In fact, scientific study after scientific study showed that DDT was failing at both roles. It triggered population explosions in insect pests who evolved resistance and whose natural enemies were killed by the spray. It poisoned birds and fish. It disrupted sex hormones in laboratory and domestic animals. It showed signs of contributing to cancer. By 1951, it had become a contaminant of human breast milk and was known to pass from mother to child.

Nevertheless, people continued using DDT until Carson's preliminary damning evidence was supplemented with more and more corroborating damning evidence, producing a great accumulation of damning evidence, and its registration was finally revoked in 1972. I find this phenomenon boundlessly fascinating. Across my desk are spread forty years of toxicological profiles, congressional testimonies, laboratory studies, field reports, and public health investigations of toxic chemicals both officially outlawed and officially permitted. Like crossing and recrossing the same field, I move back and forth between *Silent Spring* and the scientific literature that preceded it, between *Silent Spring* and the scientific literature published in the decades since. At what point does preliminary evidence of harm become definitive evidence of harm? When someone says, "We were not aware of the dangers of these chemicals back then," whom do they mean by *we*?

However banished, DDT has an ongoing presence in our lives through several routes. Its persistence in soil means that some food crops continue to bear DDT residues. Migratory songbirds carry DDT molecules in their flesh, as do many freshwater fish. DDT is a common ingredient of hazardous waste sites. It has been detected in carpet dust. Global air currents carry DDT into the North American continent from countries where its use is still permitted. And DDT periodically wells up from the deep basins of the Great Lakes.

Moreover, even after its ban, DDT has continued to be shipped abroad. Laws banning the use of particular pesticides in this country do not prohibit their export. U.S. Customs records from 1992 reveal that several million pounds of unregistered, canceled, or suspended pesticides were loaded on ships and exported from the United States that year. As of 1994, nine tons per day of domestically banned pesticides left U.S. shores for foreign lands.

Lindane, chlordane, dieldrin, aldrin, heptachlor. These names, unfamiliar to us now, are a roll call of the other pesticides Rachel Carson featured in *Silent Spring*. All are now classified as known, probable, or possible carcinogens. All are now prohibited or heavily restricted for domestic use. Many are still manufactured and exported. A chemical company in my hometown, for example, released several pounds of lindane into the air in 1992 and dumped several more pounds into the sewer system. I know this because federal right-to-know laws now make such events public information. Thus, lindane appears in the 1992 federal government's Toxics Release Inventory for Tazewell County. I was stunned to discover it there as I scanned the long computer scroll that documents industry's emissions, dumpings, and transfers of toxic chemicals. Lindane was banned for most uses in 1983, although it is still allowed in lice shampoos for humans and flea dips for dogs. Clearly, I have a more intimate relationship with lindane than I realized.

Aldrin and dieldrin were banned in 1975, although aldrin was allowed as a termite poison until 1987. Aldrin converts to dieldrin in soil and inside our tissues. Dieldrin suppresses the immune system and produces abnormal brain waves in mammals. As late as 1986, dieldrin was still turning up in milk supplies because the soils of hayfields sprayed more than a decade earlier remained contaminated. Most agricultural uses of chlordane in the United States were ended in 1980 and heptachlor in 1983. Both have been linked to leukemia and certain childhood cancers.

For those of us born in the 1940s, '50s, and '60s, the time between the widespread dissemination of these pesticides and their subsequent prohibition represents our prenatal periods, infancies, childhoods, and teenage years. We were certainly the first generation to eat synthetic pesticides in our pureed vegetables. By 1950, residue-free produce was so scarce that the Beech-Nut Packing Company began allowing detectable levels of residue in baby food.

Banned pesticides, like fugitives from justice, have not entirely disappeared. We have forgotten about them, but they are still among us. They frequent foreign ports. They languish underground. But they are beginning to surface again in the tissues of women with breast cancer, sometimes under different names—DDT is metabolized in the human body into other chemicals, including one called DDE—and sometimes along with banned industrial chemicals belonging to the same chemical clan.

Four years after DDT was banned, researchers reported that women with breast cancer had significantly higher levels of DDE and PCBs in their tumors than in the surrounding healthy tissues of their breasts. Similar but weaker trends held for lindane, heptachlor, and dieldrin. The study was small—involving only fourteen women—but the findings provocative, because DDT and PCBs were already linked to breast cancer in rodents.

Other small studies followed. Some showed an association between breast cancer and residues of pesticides or PCBs; some did not. In 1990, Finnish researchers reported that women with breast cancer had higher concentrations of a lindanelike residue in their breasts than women without breast cancer. Indeed, women whose breasts sequestered the highest levels were ten times more likely to have breast cancer than women with lower levels. Moreover, the pooled blood from women with breast cancer contained 50 percent more of this pesticide residue than the blood from women without breast cancer. Similarly, in 1992, a study of forty Connecticut women revealed that levels of PCB, DDE, and DDT in the breasts of women with breast cancer were 50 to 60 percent higher than in women who did not have breast cancer.

In 1993—seventeen years after the first pilot study—the biochemist Mary Wolff and her colleagues conducted the first carefully designed, major study on this issue. They analyzed DDE and PCB levels in the stored blood specimens of 14,290 New York City women who had attended a mammography screening clinic. Within six months, fifty-eight of these women were diagnosed with breast cancer. Wolff matched each of these fifty-eight women to control subjects—women without cancer but of the same age, same menstrual status, and so on—who had also visited the clinic. The blood samples of the women with breast cancer were then compared to their cancer-free counterparts.

On average, the blood of breast cancer patients contained 35 percent more DDE than that of healthy women. (PCB levels were only slightly higher.) The most stunning discovery was that the women with the highest DDE levels in their blood were four times more likely to have breast cancer than the women with the lowest levels. The authors concluded that residues of DDE "are strongly associated with breast cancer risk."

On the heels of the Wolff study came another by the Canadian researcher Éric Dewailly and his colleagues in Québec. Dewailly obtained breast tissue from women who had undergone biopsies for breast lumps. He chose twenty women whose lumps turned out to be cancerous and seventeen women whose lumps were benign. The removed lumps were then analyzed for chemical residues. Consistent with the findings of previous studies, the concentrations of several pesticides and industrial chemicals were moderately higher in the tissues of women with cancer than women without. When Dewailly restricted his comparison to estrogen-receptor positive tumors (that is, tumors sensitive to the presence of estrogen), the difference became more striking: DDE levels were substantially higher in women with estrogen-receptor positive cancers than in the women of the control group.

Following Wolff's and Dewailly's work came the Krieger study, which yielded a more complicated picture. The Harvard epidemiologist Nancy Krieger, then at the Kaiser Foundation in Oakland, California, examined DDE and PCB levels in blood drawn from women in the 1960s and then frozen and stored for nearly thirty years. She compared the blood from 150 women who went on to get breast cancer sometime during those intervening three decades to blood from 150 women who remained cancer free. The central question: Can exposure to DDT and PCBs many years ago predict whether a woman will contract breast cancer? Previous studies looked at DDE and PCB levels at the time of diagnosis. Hers would be the first study to take into account the lag time between exposure and onset of disease. Three racial/ethnic groups were represented—African Americans, Asian Americans, and whites. When the three groups were combined, no significant differences were found. However, when each racial group was considered separately, the results changed. Whites and especially African American women with breast cancer had significantly higher levels of DDE than women without breast cancer, even as Asian American women continued to reflect the overall pattern of no difference. More mysteriously, while African American women with breast cancer showed more past exposure to PCBs than their counterparts without breast cancer, the trend for white women went in the opposite direction: the highest levels of blood PCBs tended to occur in women *without* the disease.

The interpretation of these results—which are not inconsistent with earlier studies but which do not actually confirm them either—has sparked considerable debate. Do DDE and PCB levels in blood serum accurately mirror their levels in women's breasts? (Evidence from other studies indicates they do.) Do we know whether DDE and PCB molecules remain stable when stored for thirty years? (Persistence is certainly a well-known trait of both chemicals.) What about the red rubber tops that capped the test tubes? Could chemical contaminants have migrated into the blood and marred the chemical analysis? (A speculative concern.)

For thirty years, three hundred stoppered test tubes stood at attention in the back of a freezer, waiting to be rediscovered. Red blood. Red caps. Some of the women whose blood lay frozen in those tubes are dead of breast cancer, others have died from other causes, and others are no doubt still alive. Probably no one now living remembers those particular blood draws, but three decades later, our understanding about breast cancer and environmental contamination has become linked to the contents of these red-capped tubes.

Perhaps the image would seem less urgent if women born in the United States between 1947 and 1958 did not now have almost three times the rates of breast cancer than their great-grandmothers did when they were the same age. Or if pesticide use in the United States had not doubled since Rachel Carson wrote *Silent Spring*. But we do. And it has.

❧

Ten thousand years of tallgrass prairie have left a fainter trace on the place I call home than twenty-seven years of DDT spraying. Because it is my home, I am driven to pursue the question of the past and ongoing contamination of Illinois and its possible link to the increasing frequency of cancer there. I believe that all of us, wherever our roots, need to examine this relationship. And I think it reasonable to ask—more than three decades after *Silent Spring* alerted us to a possible problem—why so much silence still surrounds questions about cancer's connection to the environment and why so much scientific inquiry into this issue is still considered "preliminary."

From dry-cleaning fluids to DDT, harmful substances have trespassed into the landscape and have also woven themselves, in trace amounts, into the fibers of our bodies. This much we know with certainty. It is not only reasonable but essential that we should understand the lifetime effects of these incremental accumulations.

TWO

silence

The very modern Beinecke Library at Yale University is the resting place for Rachel Carson's papers. The cool, gray archival boxes that contain her correspondence, lecture notes, and personal writings must be requested one at a time from the librarian's assistant. The special room for viewing them is hushed and spacious. A wall of windows looks out over a green collegiate lawn. One enters after a ritual of giving over all personal possessions to the librarian. No ink is allowed in the viewing room—only pencils or laptop computers.

Alone in this room with the first box, I sift slowly through the pages it holds as though I were sorting botanical specimens. It is an automatic reflex, although I have not worked in a botanical herbarium for years. Herbarium sheets, onto which the delicate skeletons of dried plants are pressed, must never be flipped over like pages in a book but rather are to be laid gently in reverse order to the left of the stack one is looking through. When finished, the examiner places the sheaves, one at a time, on top of the stack to the right, and they thus assume their original position. At least, this is the method I was taught. Something about the ceremony of my current task has triggered this old behavior. I can only hope it approximates correct archival technique.

The sight of Rachel Carson's handwriting is exhilarating. I uncover a note to Carson from Jacqueline Kennedy. Deep in another file is a letter of complaint Carson sent to a music company after receiving an erroneous bill and an inferior record album. The extraordinary and the mundane lie together here.

I have come to eavesdrop, looking for no specific document but with a desire to listen to the voices behind *Silent Spring*. And while I do overhear some things, what I end up thinking about is silence.

In a nation where guarantees of free speech are carved into the heart of our legal system, we are very often baffled by those who claim they have been silenced. I myself have never feared my mail would arrive with passages blacked out by a censor's invisible hand. I have never wondered if the police would stop me on the way to class to announce that the content of my lecture was unacceptable. And yet perhaps we have all witnessed certain subtle codes of silence in operation—an unspoken agreement in the workplace or a family secret that everyone knows but does not discuss.

Rachel Carson was interested in three forms of silence. As a government scientist—she rose through the ranks of the U.S. Fish and Wildlife Service—Carson became concerned that the noise of important ecological debates carried on within federal agencies seldom reached the public. The long-running quarrel over the claim that pesticides were harmless was one she followed most closely. By virtue of her position, she had access to field reports clearly indicating that attempts to eradicate insect pests through massive chemical spraying programs had many unintended consequences for people and wildlife alike. This view, although denied vociferously by some in the government, was shared by many of Carson's colleagues. Yet the citizenry heard little of this debate. The problem was not so much that those questioning the wisdom of eradication programs were spirited away in the middle of the night but that much of their data remained soundproofed in internal documents and technical journals, that follow-up research was sorely underfunded, and that government officials turned a deaf ear to bearers of bad news.

By 1952, Carson had become a best-selling author of nature books and was able to retire from government service. However, she continued to follow the pesticide debate as it clamored through the halls of the U.S. Department of Agriculture and the National Academy of Science. Meanwhile, evidence of harm was becoming visible to many average citizens—even in the absence of public discussion. In 1958, a writer friend in Massachussets sent Carson a letter full of painful details about a mosquito control campaign that had resulted in a mass death of songbirds near her home. Those that lay scattered around her DDT-contaminated birdbath had perished in a posture of grotesque convulsion: legs drawn up to their breasts, beaks gaping open.

This letter prompted Carson to begin a comprehensive investigation of pesticides. In letters to friends about this project, she referred often to her need to speak out in defense of the natural world: "Knowing what I do, there would be no future peace for me if I kept silent." Having documented a cavalcade of problems attributable to pesticides—from blindness in fish to blood disorders in humans—she could find no magazine or periodical willing to publish her work. Carson decided to write a book.

Its title, *Silent Spring*, refers to an eerier kind of silence: the absence of bird song in a world poisoned by chemicals. Indeed, Carson argued, pesticidal warfare, waged with reckless disregard, threatens to extinguish a chorus of of living voices—those of birds, bees, frogs, crickets, coyotes, and ultimately us. On this level, *Silent Spring* can be read as an exploration of how one kind of silence breeds another, how the secrecies of government beget a weirdly quiet and lifeless world.

Through this process of silencing, the interconnectedness of all life forms is revealed. Carson studied the failed attempt to prevent the Japanese beetle from invading Iroquois County, Illinois, a rural farming community located due east of my home county. After intense and repeated pesticide bombardments by air during the mid-1950s, many insect species, sickened by the spraying, became easy prey for insect-eating birds and mammals. These creatures became poisoned in turn and, in ever-widening circles of death, went on to sicken and kill those who fed on their flesh, leaving a landscape devoid of animal life—from pheasants to barnyard cats.

Meanwhile, the targeted beetle species continued its westward advance. The protracted war against this enemy had accomplished nothing, but the residues of dieldrin remaining in the water and soil—like landmines left behind by a retreating army—guaranteed further casualties for decades to come. All for the dream of a beetleless world. The ecological tragedy of Iroquois County, said Carson, is narrated by the mute testimony of its dead ground squirrels: found with their mouths full of dirt, they had gnashed at the ground as they died.

The third kind of silence that fascinated Carson was the hushed complicity of many individual scientists who were aware of—if not directly involved in documenting—the hazards created by chemical assaults on the natural world. While dutifully publishing their research, most were reluctant about speaking out publicly, and some refused Carson's requests for more information. Writing in *Silent Spring*, Carson acknowledged the constant threat of defunding that hushed many government scientists. But she made clear in her private correspondence that she had little respect for those who knew but did not speak, a combination she saw as cowardice:

> The other day I saw a wonderful quote from [Abraham] Lincoln. . . . I told you once that if I kept silent I could never again listen to a veery's song without overwhelming self-reproach. . . . The quote is "To sin by silence when they should protest makes cowards out of men."

After *Silent Spring* was published, Carson turned her attention to the political and economic reasons behind the fearful silence of her colleagues in science. In a speech to the Women's National Press Club, she questioned the cozy relations between scientific societies and for-profit enterprises, such as chemical companies. When a scientific society acknowledges a trade organization as a "sustaining associate," Carson asked, whose voice do we hear when that society speaks—that of science or of industry?

Carson was just beginning to develop her ideas on the interlocking economic structures that bound the direction of medicine and science to the interests of industry when she herself was silenced. Leaving behind an adopted son, plans for summer fieldwork, and sketches for two more books, Rachel Carson died of breast cancer on April 14, 1964.

Sheltered from wind and waves, the Rachel Carson National Wildlife Refuge in southern Maine is essentially a salt marsh. It bears little resemblance to the rest of the Maine coastline, where the intense drama of ocean meeting rock prohibits marsh grasses from taking root. It is, therefore, a very different place from the craggy tidal pools and moonlit coves of Rachel Carson's beloved summer home farther north.

Walking along the paths of the refuge that bears her name, I realize I feel less close to Rachel Carson here than in the climate-controlled sanctum of the Beinecke Library. At the dedication site, a large plaque dutifully lists the titles of her books and then credits her for inspiring millions to greater environmental consciousness. Its brief, abstract sentences remind me how remote a figure Carson became after her death. Like Rosa Parks, Carson is a symbol, a muse, a spark that ignited a social movement, a name to be invoked before a speech. In this, she seems unknowable and unhuman.

Still, my Illinois nerve endings are stirred by the softness of the landscape here. The lay of the land feels familiar, although most of the plant species are not. Salt meadow grass knits together the higher grounds, while the lower sweeps are bound by the taller and stiffer saltwater cordgrass. The sinuous borders between them represent the reach of the tide. The trail guide boasts that these two grasses together can produce as much plant matter per acre per year as a prime midwestern cornfield. I smile. No way.

It is November 1993. I have driven here from Boston with my friend Jeannie Marshall, who patiently endures my lecture on corn productivity and then turns my attention to the weather. "Doesn't it feel like a different season?" Jeannie asks. On the dry uplands, a rich summery light pours through the oak trees that hang willfully onto their curled leaves. Like a flame, my dog streaks through the understory in pursuit of unseen life forms. Old oak leaves are a distinct shade of brown, which I am accustomed to viewing in hues of light more pale and dilute. We agree it is oddly beautiful to see them cast in such radiance.

The tidal creeks that worm their way through the stands of cordgrass confuse and delight me. I depend on surface water to reveal slope and direction, but poised here at the margin of the sea, these two concepts are subordinated to a larger force. At low tide, the creeks flow into the ocean. At high tide, the ocean flows into the creeks. The streambeds here pulse back and forth, flooding and draining, in a continual exchange of water and salt. There is no clear direction.

Which is exactly how I feel standing next to my friend: poised without direction in an uncertain but beautiful season. Hopeful yet unnerved.

Just diagnosed for a second time with a rare cancer of the spinal cord, Jeannie is in between surgery and radiation treatments. She is recovering quickly—getting well in preparation for becoming sick in an attempt to get well. She moves so nimbly along the paths looping through the refuge that I scarcely need to modify my own movements. If not for her cane, we could be mistaken for any two young day-trippers escaping from the city. But we are on an escape of another kind, and I feel protective and scan the path ahead for rocks, roots, and sinkholes.

Although our friendship is a recent one, the many parallels in our lives promote intense conversations whenever we are together. Both of us are writers in our thirties. Both of us became cancer patients in our twenties. Both of us grew up in communities with documented environmental contamination, high cancer rates, and suspicions that these two factors are related to each other. Both of us grew up in families constructed through adoption (I was adopted; Jeannie's mother was adopted), and we each have a keen curiosity about the interplay between heredity and environment in our lives.

And we have spoken at length about all of these topics. We have talked about what it means to have cancer as young women and about the relative significance of genealogy and ecology in that context. We have discussed our relationship with our doctors, our families, our hometowns, our writing, our bodies.

The depth and easiness of our talking carry us along today—through the luminous oak groves, out along the boardwalks that float over salt meadow grass, up onto the observation deck that overlooks the confluence of the Mariland River and Branch Brook, whose waters throb back and forth. It seems to me in these moments that Jeannie and I have words for everything. We have rejected the cultural taboos of the past that wrapped the topic of cancer in shrouds of silence, but we have also turned away from the happy cancer chatter that regularly arrives in our mailboxes in the form of brochures and magazines dedicated to the concepts of coping, accommodating, and adjusting to this disease. In its place, we have created a language between us that is compassionate, smart, fearless, open.

What my friend and I do not choose to talk about this afternoon are the dark days that lie ahead for her. Days of lying under the crosshairs of a proton-beam cyclotron. Fatigue, vomiting, blood tests. Continuously handing one's body over to technicians and doctors in a process that we call becoming medicalized. But between us, we have years of experience with cancer. I have no doubt that when those days arrive we will find a vocabulary for every experience.

We pause to examine some small ponded areas near the brook. These are salt pannes—low spots that hold water when the tide ebbs. Evaporation concentrates the salt to such extraordinary levels that only a few inconspicuous plants can survive. Glassworts. Sea-blite. Life thriving among bitterness.

"I like this place," I finally admit.

"I do too. It's nice to be here."

On average, breast cancer robs the woman it kills of twenty years of life. This means that in the United States, nearly one million years of women's lives are lost each year. In 1964, Rachel Carson died at age fifty-six—twenty years short of the average life expectancy for U.S. women at that time. Despite all the ways she was extraordinary, as a victim of breast cancer Carson was utterly typical.

Carson was diagnosed in 1960, in the thick of researching and writing *Silent Spring*. Her tumor spread to her lymph nodes and to her bones, eventually including her spine, pelvis, and shoulder. She continued writing, even though surgery left her exhausted and radiation treatments, nauseated. Other ailments—joint and heart problems that were exacerbated, if not caused, by the radiation—brought crippling and immobility. The tumors in her cervical vertebrae caused her writing hand to go numb.

Carson lived for eighteen months after finishing *Silent Spring*—long enough to smoke out a hornet's nest of ridicule and invective from the chemical industry, as well as to receive every imaginable award from the world of arts, letters, and science. Privately, Carson expressed relief and satisfaction at having lived to see *Silent Spring* complete—a reaction many of Carson's commentators and colleagues have repeatedly underscored.

But there is another story embedded in the remaining fragments of Carson's private writings. Far from viewing *Silent Spring* as her crowning achievement, Carson ached to go on to new projects as well as to seize the opportunities that her success now afforded. She did not go gently or gratefully into any good night. As her letters reveal, she died hoping for another remission, another field season, more time. And in this desire, Carson appears before us again as a typical woman with breast cancer.

From a letter to her dearest friend, Dorothy Freeman, in November 1963:

> There is still so much I want to *do*, and it is hard to accept that in all probability, I must leave most of it undone. And just when I have attained the power to achieve so much I feel is important! Strange, isn't it?

And a few months later:

> But in spite of the blow yesterday, darling, [presumably, news of more cancer] I am able to feel that another reprieve can perhaps be won. . . . Now it really seems possible there might be another summer.

There was not.

The winter of 1994 let go of Boston during the second week of March. Over a hundred inches of snow had fallen since December, and most of it lay in towering mounds over every inch of grass and concrete that was not a passage for car traffic or an entrance to a building. Now the ice piles were finally melting, and everything that had been lost or abandoned began to surface: mittens, shovels, coat hangers, trash cans, lumber, laundry baskets, entire automobiles. Stratified layers of sand, cat litter, and gravel, which had been trapped at various depths, redeposited themselves in swirling alluvial fans along the sidewalks as rivulets of meltwater streamed toward the storm sewers.

Jeannie and I move through this landscape on our way from the Massachussetts General Hospital to her apartment in the North End. Neither of us speaks. The sound of our boots on the gravelly outwash seems deafening. Jeannie is not using a cane today, and we are walking even faster than we did four months ago in the salt marsh. In my mind's eye, I am tossing all obstacles out of our way—chunks of ice, orange traffic cones, parked cars, cement barricades. I am aiming a wrecking ball at every building.

Neither of us can believe what we have just heard. After eight miserable weeks of radiation treatments to the tumor in her lower back, the original tumor in her neck—successfully removed and treated six years ago—has returned. "Massive recurrence," to quote the neurologist who had just received the scans from the radiologist.

In fact, he said these words to us as soon as we walked into his office and closed the door. We were still standing in our winter coats and had not yet found our chairs. "Massive recurrence." I struggled with my buttons, my scarf, the zipper to my book bag. My hands refused to work correctly. It had become my job in these settings to serve as the scribe and, as such, to provide complete documentation of conversations between patient and doctor.

This ritual could not withstand the current assault. I am a crack note-taker, but my hands did not want to write the words being spoken. All my attention was trained on overriding my desire to lay down the pen. The doctor spoke quickly and relentlessly as he described the tissues that were being "destroyed" or "strangled" by the chordoma's advance. He was clearly upset but seemed unable to blend his despair with a demonstration of compassion or hope.

Jeannie remained calm. She asked him to conduct a neurological exam; her symptoms, after all, were improving. Her body seemed to be telling a different story. He refused. What would be the point? The scans told the whole story. He asked her to look at them. She refused. They each accused the other of not listening. I focused on writing faster. It was a battle of narrative. Which told the true story? the radiologist's report? or Jeannie's body? Finally, the meeting ended.

"Don't shoot the messenger," he said flatly as we were once again standing and struggling with our coats.

Now we are back in Jeannie's apartment. A garbage truck backing down the street sets off a car alarm. I imagine setting fire to them both. Jeannie lies on the bed, saying nothing. I make tea.

Say something, I order myself. The words I have just transcribed in the doctor's office are the same ones I have dreaded since my own diagnosis. Now I have heard them spoken—by a doctor who was looking into the eyes of the person sitting next to me. Not mine. Not me.

Say something.

On the day of my diagnosis, I was hospitalized and friends from college came to visit. They politely stepped into the hallway when the doctor came in. He gently told me the results of the pathology reports and the treatment plan he had in mind. We sat together for a while. After he left, my friends gingerly reentered the room. They were trying to be appropriate.

"I have cancer."

There was silence—and then some kind of awkward talking, but no one really acknowledged what I had said, including myself. Later, I was furious with all of us.

Say something.

But what? I sit down at Jeannie's kitchen table and begin to review the notes I have taken to make sure they are legible and complete. Were these the words that were really said? Can their meanings be trusted? Perhaps we had simply entered an unfamiliar culture where the phrase "massive recurrence" actually means "hello, have a seat," and "don't shoot the messenger" is a way of saying "so long, take care."

You are not saying anything.

I think back to the sunlit oak grove and the salt pannes where language was so easy. How sure I was then that I could be depended on to push any situation, no matter how dire, into the bright daylight of human speech. I think back to Rachel Carson. Tumors in her cervical vertebrae caused loss of functioning in her right hand, the writing hand. Jeannie is also right-handed. It is her left hand that is becoming weak.

In the four years Rachel Carson struggled with breast cancer, she worked to break silence in the public arena. Yet in her private life, she created at least two kinds of silence. One was permeable; one, absolute.

The former kind was a sort of drapery Rachel periodically pulled between herself and her confidante, Dorothy Freeman. In some of her letters to Dorothy, Rachel described the progress of her disease in detailed medical terms. But in others she spoke only in code, referring elliptically to "menacing shadows." Rachel often refrained from divulging bad news, downplayed the miseries of treatment, and stated her belief that the expression of fearful thoughts would only make them loom larger.

Reading again the collected letters between these two friends, I see an elaborate dance of silence. At times, Dorothy seemed relieved at the abstentions and forebearances, even seeming to encourage Rachel to keep her own counsel. Dorothy did not share her correspondent's taste for writing about cancer in detached, medical tones. She refers not to Rachel's radical mastectomy but to her "hurt side."

And yet at other times, Dorothy seemed to feel shut out by Rachel's silences. Both correspondents entreated the other not to censor her thoughts or feelings. Both correspondents also admitted they were not fully disclosing their own secret fears, out of a need to protect the other. Rachel sometimes pulled back the curtain and confided a darker story—one that admitted to pain and despair. Sometimes she followed these communications with retractions and apologies. And sometimes the letters containing these dark confessions were, upon request, destroyed.

Confessing and recanting. Withholding and divulging. This mesh of conflicting impulses is part of a familiar script that is enacted again and again between cancer patients and those who love them. And in this familiarity, Carson emerges once more, poignantly, as an ordinary woman.

The second kind of silence was a fortress of secrecy Rachel constructed around her own diagnosis, a secrecy she expected Dorothy to collude with her in maintaining. Rachel strictly forbade any discussion, public or private, about her illness. This decision was intended to retain the appearance of scientific objectivity as she was documenting the human cost of environmental contamination. She also wished to yield her enemies in industry no further ground from which to launch their personal attacks.

Accordingly, Rachel instructed Dorothy to say nothing of her condition to their mutual friends and aquaintances, lest rumors take root. If need be, Dorothy was to lie. "Say you heard from me recently and I said I was fine," she told Dorothy to tell her neighbors in Maine. "Say . . . *that you never saw me look better.* Please say that."

What personal price each of these women paid for upholding this code of silence is impossible to know. Being sworn to secrecy can be a terrible burden. Anticipating the unintentional slip of the tongue that could ruin one's career must have been equally crushing. Against this backdrop of agreed-upon silence is the fact that Carson's state of health should have been obvious to anyone who cared to look at her. But not seeing is another form of silence.

As soon as *Silent Spring* was published, Carson was thrust into the national spotlight. She spoke in front of Congress, at the National Press Club, and on national television. In the photographs and old film clips documenting these occasions, she looks for all intents and purposes like a woman in treatment for cancer. She wears an unfortunate black wig. Her face and neck exhibit the distorting puffiness characteristic of radiation. She holds herself in the ginger, upright manner of one who has undergone surgery. The alteration in her appearance that followed her cancer diagnosis is dramatic.

The newspaper clippings in the Beinecke Library that trace her various public appearances in the waning days of her life are full of elaborate descriptions of what type of elegant suit Miss Carson chose to wear and how delightfully she comported herself. The accompanying pictures tell a different story. But it is a story read in silence by a woman from a future generation who knows how it will end.

❧

Thanksgiving morning is sunny and mild. Jeannie and I decide to walk to Waterfront Park overlooking Boston Harbor. It is now more than a year since our buoyant walk through the wildlife refuge. Jeannie has just finished another round of radiation treatment, and because her balance has been affected, our pace is much slower. Orange tail swishing, my dog circles patiently, herding us toward the water. Somehow, Jeannie has managed to finish writing two articles, one about the search for cancer genes and another on breast cancer prevention for a British medical text. Feeling triumphant, she is in the mood to talk about cancer—but not her own.

"You remind me of Rachel Carson," I laugh.

We talk all the way to the ocean and back.

❧

Silent Spring is remembered for the birds. When I ask people to name words, phrases, or images that Rachel Carson's book evokes for them, "thin eggshells" is among the most frequent responses. Yet this consequence of pesticide exposure—bird eggs so fragile they crush under the airy weight of their own brooding parents—is scarcely mentioned in *Silent Spring*. Perhaps we like to equate Carson with eggshell thinning because it is a problem that largely fixed itself after DDT and a handful of other pesticides were finally restricted for domestic use. In this way, Carson's predictions of disaster can be simultaneously viewed as both prophetic and successfully averted. A comfortable reckoning.

Of course, the fate of birds and other innocents caught in the chemical crossfire certainly was a central concern of *Silent Spring*. As proof of harm, their deaths were starkly visible. Who can deny the ground squirrels' cold little mouths packed with dirt? Or shrug off the pitiful sight of songbirds writhing in the grass? But *Silent Spring* makes clear that this kind of evidence, however immediate and tangible, is only one part of a much larger assemblage that also includes human cancer. Even while hiding the image of herself as a cancer patient, Carson provided many others: from farmers with bone marrow degeneration to spray-gun-toting housewives stricken with leukemia.

Making visible the links between cancer and environmental contamination was challenging for Carson, and the task continues to be daunting. However agonizing their deaths, cancer patients do not collapse around the birdbath. Decades can transpire between the time of exposure to cancer-causing agents and the first outward symptoms of disease. When birds drop out of the sky in great numbers, we ask why. When someone we love is diagnosed with cancer, questions of cause are often of less immediate relevance than questions about treatment. Questions about the past are subordinated to questions about the suddenly uncertain future.

Based on all the data available to her in 1962, Carson laid out five lines of evidence linking cancer to environmental causes. While any one alone would be insufficient proof, when viewed all together, Carson asserted, a startling picture emerges that we ignore at our peril. First, although some cancer-producing substances—called carcinogens—are naturally occurring and have existed since life began, twentieth-century industrial activities have created countless such substances against which we have no naturally occurring means of protection.

Second, since the arrival of the atomic and chemical age that followed World War II, everyone—not just industrial workers—has been exposed to these carcinogens from the moment of conception until death. Industry manufactures carcinogens in such large quantities and in such diverse array that they are no longer confined to the workplace. They have seeped into the general environment, where we all come into intimate and daily contact with them.

Third, cancer is striking the general population with increasing frequency. At the time of Carson's writing, the postwar chemical era was less than two decades old—less than the time required for many cancers to manifest themselves. Carson predicted that the full maturation of "whatever seeds of malignancy have been sown" by the new lethal agents of the chemical age would occur in the years to come. She also believed that the first signs of catastrophe were already visible. At the end of the 1950s, death certificates showed that a far greater proportion of people were dying of cancer than had been true at the turn of the century. Most ominously, children's cancers, once a medical rarity, were becoming commonplace—as revealed both by vital statistics and by doctors' observations.

Carson's fourth line of evidence came from animals. Experimental tests were beginning to reveal that low doses of many pesticidal chemicals in common use caused cancer in laboratory mice, rats, and dogs. Moreover, many animals inhabiting contaminated environments develop malignant tumors; *Silent Spring* not only documents acute poisonings of songbirds but also reports on cases of sheep with nasal tumors. These incidents supported the circumstantial evidence from human populations.

Finally, Carson argued, the unseen inner workings of the cell itself corroborate the story. At the time of *Silent Spring*'s publication, the mechanisms responsible for basic cellular processes such as energy production and regulation of cell division were just beginning to be elucidated. The role and structure of the twisting DNA molecule had been discovered only recently. From the glimmers she was able to gather from widely scattered studies, Carson spotlighted three properties that she believed would ultimately explain why these new chemicals were associated with cancer: they were able to damage chromosomes and thereby cause genetic mutations (a property shared with radiation, which had already been shown to cause cancer); they were able to mimic and disrupt sex hormones (high estrogen levels were already being correlated with high cancer rates); and they were able to alter the enzyme-directed processes of metabolism (by which we break apart molecules to generate energy and synthesize new substances). Carson predicted that future studies on the mysterious transformation of healthy cells into malignant ones would reveal that the roads leading to the formation of cancer are the same pathways that pesticides and other related chemical contaminants operate along once they enter the interior spaces of the human body.

Like the assembling of a prehistoric animal's skeleton, this careful piecing together of evidence can never furnish final or absolute answers. There will always be a few missing parts, first because experimenting on human beings is not, thankfully, considered ethically acceptable. Human carcinogens must, therefore, be identified through inference. One set of clues is provided by observations of people who have been inadvertently exposed to substances suspected of having cancer-causing tendencies. But often these people have been exposed to unknown quantities over unknown periods of time. Observations of laboratory animals exposed to known quantities of possible carcinogens supply a second set of clues. But different animals can vary in their vulnerability to certain kinds of cancers and in their sensitivity to certain kinds of chemicals. Which species should serve as our surrogates in these studies? Rats? Mice? Fish? Dogs? Which species' lymph nodes, bone marrow, brain tissue, prostate glands, bladders, breasts, livers, and spinal cords behave most like those in humans when exposed to particular substances?

Another reason for scientific uncertainty is that the widespread introduction of suspected chemical carcinogens into the human environment is itself a kind of uncontrolled experiment. There remains no unexposed control population to whom the cancer rates of exposed people can be compared. Moreover, the exposures themselves are uncontrolled and multiple. Each of us is exposed repeatedly to minute amounts of many different carcinogens and to any one carcinogen through many different routes. From a scientific point of view, such combinations are especially dangerous because they have the capacity to do great harm while yielding meaningless data. Science loves order, simplicity, the manipulation of a single variable against a background of constancy. The tools of science do not work well when everything is changing all at once.

It is March 1995. Winter and spring have hung together in the air for weeks, neither yielding to the other. On the phone, Jeannie is trying to describe to me a new sensation she feels across the skin of her chest. It is vague and formless. There are no real words for it. I am attempting to understand how this symptom fits together with a few other recent problems she has reported. Morning vertigo. A funny feeling when she swallows. What picture is emerging here? What does her doctor say? She turns back my questions.

"Let's talk about the chapter you're writing now. What is it called?"

"Silence."

"Let's talk about that."

THREE

time

Like a jury's verdict or an adoption decree, a cancer diagnosis is an authoritative pronouncement, one with the power to change your identity. It sends you into an unfamiliar country where all the rules of human conduct are alien. In this new territory, you disrobe in front of strangers who are allowed to touch you. You submit to bodily invasions. You agree to the removal of body parts. You agree to be poisoned. You have become a cancer patient.

Most of the traits and skills you bring with you from your native life are irrelevant, while strange new attributes suddenly matter. Beautiful hair is irrelevant. Prominent veins along the soft skin at the fold of your arm are highly prized. The ability to cook a delicious meal in thirty minutes is irrelevant. The ability to lie completely motionless on a hard platform for half an hour while your bones are scanned for signs of tumor is, conversely, quite useful.

Whether it happens at a hospital bedside, in a doctor's office, or on the phone, most of us remember the event of our diagnosis with a mixture of photographic recall and amnesia. We may be able to describe every word spoken, the arrangement of photographs on the doctor's desk, the exact color of the office draperies—but have no memory of how we got home that day. Or we may remember nothing that was said but everything about the bus ride. The scene I happen to remember most vividly—and this must have occurred weeks after my discharge from the hospital—is unlocking my door and discovering that my roommate had moved out. She did not want to live with a cancer patient. This was my redefining moment. Fifteen years later, the sight of a bare mattress can still cause me to burst into tears.

In 1995, an estimated 1.2 million people in the United States—thirty-four hundred people a day—were told they had cancer. Each of these diagnoses is a border crossing, the beginning of an unplanned and unchosen journey. There is a story behind each one.

❧

These diagnoses also form a collective, statistical story. When all the diagnoses of years past and present are tallied, an ongoing narrative emerges that tells us how the incidence of cancer has been and is changing. Changes in cancer incidence, in turn, provide key clues about the possible causes of cancer. For example, if heredity is suspected as the main cause of a certain kind of cancer, we would not expect to see its incidence rise rapidly over the course of a few human generations because genes cannot increase their frequency in the population that quickly. Or if a particular environmental carcinogen is suspected, we can see if a rise in incidence corresponds to the introduction of such substances into the workplace or the general environment (taking into account the lag time between exposure and onset of disease). Such an association does not constitute absolute proof, but it gives us ground to launch additional inquiries.

The work of compiling statistics on cancer incidence is carried out at a network of cancer registries, which exist in the United States at both the state and the federal levels. Theoretically, for each new cancer diagnosis, a report is sent to a registry. How a diagnosed person has experienced, reacted to, coped with, remembered, or repressed this stunning event are aspects not included in this accounting, of course. What each report does contain is a coded description of the type of cancer; the stage to which it has advanced; and the geographic region, age, sex, and ethnicity of the newly diagnosed person.

This incoming information is then processed, analyzed, audited, graphed, and disseminated by teams of statisticians. In and of itself, a head count is not very useful. The prevalence of cancer is higher now than it was a century ago, in part because there are simply more people now. There are also proportionally more older people alive now than ever before, and the aged tend to get more cancers than the young. Between 1970 and 1990, for example, the U.S. population increased by 22 percent, and the number of people over sixty-five increased by 55 percent. To eliminate the effects of the changing size and age structure of the population, cancer registries standardize the data. One way of doing this is to calculate a cancer incidence rate, which is traditionally expressed as the number of new cases of cancer for every 100,000 people per year. For example, in 1982, 90 out of every 100,000 women living in the state of Massachusetts were diagnosed with breast cancer. By 1990, the incidence rose to 112 out of 100,000.

These numbers are also age-adjusted. That is, the data from all the differently aged people from any given year are weighted to match the age distribution of a particular census year. Thus standardized, the statistics from various years can be compared to each other. In this way, we know that the 24 percent rise in breast cancer in Massachusetts that occurred between 1982 and 1990 did not happen because the population of New England women was aging. Alternatively, cancer registry data can be made age-specific: the percentage of forty-five- to forty-nine-year-olds contracting breast cancer can, for example, be compared with the percentage from a decade ago.

I have often wondered about the daily lives of tumor registrars, those souls responsible for keeping count of cancer's casualties. How strange it must be to monitor the thousands of cancer reports that flow into the registries every day in the form of paper files or electronic transfers. Surely I would want to pluck each one from the current and imagine the life behind the name. A seventy-five-year-old black woman from an urban area with advanced-stage breast cancer . . . or a forty-five-year-old white man from a farming community with chronic lymphocytic leukemia . . . or a seven-year-old girl with a brain tumor. I would long to sit down and talk with each one. "What has happened to you since your diagnosis? Are you getting good care? Are you surrounded by people who love you?"

As a group, tumor registrars seem like an affable lot—happy to converse about their work. Susan Gershman is the director of the Massachusetts Cancer Registry. Speaking to the public at a small, suburban library one Saturday afternoon, she was cool, well organized, and articulate as she stood in the tiny spotlight of the overhead projector, illuminated by her data. People in the audience took notes. Later, during the coffee and doughnut reception, she mentioned casually that her mother and father had both died of cancer when they were young, and I knew that she must bring a double perspective to her work.

Cancer registries publish their findings in thick annual volumes replete with tables and graphs, much like sports almanacs. My own reaction to these reports follows a particular evolution. At first examination, my eyes disassemble the data. In a graph displaying the age-adjusted rates for ovarian cancer, for example, I initially focus on the points rather than the lines that connect them. I wonder at the individual women whose lives are contained by the little black circles and gray squares that float in a white field of mathematical space. Gradually, as when I am looking at a picture that contains a hidden pattern, another way of seeing emerges from the page. Years of biological training kick in, and my eyes automatically begin to trace the slope of the lines, check the coordinates, imagine how the data might appear if displayed logarithmically.

In many ways, tracking the changing patterns of cancer incidence is not unlike tracking the patterns of ecological change. The statistical methods are certainly very similiar—as are the vexing problems.

I once compiled old and current species inventories in order to monitor gradual changes in the composition of a Minnesota forest over several decades. During this time, some species became more common and others more rare. Sometimes I literally could not see the forest for the trees. The graphs constructed from my data showed clear trends often not apparent to me as I walked the deer paths that meandered among the pillars of the ancient canopy pines and through the green tangle of shrubs and saplings below. Without an exact count, I tended to overestimate the presence of rare plants because my delight at discovering them was more memorable than my efforts to note the existence of their more common neighbors. Perception can be misleading.

But I also had reasons to distrust parts of my data. To study time trends over half a century, one must rely on census counts conducted by many previous researchers, including some no longer living. If their system of coding and classifying differed significantly from mine, or if any one of us consistently misidentified certain species, then the changes indicated by my graphs were artifacts of our different techniques rather than reflections of a real biological shift. The seeming disappearance of a species that then suddenly reappeared in abundance five years later was a likely indication of a methodological snafu.

Cancer registry data are cursed with similar problems. We need these data because perception can mislead. It may seem to us that more and more people are getting brain tumors or that breast cancer is striking women at increasingly younger ages, but what do the numbers actually show? Perhaps people with cancer are now simply more outspoken than their predecessors. The numbers, on the other hand, can also deceive. Earlier detection, changes in the rate of misdiagnosis, and alterations in coding and classifying tumor types mean that apparent rises or falls in incidence rates can be artificial. How to quantify and correct such problems is a recurring question at tumor registrars' conferences and in publications such as *Cancer Registry News.*

Breast cancer incidence, for example, rose by nearly 25 percent in the United States between 1973 and 1991. During that time, the introduction of mammography changed the way many U.S. women were diagnosed with the disease, presumably because malignancies could be identified before being felt as a lump. How much of this rise can be explained by the increased use of mammograms? To answer this question, statisticians first look to see whether breast cancer incidence began to rise at the same time mammography became widely available. An internal audit of the data can also show whether groups of women with the highest rates of cancer are those receiving the most mammograms. And, since mammograms purportedly detect cancer earlier, statisticians can check whether the diagnosis of small breast tumors has been increasing faster than the diagnosis of large, advanced ones.

While still a matter of some debate, the most widely accepted estimate is that between 25 and 40 percent of the recent upsurge in breast cancer incidence is attributable to earlier detection. Underlying this acceleration exists still a gradual, steady, and long-term increase in breast cancer incidence that has just recently begun to level off. This slow rise—between 1 and 2 percent each year since 1940—predates the introduction of mammograms as a common diagnostic tool. Moreover, the groups of women in whom breast cancer incidence is ascending most swiftly—blacks and the elderly—are among those least served by mammography. Between 1973 and 1991, the incidence of breast cancer in females over sixty-five in the United States rose nearly 40 percent, while the incidence of breast cancer in black females of all ages rose more than 30 percent. Therefore, the majority of the increase in breast cancer cannot be explained by mammograms.

This kind of analysis is possible only when many years of data are available. Unfortunately, many state cancer registries are new; they cannot look back across fifty years as I could with my tree inventories. The Illinois State Cancer Registry was created in 1985. My own diagnosis, which took place in 1979, is therefore not part of the collective story of cancer in Illinois. Unless I die from the disease, I will never be officially counted among those touched by cancer. The first year of reliable data in the Illinois State Cancer Registry is 1986. Moreover, like many state registries, Illinois's is about five years behind in analyzing and publishing its data. Currently, therefore, Illinois residents have only a four-year picture of cancer incidence in their home state. Studying these time trends is like watching four minutes of a feature-length movie and trying to figure out the whole story.

Regional comparisons are often difficult because cancer registries in neighboring states can vary wildly in their length of operations. For example, Connecticut has the oldest functioning registry, one started in 1941. The Connecticut Tumor Registry provides one of the only truly long-term views of U.S. cancer incidence. Massachusetts, on the other hand, established its cancer registry in 1982. Nearby Vermont is one of ten states that had no cancer registry at all until 1992, when Congress established the National Program of Cancer Registries.

This patchwork of state-based registries is afflicted with another problem that we who count plants never have to worry about. People, unlike trees, move. Lifelong residents of one state, for example, may migrate to another upon retirement and become statistics in their new community. Without a comprehensive national cancer registry—which the United States does not have—state registries must rely on an elaborate system of data exchange. This is especially crucial for my elongated home state of Illinois, which shares a border with five other states. When faced with a serious health problem, many rural folk in the central and southern counties wind up being diagnosed across the Mississippi and Wabash Rivers because they would rather travel to cities in Iowa, Missouri, Indiana, or Kentucky than make the long trek north to Chicago. Illinois recently began trading registry data with its neighbors, thereby considerably boosting cancer incidence figures in its many east and west border counties.

Five state registries also contribute data to the federal cancer registry. The so-called SEER Program (Surveillance, Epidemiology, and End Results), overseen by the National Cancer Institute, does not attempt to record all cases of cancer in the country, but instead samples about 14 percent of the populace. SEER is a child of the War on Cancer as declared by President Richard Nixon and codified as the National Cancer Act of 1971. SEER has been collecting cancer diagnoses since 1973 and currently represents the states of Connecticut, Hawaii, Iowa, New Mexico, and Utah, as well as five specific metropolitan areas: Atlanta, Detroit, San Francisco–Oakland, Seattle, and Los Angeles. Everyone living in one of these states or cities who is diagnosed with cancer becomes a bit of data in the SEER Program registry, and their tumors stand in for all of ours.

Without a nationwide registry, no one can know exactly how many new cases of cancer are diagnosed in the United States every year. Instead, such numbers are estimated by applying rates from the SEER registry to the population projection for any particular year. To generate estimates before 1973, statisticians combine data from older individual state and city registries across the country. In this way, we now have reasonably reliable incidence figures going back to 1950.

Incidence data were not available to Rachel Carson when she first documented what she believed was the beginnings of a cancer epidemic. Instead, Carson focused on rising death rates from cancer. She was most disturbed by evidence that childhood cancer had jumped from the realm of medical rarity to the most common disease killer of American schoolchildren within a few decades.

Some researchers believe that mortality rates—which are also adjusted for age and population size—are still a more reliable indicator than incidence because they are less affected by changes in diagnostic technique. Death, after all, is certain and absolute. Moreover, causes of death, duly noted in all states of the union, have been tallied for far longer than tumors have been registered. We have a much deeper and wider view when we examine cancer trends over time using information gleaned from death certificates.

But mortality is also an imperfect measure of the prevalence of cancer. Not everyone diagnosed with cancer, thankfully, goes on to die from it. If treatment improves, mortality can decline even as incidence rises. This is certainly the case for childhood cancers, which, according to SEER data, jumped in incidence by 10.2 percent between 1973 and 1991 even as the death rate fell by almost 50 percent. Long-term trends show that childhood cancers have risen by one-third since 1950. Using mortality to measure the occurrence of cancer in children today would create a falsely rosy picture. Heroic measures may be saving more children from death, but every year more children are diagnosed with cancer than the year before. Increases are most apparent for leukemia and brain tumors. At present, eight thousand children are diagnosed with cancer each year; one in every four hundred Americans can expect to develop cancer before age fifteen.

Cancer among children provides a particularly intimate glimpse into the possible routes of exposure to contaminants in the general environment and their possible significance for rising cancer rates among adults. The lifestyle of toddlers has not changed much over the past half century. Young children do not smoke, drink alcohol, or hold stressful jobs. Children do, however, receive a greater dose of whatever chemicals are present in air, food, and water because, pound for pound, they breathe, eat, and drink more than adults do. In proportion to their body weight, children drink 2.5 times more water, eat 3 to 4 times more food, and breathe 2 times more air. They are also affected by parental exposures before conception, as well as by exposures in the womb and in breast milk.

The night before Jeannie's death, I dreamed I traveled on a large boat with many other people. No shorelines were visible. Someone suggested I walk out onto the deck and get some sun. It's too hot, I said. But I walked out anyway and discovered the weather very pleasant. Someone suggested I go for a swim. Too dangerous, I said. But I dove in, and the water was cool and crystalline. Dolphins circled me protectively. Back in the boat, I asked, Where are we? And someone smiled and handed me a map.

Driving across the Charles River to the hospital the next morning, I took the dream as a sign that I had accepted what I understood now to be imminent. But by the time I crossed the river again that night, I knew I had not and never would.

I wanted time to stop. I wanted all the clocks unplugged and the calendars nailed flat to the walls. It was April. I wanted no leaves to emerge from the buds that blurred the outlines of the trees.

Time had become such a strange commodity in the preceding month. On the surface, it had seemed to speed up as the vague progression of Jeannie's various symptoms had suddenly accelerated. One day she found she could no longer type. A week later she could not turn doorknobs. The next week, buttons were impossible. Each loss was profound and irrevocable—the ability to write, to walk through a doorway, to undress.

But under the quick surface, in the deep water at the center of every hour and every moment, time was slowing down. Each meal, each conversation, each walk from one room to another unfolded with such deliberateness that an afternoon spent in Jeannie's apartment was the equivalent of a week.

"You understand this is a terminal event." A doctor's voice on the magnetic tape of my answering machine. The dazed drive to the intensive care unit. Each heartbeat visible as data on a video screen. Slow drippings in tubes. An endless night. A blue-black dawn. A nurse's voice, as though from a distant room: "Okay. These are her last breaths now."

The whole concept of time was unbearable. I wanted to be back in Illinois in the middle of winter. I wanted to walk across frozen fields. No ocean. No leaves. No boats. She was gone.

All types combined, the incidence of cancer in the United States rose 49.3 percent between 1950 and 1991. This is the longest reliable view we have available. If lung cancer is excluded, overall incidence still rose by 35 percent. Or, to express these figures in another way: at midcentury a cancer diagnosis was the expected fate of about 25 percent of Americans—a ratio Carson found so shocking that it inspired the title of one of her chapters—while today, about 40 percent of us (38.3 percent of women and 48.2 percent of men) will contract the disease sometime within our lifespans. Cancer is now the second leading cause of death overall, and the leading cause of death among Americans aged thirty-five to sixty-four.

More of the overall upsurge has occurred in the past two decades than in the previous two, and increases in cancer incidence are seen in all age groups—from infants to the elderly. If we exclude cancer of the lung and restrict our view to the period covered by SEER, overall incidence rose 20.6 percent between 1973 and 1991, while mortality declined 2.8 percent.

Adding lung cancer to the picture, overall cancer mortality *rose* by 6.9 percent from 1973 until 1991—a difference that testifies to the deadly nature of this disease. Happily, the decline in smoking is finally affecting the cancer death rate. In a recent study of cancer mortality rates from 1991 to 1995, researchers found a small but decisive decline in overall cancer mortality (about 3 percent) during this period. The single largest factor behind this decline is a decrease in lung cancer deaths.

One-fourth of all cancer deaths are from lung cancer. Because the fatality rate is so high, lung cancer incidence and lung cancer mortality are very nearly the same statistic, and, in the United States, both closely mirror historical patterns of cigarette consumption. (Among American women, who began smoking in large numbers later in the century than did men, lung cancer mortality is still rising.) Overall, approximately 87 percent of the deaths from lung cancer can be attributed to cigarette smoking.

This also means, of course, that 13 percent of all lung cancer deaths occur among people who do not smoke. Thus, although smoking dominates the lung cancer picture, additional mysteries need sleuthing here. And, while smoking remains the largest single known preventable cause of cancer, the majority of cancers cannot be traced back to cigarettes. Indeed, many of the cancers now exhibiting swift rates of increase—cancers of the brain, bone marrow, lymph nodes, skin, and testicles, for example—are not related to smoking. Testicular cancer is now the most common cancer to strike men in their twenties and thirties. Among young men both here and in Europe, it has doubled in frequency during the past two decades. These increases cannot be attributed to improved diagnostic practices. Brain cancer rates have risen particularly among the elderly. Between 1973 and 1991, brain cancers among all Americans rose 25 percent. Those over sixty-five suffered a 54 percent rise.

Mortality and incidence do not always track each other. No cancers are increasing in mortality while decreasing in incidence, but several cancers have increased in incidence even as their death rates have declined due to more effective treatments. According to SEER data, these include cancers of the ovary, testicle, colon and rectum, bladder, and thyroid. There are eight cancers whose incidence and mortality are both on the decline: those of the stomach, pancreas, larynx, mouth and pharynx, cervix, and uterus, as well as Hodgkin's disease and leukemia. Stomach cancer has been declining for decades, probably owing to improvements in food handling and the increased consumption of fresh foods made possible when refrigeration replaced more toxic methods of food preservation, such as smoking, salting, and pickling. Pap smears have been credited with bringing down the incidence of cervical cancer because precancerous lesions can be detected and cut out before they are transformed into invasive tumors.

However, these modest gains are swamped by the cancers that show both increasing incidence and increasing mortality: cancers of the brain, liver, breast, kidney, prostate, esophagus, skin (melanoma), bone marrow (multiple myeloma), and lymph (non-Hodgkin's lymphoma) have all escalated over the past twenty years and show long-term increases that can be traced back at least forty years. In recent years, breast cancer mortality among white women has begun to slow down, declining 6.8 percent from 1989 to 1993. However, the death rate is still higher than it was when Rachel Carson died of the disease in 1964, and it is still rising for black women. Moreover, breast cancer incidence rates are still rising for localized disease even as they are falling for more advanced-stage diagnoses (a shift probably indicating that breast cancer is being detected and treated earlier); the proportion of women developing the disease at all remains at the highest level ever recorded.

"Explanations for these increases do not exist," according to Philip Landrigan, a pediatrician and leading public health researcher. Medical literature is accustomed to summations more temperate and indirect, but this one has been echoed again and again in recent research papers on trends in cancer rates. In a 1995 assessment of the situation, a research team at the National Cancer Institute similarly concluded, "Some trends remain unexplained . . . and may reflect changing exposures to carcinogens yet to be identified and clarified."

Clarification about carcinogens, Landrigan believes, requires an environmental line of inquiry:

> The possible contribution to recent cancer trends of the substantial worldwide increases in chemical production that have occurred since World War II (and the resulting increases in

> human exposure to toxic chemicals in the environment) has not been adequately assessed. It needs to be systematically evaluated.

I have read the preceding two sentences many times. Most of my life spans the time between Carson's call for a systematic evaluation of the contribution of toxic chemicals to increased human cancers and Landrigan's repetition of this call. Both give one pause.

I am struck also by the symmetry between Landrigan's recommended course of action and an observation made thirty years earlier by two senior scientists at the National Cancer Institute, Wilhelm Hueper and W. C. Conway: "Cancers of all types and all causes display even under already existing conditions, all the characteristics of an epidemic in slow motion." This unfolding crisis, they asserted, was being fueled by "increasing contamination of the human environment with chemical and physical carcinogens and with chemicals supporting and potentiating their action." And yet the possible relationship between cancer and what Hueper and Conway called "the growing chemicalization of the human economy" has not been pursued in any systematic, exhaustive way.

The environment, it seems, keeps falling off the cancer screen. The circumstances surrounding the birth of the Illinois State Cancer Registry is a case in point. The registry came into being when the Illinois Health and Hazardous Substances Registry Act was signed by the governor in September 1984. As implied by its name, this state law was intended to "monitor the health effects among the citizens of Illinois related to exposures to hazardous substances in the work place and in the environment." Accordingly, the registry system was to collect information not only on the incidence of cancer among the Illinois populace but also on their "exposure to hazardous substances, including hazardous nuclear material," thus prompting public health studies that would relate "measurable health outcomes to environmental data to help identify contributing factors in the occurrence of disease."

The cancer registry was funded. The hazardous substances registry was not.

Like a thriving child with a stillborn twin, the Illinois State Cancer Registry dutifully acquires information on health outcomes, but this activity now goes on independently of any attempt to correlate health with exposure to hazardous substances.

❧

Two months pass before I visit the cemetery. It is June. Four days of stormy weather have pelted the last of New England's rhododendron blossoms into the grass. The just-awakening roses, however, are luminescent in the streaming rain. In fact, their buds seem to be opening before my eyes.

Time still seems speeded up, as in an old movie when the wind tears the pages from the calendar and the characters leap forward into another season. Cars drive too fast. People walk too fast. Food even seems to cook too fast. I have learned to avoid quickness—like dashing out to the post office before it closes—because sudden movements seem to rush time forward even faster. I am hoping an afternoon in the cemetery will slow the world down again.

I realize immediately that I have no idea where her gravesite is. When last here, I had noticed nothing but the flower-swathed casket and the mound of dirt draped in green plastic. There was some kind of old, severely pruned tree nearby, but I can't recall the species. In my mind, I can see the round bull's eyes of its sawed-off limbs and the humped roots that had pushed away the hurricane fence behind it. I scan the fence line. A row of old basswoods runs along the far side. Finally, I see it: nearly at the end and standing exactly as in my memory. So, it is a basswood. The tree leads me to the rectangle of earth I am looking for. At the top is a plastic plaque coated with wet petals, seed coats, leaf bits, and stems. JeanMarie Marshall. 1958–1995. Finally, everything seems still enough.

❧

Studying cancer time trends is like ascending a glacial moraine in central Illinois. The rise is gradual, steady, and real. What seems imperceptible from the ground—percentage changes that unfold over miles or over decades—is plainly revealed by graphs of the data. In regard to cancer incidence in the United States, we are, in fact, walking on a sloping landscape.

The failure to evaluate systematically the relationship between rising cancer rates and rising exposures to environmental carcinogens is beginning to receive attention. The National Cancer Advisory Board stated bluntly in its fall 1994 report to Congress that a lack of appreciation for environmental and food source contaminants has frustrated cancer prevention efforts. Further asserting that the government has a responsibility to identify and prevent environmental hazards, the board called for a coordinated investigation of industrial chemicals and pesticides as causes of cancer.

Recent analyses of cancer registry data have made the need for such a coordinated investigation more urgent. Using data unavailable to Rachel Carson, the public health researcher Devra Davis and her colleagues have analyzed U.S. cancer patterns in a novel and revealing way. Rather than simply look at changes in cancer rates over calendar time, Davis grouped people according to year of birth, as well as year of diagnosis, and explored how cancer has affected successive generations. Because data on nonwhites in the early years of the SEER Program are unreliable, she restricted her view to U.S. whites and separated cancers generally believed to be associated with smoking from those not known to be so associated.

Davis found that cancer not tied to smoking has increased steadily down the generations. U.S. white women born in the 1940s have had 30 percent more non-smoking-related cancers than did women of their grandmothers' generation (women born between 1888 and 1897). Among men, the differences were even starker. White men born in the 1940s have had more than twice as much non-tobacco-related cancer than their grandfathers did at the same age. "What this is telling us," Davis says, "is that there is something going on here in addition to smoking, and we need to figure out what that is."

The grandparents of those born in the 1940s are mostly all dead now. Of all the worries they carried for their baby-boom grandchildren—those riotous offspring who opened the original generation gap—cancer, as I recall, was not high on their list. I say this as a lifelong observer of this birth cohort. At one point in my childhood, it seemed that the entire generation born in the decade before mine might die young. By eleven, we all wore metal bracelets engraved with the names of those officially missing in action in Vietnam.

About the rest, we heard various dire predictions from adults: perhaps they would all be felled by police truncheons or end up crazed and deafened from rock and roll. But I never heard anyone's grandmother predict that those born in the 1940s would surely undergo chemotherapy regimens in record numbers or that a cancer diagnosis would become as significant a generational marker as patchouli oil.

Nothing slows time down as much as waiting for lab reports. This time I am the patient. In the interior waiting room, dressed in a wraparound smock identical to the ones worn by every other human being who has entered this room, I try to conjure Jeannie out of thin air. Of the ample supply of magazines provided us here, she would choose *Vanity Fair.* Of this, I feel certain. During these moments of waiting, which celebrity interview would she, in her unflagging attempt to bring me up to snuff on popular culture, read aloud to me? And when I drifted into anxious thinking, what clever thing would she say to keep me from floating off too far?

Last summer she waited with me for hours at the ultrasound clinic.

"They had a hard time seeing what they wanted to see," I reported back to her as we finally walked out the door. "And then one of the technicians looked at the image in the monitor and whistled."

She laughed. "You know that ranks right up there with 'Hey, nice tits!'"

My name is called and I follow the doctor down the corridor to her office. Like a defendant studying the faces of the jurors as they file back into the courtroom, I try to read her expression.

It seems my situation today is mostly good, but a little bit ambiguous. The specialists have conferred and would like to recommend I undergo a new type of test, which the doctor explains in clear detail.

"I know this isn't what you wanted to hear," she says, with genuine compassion. "But you don't need me to be your best friend right now."

Time lurches forward again. Where is she?

The rise in cancer incidence over calendar time is one line of evidence that implicates environmental factors. The increase in cancer incidence among successive generations is another. A third line of evidence comes from a close consideration of the cancers that exhibit particularly rapid rates of increase. If we restrict our view to these cancers, what patterns emerge? Who gets these cancers and what do we know about their possible causes?

After lung cancer in women, the three cancers ascending most swiftly in the United States are melanoma of the skin, non-Hodgkin's lymphoma, and multiple myeloma. These are not the most prevalent cancers—breast cancer remains the most frequently diagnosed cancer in women, for example—but these are the ones galloping forward at the fastest rate.

Melanoma accounts for only 5 percent of all skin cancers, but it is the most dangerous kind, accounting for 75 percent of skin cancer deaths. The U.S. incidence of melanoma rose nearly 350 percent between 1950 and 1991, and mortality rose by 157 percent. Between 1982 and 1989 alone, melanoma incidence jumped 83 percent. Each year, about 4 percent more people contract melanoma than the year before, and the average age at diagnosis is going down. The more common basal cell and squamous cell skin cancers are also on the rise. But because they rarely spread to other parts of the body and are seldom life-threatening, these skin cancers are not even included in cancer registry data. Only melanoma diagnoses are recorded in U.S. registries.

A melanoma is a cancer that begins in a melanocyte, a cell type that surely serves as the excuse for more wars, social strife, injustice, and oppression than any other human tissue. Melanocytes are the pigment-producing cells of the skin. Those who ponder the origins of racism would do well to consider the humble biology of the melanocyte. Comprising only about 8 percent of all skin cells, the melanocytes appear in microscopic cross section as dark, delicate shrubbery. They are surrounded by Langerhans cells, which migrate up from the bone marrow and play a role in immunity, and by keratinocytes, layers of flat stepping stones that comprise 90 percent of our epidermis and produce a waterproofing protein. The melanocytes' slender branches extend between and around the keratinocytes and deliver to them the molecules of melanin they cannot produce on their own. Once inside the keratinocytes, the blackish-brown granules float to the surface and form a sunlight-absorbing cloak that lies over the fragile chromosomes inside the nuclei. Exposure to ultraviolet radiation—that high-energy wavelength lying just below violet in the spectrum of visible light—stimulates the melanocytes to make more melanin and in darker shades. More grains of melanin are dispatched to the keratinocytes. Therefore, we tan.

Everyone, regardless of race, has approximately the same number of melanocytes. Differences in skin color represent differences in the amount of melanin produced. Not everyone, however, has an equal chance of contracting melanoma. Incidence among whites is ten times higher than among blacks. Among white men, the disease more often originates in a melanocyte located somewhere on the trunk of the body; among white women, on the lower leg. When a melanocyte becomes cancerous—multiplying out of control and, if undetected, seeding itself in deeper and more distant parts of the body—its pigment-producing activities do not stop. The dark interior spaces of the body, where rays of sunlight never penetrate, thus become filled with black tumors that go on crazily producing molecules of light-shielding melanin with no companion cells to receive them.

Melanomas are clearly associated with exposure to ultraviolet radiation (albeit in a complicated way that is a matter of some debate), and here is where individual behavior and changes in the global environment come together. Basal and squamous cell cancers, which arise from keratinocytes, appear to increase in proportion to one's cumulative lifetime exposure to sunlight. Melanomas, by contrast, are thought to be initiated by acute exposures, such as a bad sunburn in childhood. In essence, the cells designed to protect us from the chromosome-breaking effects of the sun are themselves damaged by an overdose of the very element they strive to shield us from. Decades later, another insult of some kind causes wild cell divisions within the damaged melanocyte to commence. A melanoma forms. A borderline is crossed. This second event may be more sunlight, but it may also include exposure to certain chemicals. Excess rates of melanoma are found in rubber and plastics workers, as well as in those employed in electronics and metal industries.

The accelerating incidence of melanoma means exposure to ultraviolet radiation is probably increasing. This could be happening for two reasons. First, more people are spending more time in the sun. Second, the sunlight to which we are exposed contains more ultraviolet rays. Since the 1974 discovery that earth's ultraviolet-shielding ozone layer is thinning, a growing group of physicians and climatologists have come to believe both forces are at work, especially in raising the risk for future melanomas. The U.S. Environmental Protection Agency (EPA) projects that tens of thousands of additional fatal skin cancers will result from the 5 percent loss of ozone that has already occurred in the stratosphere above North America. Individual behavior also plays a role. Melanomas have been on the rise for many decades—since Coco Chanel first popularized the suntan in the 1930s, according to some researchers. However, the *worldwide* increase in melanoma incidence points to a role for ecological factors. A recent study published in the *Journal of the American Academy of Dermatology* observes:

> Because of the worldwide increase in melanoma incidence, global factors need to be considered as potentially involved. Stratospheric ozone depletion, allowing more intense UV light to reach the earth's surface, may, in part, be responsible.

Ultraviolet —or UV—light is a strange energy. It is responsible for creating the very layer of ozone that defends us from it. As UV rays stream into the stratosphere, they cleave oxygen molecules in half, creating free atoms of oxygen, which then react with intact oxygen pairs to form little triangles called ozone. Ozone in turn can absorb UV rays, preventing their further passage down miles of deep air to the earth's surface. The ozone layer intercepts some, but not all, of the UV rays beamed at us from the sun.

Chemicals responsible for destroying the layer of ozone that resides twelve to thirty miles above us include the now notorious chlorofluorocarbons (CFCs). They are unlikely culprits. Not carcinogenic or even toxic, CFCs belong to a big family of synthetic, organic, chlorinated chemicals whose other members—DDT and

PCBs, for example—certainly are. But however harmless at ground level, a CFC molecule behaves quite differently when wind currents sweep it into the ozone layer. Ultraviolet rays split the CFC molecule apart, releasing a chlorine atom that quickly reacts with a molecule of ozone. The triangle of oxygens falls apart as the chlorine temporarily binds with one member of the triad. An unstable union, the chlorine atom soon shakes free of its oxygen partner and goes on to react with and destroy other ozone molecules. Before the chlorine is finally enveloped by a raindrop and redeposited on the earth's surface, it may break apart some 100,000 ozone molecules.

Less ozone allows more UV rays to beam their way through the atmosphere and down to us. Some will be halted by the veil of melanin spread across our skin for that purpose. But if its absorptive capacity is exceeded—which happens easily to the fair-skinned among us—some rays will penetrate further inside the skin cells until they are absorbed by the DNA strands themselves. If this occurs inside a melanocyte, the resulting genetic damage can place this cell on the pathway toward melanoma. In this way, noncarcinogenic CFCs contribute to rising cancer incidence by intensifying incoming sunlight, thereby making it more carcinogenic.

Non-Hodgkin's lymphoma strikes at another tissue designed to protect us from harmful invasions: the knobby lymph nodes clustered in our throats, armpits, groins, and elsewhere. Our tonsils, the most accessible example, represent a constellation of lymph nodes wrapped in a mucous membrane.

The watery fluid that fills the microscopic spaces between all of our cells is, for all intents, lymph. It does not receive that name, however, until it flows from those spaces, like rainwater from a field, into the creekbeds called lymphatic vessels. The origin of all this fluid is the bloodstream, and when held within that system, it is known as plasma. Each day, about three quarts of blood plasma leak out of the capillaries, swirl around freely, and then drain into the lymph vessels. Eventually, lymph becomes plasma again when it is poured back into blood just at the point where the jugular vein joins the subclavian in their return to the heart. Several tasks are accomplished during the ceaseless transformation of lymph to plasma and plasma to lymph.

The identification and destruction of foreign substances is one of them. Lymph nodes, scattered along the lymph vessels at various intervals, are honeycombed with a diverse array of cell types specialized for immune response. As the fluid is channeled through the nodes' intricate meshwork, alien life forms are trapped and killed. Lymph nodes can also send immune-responsive cells forth to circulate in other territories of the body.

Because the lymph system also serves as a highway for runaway cancer cells of all kinds, lymph nodes are a significant feature in the cancer landscape. Breast cancers very often spread to nearby lymph nodes, for example. Breast cancer patients are quickly categorized as node positive or node negative, a distinction that depends on whether breast cancer cells, shed from the orginal tumor, have lodged themselves in the lymph nodes beaded between the arm and the trunk of the body. Their presence there indicates the disease has likely dispersed to other, more distant locations.

As a way of measuring the extent of this cancer diaspora, node-positive women are further classified by the number of nodes containing breast cancer cells: 1 to 4 is one kind of identity; 11 to 17 is quite another. "How many nodes?" is very often the first question women in breast cancer support groups ask each other.

But a lymphoma is a different condition. In this case, the tumors derive from lymph tissue itself, not from immigrant cells that have floated in from someplace else. Lymphoma can arise inside a node, or, because lymph tissue is diffused throughout the body, it can originate almost anywhere elsewhere—in the spleen, for example, or even in the skin. Non-Hodgkin's lymphoma is therefore a collection of diseases, in contrast to the very specific and highly curable lymphoma called Hodgkin's disease.

While the incidence of Hodgkin's disease has declined modestly over the past two decades, non-Hodgkin's lymphoma has shot up—approximately tripling in incidence since 1950. This increase is evident in both sexes and within all age groups except the very young. Non-Hodgkin's lymphoma is also far less curable than Hodgkin's disease. Jackie Kennedy Onassis was killed by one of its most malignant incarnations.

AIDS has contributed to some, but not all, of the increase in non-Hodgkin's lymphoma. A small but significant percentage of AIDS patients are diagnosed with lymphoma, which for many causes death. However, the steady upward momentum of non-Hodgkin's lymphoma incidence in the United States was already under way decades before the AIDS epidemic sank its teeth in.

Lymphomas do seem to be consistently associated with exposure to synthetic chemicals, especially a class of pesticides known as phenoxy herbicides. These synthetic chemicals were born in 1942 as part of a never-implemented plan by the U.S. military to destroy rice fields in Japan. The most famous phenoxy is a mixture of two chemicals, 2,4,5-trichlorophenoxyacetic acid (2,4,5-T) and 2,4-dichlorophenoxyacetic acid (2,4-D). This combination is called Agent Orange, and it was used between 1962 and 1970 by U.S. troops to clear brush, destroy crops, and defoliate rainforests in Vietnam. The military career of phenoxy herbicides was thus revived.

Linked to miscarriages and contaminated with dioxin, 2,4,5-T was eventually outlawed. By contrast, 2,4-D went on to become one of the most popular weed killers in lawns, gardens, and golf courses, as well as in farm fields and timber stands. It has been marketed under a schizophrenic collection of trade names: Ded-Weed, Lawn-Keep, Weedone, Plantgard, Miracle, Demise.

Evidence for an association between phenoxy herbicides and non-Hodgkin's lymphoma comes from several corners. Vietnam veterans have high rates of non-Hodgkin's lymphoma. So do farmers in Canada, Kansas, and Nebraska who use 2,4-D. Studies show that the risk of lymphoma to farmers rises with the number of days per year of use, the number of acres sprayed, and the length of time they wear their "application garments" before changing clothes. In Sweden, exposure to phenoxy herbicides was shown to raise one's risk of contracting lymphomas sixfold. In a comprehensive review of the topic, the National Cancer Institute scientists Sheila Hoar Zahm and Aaron Blair concluded:

> NHL is associated with pesticide use, particularly phenoxy herbicides. Exposure to phenoxy herbicides is widespread in the agricultural and general populations. The use has increased dramatically preceding and during the time period in which the

> incidence of NHL has increased, which could explain at least part of the rising incidence.

Similarly, an 812-page study of herbicide-exposed Vietnam veterans conducted by the Institute of Medicine offered the following terse opinion:

> Evidence is sufficient to conclude that there is positive association. That is, a positive association has been observed between herbicides and the outcome [non-Hodgkin's lymphoma] in studies in which chance, bias, and confounding could be ruled out with reasonable confidence.

Dogs also acquire lymphoma. One recent study showed that pet dogs living in households whose lawns were treated with 2,4-D were significantly more likely to be diagnosed with canine lymphoma than dogs whose owners did not use weed killers. Risk rose with number of applications: the incidence of lymphoma doubled among pet dogs whose owners applied lawn chemicals at least four times per year.

From jungle warfare to suburban dandelions: our ongoing war against plants is now waged on a domestic grid of tiny battlefields. One in ten single-family American households now uses commercial lawn care services, and one in five applies the chemicals themselves. The evidence linking phenoxy compounds to non-Hodgkin's lymphoma is preliminary. No one knows exactly how traces of weed killer find their way into our extracellular fluid as it is funneled back and forth between blood and lymph. Absorption through the skin is considered the most likely route of exposure. No one has explicated the exact mechanism by which these chemicals might alter the cells inside the far-flung network of nodes, canals, and lymph tissues-at-large and thereby set the stage for a lymphoma. No one knows whether phenoxys require interactions with other agents to work their damage nor what proportion of the current rise in lymphoma might be attributed to phenoxy exposure.

Most of us are probably far less exposed to phenoxy herbicides than soldiers, farmers, or even our own beloved dogs who use our lawns for their bedrooms. Nevertheless, the presence of disease in these specific groups is a clue to which we, as readers of a complicated mystery, need to pay attention when trying to determine why non-Hodgkin's lymphoma casts an ever-longer shadow among us all.

Bone marrow is the mother of lymphocytes, the immune cells that inhabit the lymph nodes. Multiple myeloma is cancer of the cells inside the bone marrow that give rise to a particular type of lymphocyte called plasma cells. Its main symptom is horrible pain. As the tumors grow, blood, lymph, and bone marrow are filled with an excess of abnormal plasma cells, which then churn out an excess of abnormal antibodies. The bones themselves, riddled with lesions, begin to fracture. Calcium spills into the bloodstream. Although multiple myeloma is thought to begin with a single mutation in a single cell, tumors created by aggregations of plasma cells are usually diffusely present throughout the bone marrow by the time of diagnosis. The skull is often severely affected.

As with non-Hodgkin's lymphoma, the incidence of multiple myeloma in the United States has approximately tripled since 1950, and the mortality rate is not far behind. As a sign of its rise from obscurity, some cancer newsletters now run announcements for multiple myeloma support groups. (In San Francisco, the third Saturday of each month at the Women's Cancer Resource Center is multiple myeloma day.)

There is much less to say about myeloma than about the other contenders for Most Swiftly Moving Cancer. Much less is known. It tends to stalk the elderly, and blacks are at higher risks than whites—but for unclear reasons. Because the presence of abnormal antibodies in the blood and urine provides a definitive diagnosis, the registration of multiple myeloma is considered to be particularly accurate.

Exposure to ionizing radiation is recognized as one probable cause. U.S. radiologists exhibit excess rates of multiple myeloma, as do survivors of the 1945 atomic bomb blasts in Japan. Some evidence suggests that workers in the nuclear industry also have increased risks for myeloma.

In contrast to solar radiation, such as UV light, ionizing radiation has energy sufficient to penetrate the skin's surface, stream through the soft tissues, and in some cases, enter the bones themselves. Released when atoms are split, ionizing radiation is so called because it alters the molecules through which it passes, knocking away their electrons and creating electrically charged particles, or ions. Because of this property, ionizing radiation is classified as a known human carcinogen at any exposure level. When the atomic modifications induced by radiation involve molecules of DNA, tucked as they are into the nucleus of every cell, cancer-inducing mutations can result. Alternatively, radiation can create ions of surrounding atoms, which then bind with DNA to create mutations. In either case, our chromosomes are equipped with DNA repair mechanisms designed to detect and correct such problems, but it is a system that can be overwhelmed and overpowered. Multiple myeloma is one of the most recent cancers to be linked with exposure to radiation; the lag time between exposure and diagnosis is much longer than that for other cancers caused by irradiation of the bone marrow.

Multiple myeloma is also associated with exposure to a variety of chemicals—metals, rubber, paint, industrial solvents, and petroleum. Farmers and agricultural workers exposed to pesticides and herbicides have higher rates of multiple myeloma than the general population. Multiple myeloma is on the rise in all major industrialized countries. But the parallel increase among both sexes argues against a purely occupational cause. According to one researcher who has examined multinational mortality trends, the patterns of multiple myeloma among generational cohorts suggest a general environmental exposure of some kind, common to all industrialized countries, which would have begun increasing at the turn of the twentieth century.

Other researchers urge an investigation of one very specific industrial chemical: benzene. Consisting of a simple ring of six carbon atoms, benzene is used as a solvent in which other petrochemicals are dissolved, as an additive to gasoline, and as a raw material for the creation of synthetic materials including certain foams, plastics, and pesticides. It is a ubiquitous pollutant of outdoor and indoor air and a common contaminant of drinking water. Benzene can pass through the waterproofed layer of our skin and thus seep into blood upon direct contact; it also evaporates quickly and can be easily inhaled.

Benzene is a suspect in myeloma because it is a known offender in a related crime, namely, leukemia. A proven bone marrow toxin, benzene alters the cells of the marrow that give rise to leukocytes, or white blood cells. Could the same toxin also preside over alterations in the marrow's production of plasma cells? According to the U.S.

Agency for Toxic Substances and Disease Registry, "Although this is plausible, no scientific proof of a causal relationship exists." The question thus becomes, Is anyone looking?

Bone marrow. Lymph nodes. Skin. From the body's dark tunnels to its sunlit surface, cancers of all kinds are presenting themselves with increasing frequency. Melanoma, lymphoma, and multiple myeloma are simply traveling at especially high velocities.

A month before her death, Jeannie initiated a massive housecleaning project. She reorganized all her files, returned books, gave away clothing. Waiting for me on her kitchen table one morning was a stack of medical papers, department of public health reports, press releases, and newspaper clippings. They were her collection of articles about the cluster of cancer cases in southeastern Massachusetts, where she grew up.

"I thought you might want them for your research."

"You don't want to keep these?"

"You take them."

Eighteen months after Jeannie's death, I finally read them—prompted by the release of a new study confirming the patterns documented by the previous ones. Jeannie's cancer is not included in any of these these studies, which concern sharply rising leukemia rates in five neighboring towns during the 1980s and their possible relationship to documented radioactive releases at the Pilgrim nuclear power plant—the result of a fuel rod problem—ten years earlier. While no firm cause-and-effect relationship has been established, meteorological data indicate that coastal winds may have trapped the airborne radioactive isotopes and recycled them within a five-town area. "Individuals with the highest potential for exposure to Pilgrim emissions . . . had almost four times the risk of leukemia as compared with those having the lowest potential for exposure."

Although one of the towns is her own, Jeannie's cancer was far too rare for the case-control comparisons made here. Her cancer has no known cause, and cancer registries do not track its incidence. I will not find her here.

FOUR

space

Pekin is the judicial seat of Tazewell County, Illinois. It is situated across the Illinois River and a few miles downstream from Peoria. Just outside of Pekin's city limits, about two miles west of the house I grew up in, is the unincorporated subdivision of Normandale. The community was created in 1926 to provide housing for factory workers, and its streets are named for the original prewar products that the residents who slept here at night toiled by day to create: Karo Street (after the syrup), Quaker Street (after the paper mill's round oatmeal boxes), Fleischmann Street (after the yeast).

Normandale is home to 480 people, a popular supper club, a beautiful brick church, and a root-beer stand where I hung out with my best friend in the summers after we learned to drive. Eating onion rings in her father's car, Gail Williamson and I debated the merits of German versus Latin, big universities versus small colleges, sex versus celibacy. In this parking lot, we decided to settle for nothing short of everything. Gail would go to medical school *and* play the violin. I would go to graduate school *and* write poetry. Our present and future boyfriends would just have to understand. Of course. And they also would have to play the guitar.

Normandale is situated on a triangular wedge of land near Dead Lake, a dumping pond for industrial wastes near the river's east bank. It is flanked on two sides by industry: a foundry, a grain-processing plant, a couple of chemical companies, a coal-burning power plant, and an ethanol distillery. Its third side is bounded by a landfill that operated without state permits until the Illinois Pollution Control Board shut it down in 1988. Mysterious dumpings still go on. Twenty rusted barrels leaking an unknown tarry substance were discovered recently along the blacktop just south of town. This is also Normandale.

The distribution of cancer across space, like its trajectory through time, reveals key clues about its possible causes. For example, if ethnicity played a major role in determining cancer risk, then immigrants should retain the cancer incidence of their homelands. Conversely, if the cancer rates of immigrants come to approximate those of their host country (and this is, in fact, the case), then we have good reason to suspect that environmental agents are at work. If cancer rates are elevated in certain geographic areas—within cities, for example, or in areas of intensive agriculture—we have further leads to pursue. If high rates of cancer follow the course of a river or the path of the prevailing wind or are clustered around a drinking-water well or a certain industrial site, then we have very strong clues indeed.

Paradoxically, the closer we stare at the map of cancer, the more unclear the picture becomes. On the largest scale, when cancer registry data from many nations are pooled and we are looking across whole continents, distinct areas of high and low cancer rates are clearly visible. As we narrow our view to one regional area—a single county or town or, like Normandale, a particular subdivision within a town—our power to discriminate differences decreases. Recall that cancer rates are based on the number of people annually diagnosed for each 100,000 people. Determining whether a cancer cluster exists in small community of only a few thousand or a few hundred inhabitants is statistically difficult work, and it is at this level where the fiercest arguments fly.

At a global level, fewer arguments arise. The spatial features of cancer's occurrence around the globe clearly belie the notion that cancer is a random misfortune. Industrialized countries have disproportionately more cancers than countries with little or no industry (after adjusting for age and population size). One-half of all the world's cancers occur among people living in industrialized countries, even though we are only one-fifth of the world's population. Closely tracking industrialization are breast cancer rates, which are highest in North America and northern Europe, intermediate in southern Europe and Latin America, and lowest in Asia and Africa. Breast cancer rates are thirty times higher in the United States than in parts of Africa, for example. Breast cancer incidence in the United States is five times higher than it is in Japan, but this gap is rapidly narrowing. Of all the world's nations, Japan has the most rapidly rising rate of breast cancer.

Among the nations of the developed world, similar time trends are in motion for a number of major cancers. Mortality rates of breast cancer and prostate cancer are rising in almost all industrialized countries. The accelerating U.S. rates of brain cancer, kidney cancer, multiple myeloma, non-Hodgkin's lymphoma, and melanoma are replicated in France, West Germany, England, Japan, and Italy. Increased access to health care, improved diagnostic techniques, and a greater cultural willingness to write the word *cancer* on a death certificate account for some, but not all, of these increases.

Especially alarming is the rapid rise of brain cancer mortality among those over sixty-five years old—a pattern mirrored throughout the industrialized world. The brain is a tricky site to be accurate about because, as with the lymph nodes, so many other types of cancers arrive there from other places. Breast cancer patients can have tumors inside their skulls, for example, but these are not *brain* cancers insofar as the tumors are derived from breast, and not brain, tissue. Medical imaging technology introduced in the 1980s profoundly altered the ways in which brain cancers are diagnosed. However, the greatest rise in brain cancer occurred well before this time and has continued to increase, even though use of the new methodologies has leveled off. Moreover, brain cancer incidence is also rising in countries where enthusiasm or budgets for these new imaging technologies are much less.

These kinds of international comparisons are made possible because of a little-known office of the World Health Organization. Located in Lyon, France, the International Agency for Research on Cancer is charged with the daunting job of monitoring cancer incidence around the world. It does so by collecting registry data from as many countries as possible. The United States, for example, sends its Surveillance, Epidemiology, and End Results Program data on to Lyon. The World Health Organization also collects and analyzes cancer mortality data gleaned from death certificates in seventy different countries. From these data, the organization concluded that at least 80 percent of all cancer is attributable to environmental influences.

This is a stunning statistic—and not one often cited in popular cancer literature in the United States. It was derived by subtracting the rates of the countries with the least cancer from the rates of the countries with the most. The lowest rates are presumed to represent the basement level for cancer—tumors caused by spontaneous mutations, heredity, cosmic radiation, or other factors impossible to avoid. The highest rates represent all these cancers plus the ones attributable to a consortium of extrinsic causes, including tobacco smoke. The difference between these two numbers is thought to represent the environment's contribution to cancer.

A number of researchers have asked what *environment* really means in this context. Certainly, it does *not* mean what ecologists use the word to mean. To an ecologist, the term specifically refers to the physical world in which an organism lives; everything outside an animal's own skin is its environment. To a geneticist for whom the chromosome is the object of interest, anything beyond the cell membrane is the outside world. Circulating sex hormones, vitamins, fat globules, caffeine molecules, viruses—any element not a gene is part of the environment.

These are not contradictory definitions. What we drink, inhale, and find to eat in the environment external to ourselves quickly becomes our internal environment. Definitions do become relevant, however, when we ponder what specific extrinsic causes might be behind these 80 percent of all cancers. The traditional answer is that they include both environment and lifestyle. In this dichotomy, *environment* is used to label everything we interact with or consume that is not freely chosen and *lifestyle* to refer to that which we choose to consume: breathing air as opposed to eating dessert, or drinking water as opposed to dipping snuff.

Which is more significant in explaining geographical differences in cancer rates? Here is where the waters grow murky. Our diets are—at least for some of us—a freely chosen aspect of our lifestyle. On the other hand, many environmental carcinogens, such as PCBs and pesticide residues, wind up in the food we eat. These are not freely chosen. Therefore, our diets have one foot each in lifestyle and environment. Coffee consumption, at first glance, seems like a classic lifestyle choice. But the coffee we drink includes the water we pour through the beans—and this may be the same water used for showering and cooking food. If our tap water contains, say, traces of weed killer and dry-cleaning fluids, we are being exposed to environmental carcinogens through multiple pathways and through no individual choice of our own, even as we freely determine our own bathing, cooking, and coffee-drinking habits.

However culturally distinct immigrants may remain in their adopted country, their cancer rates assimilate. According to the International Agency for Research on Cancer, "The most important single conclusion to derive from migrant studies is that, for a group as a whole, it is the new 'environment' that determines cancer risk and not the genetic component associated with the ethnic stock of the migrants." The quotation marks around that slippery word *environment* acknowledge its many meanings.

Migrants to Australia, Canada, Israel, and the United States all illustrate this pattern. Consider Jewish women who migrate from North Africa, where breast cancer is rare, to Israel, a nation with high incidence. Initially, their breast cancer risk is one-half that of their Israeli counterparts. But risk rises rapidly with duration of stay: within thirty years, African-born and Israeli-born Jews show identical breast cancer rates. Jewish women from the Middle East and Asia also increase their risk of breast cancer upon arrival in Israel, although the pace at which they do so is considerably slower.

Likewise, in the United States, the breast cancer rates of European, Chinese, and Japanese women immigrants all eventually rise to conform to the U.S. rate, but they do so at different speeds. Polish women assume U.S. rates of breast cancer quickly. Japanese women migrating to the U.S. mainland require two generations to achieve our breast cancer rate. First-generation Japanese immigrants show a rate intermediate between that for Japan and the United States; their daughters, however, reflect the U.S. rates completely.

Happily, the reverse is also true. Women moving to a new country with lower breast cancer rates experience a decline in their chances of contracting the disease—as when, for example, an English woman immigrates to Australia.

These results lead us back to the Möbius strip of lifestyle and environment. Both change simultaneously when someone moves from one part of the world to another. At present, no one understands precisely how these changes interact to create the patterns described above.

By 1991, half the homes on Karo Street had a cancer patient residing there. It also seemed to some residents that Normandale's children were unusually susceptible to eye and ear infections. In one neighborhood, fourteen residents were diagnosed with cancer over a ten-year period. These numbers were calculated by the people themselves and presented to the health department and the local newspaper. A citizens' group was organized and a letter dispatched to the Tazewell County Health Department requesting an investigation of cancer incidence in their community.

Those quoted in the newspaper mentioned neighbors who died of cancer, as well as those who had moved away out of fear of it.

"Oh, but we've lost so many," said one.

Because we have no nationwide cancer registry, we also have no definitive geography of cancer incidence in the United States. The National Cancer Institute has, however, published the two-volume *Atlas of U.S. Cancer Mortality among Whites and Nonwhites: 1950–1980*—in essence, cancer death maps. Areas of high mortality are colored in scarlet, fuchsia, and orange; areas of unusually low mortality are painted deep blue. At first glance, they look like pieces of a jigsaw puzzle.

Death from cancer is not randomly distributed in the United States. Shades of red consistently light up the northeast coast, the Great Lakes area, and the mouth of the Mississippi River. For all cancers combined, these are the areas of highest mortality; they are also the areas of the most intense industrial activity. The trend maps show that *rates* of increase, on the other hand, are actually higher in the parts of the country with lower mortality, indicating that cancer deaths are tending to become more geographically uniform as time passes, possibly due to the growing urbanization of formerly rural areas, the increasing mobility of the population, and the rising use of pesticides. In one 1988 study, researchers found significant associations between agricultural chemical use and cancer mortality in 1,497 U.S. rural counties.

A mapping project carried out by researchers at the public service organization Public Data Access, Inc., confirms these patterns. These investigators not only charted county mortality data but also created an atlas of various environmental measures, such as hazardous waste incinerators, pesticide usage, and workplace toxins. Deepening shades of gray indicate counties with increasing levels of contamination. Investigators found a close overlap between cancer mortality and environmental contamination. Concentrations of industrial toxins were higher in the top-ranked cancer counties than in the rest of the country.

When examining either of these atlases, a reader must keep in mind that these maps display cancer *deaths*, not cancer diagnoses. Counties with higher levels of contamination may also have worse health care: cancer patients from more polluted, more pesticide-saturated counties may be dying at faster rates simply because they are receiving poorer treatment. On the other hand, the rates of death from other causes—such as cardiovascular or infectious diseases—are not as closely linked to environmental contamination as cancer is. Disparities in health care, then, cannot account for all of the differences in the geographic distribution of cancer deaths.

The patterns of certain cancers depart from the overall picture. The death rates from multiple myeloma are highest not along industrial corridors but in rural farming areas. Also found in the central agricultural region of the United States are high rates of leukemia and lymphoma. All three of these cancers have been associated with pesticide exposure. In contrast, southeastern and south-central regions predominate in the maps of melanoma, a pattern consistent with increasing exposure to sunlight.

Deaths from breast cancer also follow a north–south gradient, but with rates higher in the North. Rates are especially high in heavily industrialized areas of the Northeast. These regional differences, however, have been lessening over time. Today, breast cancer mortality is rising most quickly in areas of the South, including Appalachia.

Another way of mapping cancer is to examine how it distributes itself among people of various occupations. Just as cancer is not scattered uniformly across the physical landscape, neither does it afflict with an even hand the landscape of work. Understanding occupational cancers is important not only because people spend so many hours of their lives in the workplace but also because it yields critical clues about cancers beyond the factory wall and the office door. Released into air or water, hauled away as toxic waste, or mixed into consumer products, most cancer-causing agents in the workplace ultimately become part of the general environment in which we all live. Workplace carcinogens are largely identical to those agents that cause cancer in the general population. Indeed, the majority of substances now classified as known human carcinogens by the International Agency for Research on Cancer were first identified in studies of workers. In spite of the crucial role that workers have played in our understanding of the disease, cancer registries in the United States are not funded to collect occupational histories. Therefore, we are once again left to work with mortality data derived from death certificates rather than registry-derived incidence data when taking up the question of cancer and work.

People in at least sixty different occupations have elevated death rates from cancer. One of these is farming.

Farmers from industrialized countries around the world exhibit consistently higher rates of many of the same cancers that are also on the rise among the general population. Farmers, in other words, die more often from the same types of tumors that are also afflicting, with increasing frequency, the rest of us. These include multiple myeloma, melanoma, and prostate cancer. Farmers also suffer from rates of non-Hodgkin's lymphoma and brain cancers higher than those of the general population—although these excesses are more modest. In spite of lower overall mortality and lower rates of heart disease, farmers also die significantly more often than the general public from Hodgkin's disease, leukemia, and cancers of the lip and stomach. Likewise, migrant farmworkers suffer excess rates of multiple myeloma, as well as of stomach, prostate, and testicular cancers.

Elevated cancer rates are also found among painters, welders, asbestos workers, plastics manufacturers, dye and fabric makers, firefighters, miners, printers, and radiation workers. People who work in a number of so-called professional jobs are also at higher risk: for example, chemists, chemical engineers, dentists and dental assistants, and—perhaps most ironically—chemotherapy nurses. Many of the chemicals used to treat cancer are themselves carcinogenic, as the high rate of adult cancers among childhood leukemia survivors attests. Thus, we should not be surprised that those who work daily with these substances in an attempt to save others' lives themselves succumb in numbers higher than average.

The children of adults who work in specific occupations also have higher rates of cancer. Childhood brain cancers and leukemias are consistently associated with parental exposure to paint, petroleum products, solvents, and pesticides. Some exposures may occur before birth. Children can also be exposed when these materials are carried into the home on their parents' clothes and shoes, through breast milk (which can be contaminated directly or through maternal contact with the father's clothing), or even through exhaled air: because solvents are, in part, cleared by the lungs, parents can expose their children to carcinogens simply by breathing on them. In this way, a father's homecoming kiss and work-clothed embrace can contaminate his child.

Another clue that exposures to carcinogens in the workplace may be significantly affecting our health is the fact that cancer mortality is not evenly balanced between the sexes. In many industrialized countries and for many types of cancers, the death rate is rising faster among men than among women. (The exception is U.S. deaths from lung cancer.) Because proportionally more men have worked in jobs that are known to entail exposure to chemical carcinogens, some researchers contend that occupational exposures override lifestyle factors in contributing to cancer risk. Because mortality rates are at least several years old when they become available and because cancer itself requires ten to thirty years to develop and its victims usually require several more years to die after the diagnosis is made, today's mortality data probably reflect occupational exposures during the mid-1900s—a time when gender divisions among job types were more rigid. Certainly, a subgroup of traditionally female jobs have also been associated with certain kinds of cancer. Studies show higher rates of bladder and salivary gland cancers among hairdressers, for example.

As more women move into the workforce and obtain jobs traditionally held by men, these differences may fade. In 1993, an international conference on women's occupational cancers attempted to synthesize what is known about cancer rates among working women of various professions. Researchers found, for example, elevated rates of pancreatic cancer among women autoworkers assigned to the paint, plastic, and trim departments.

Researchers also discovered that old enemies sometimes reveal new dangers when women workers are the objects of study.

Consider vinyl chloride, which is used in the manufacture of a substance familiar to us all: polyvinyl chloride, otherwise known as PVC or simply vinyl. Credit cards are made of PVC, as are garden hoses, lawn furniture, floor coverings, children's toys, and food-packaging materials. PVC, in turn, is made of many vinyl chloride molecules all bonded together. Vinyl chloride, a sweet-smelling gas at room temperature, has long been classified as a known human carcinogen. Its cancer-causing properties were discovered when high numbers of male vinyl chloride workers began contracting angiosarcoma, a rare cancer that causes tumors to grow inside the liver's blood vessels. The incidence among vinyl chloride workers was found to be three thousand times higher than among the general population. Animal studies, as well as further studies of male workers, also revealed the ability of vinyl chloride to contribute to lung and brain cancers. In response to these results, allowable workplace air levels of vinyl chloride were drastically reduced. But it was not until researchers also studied female workers that vinyl chloride's potential as a breast carcinogen was uncovered. In a 1977 study, women who breathed vinyl chloride vapors on the job had elevated death rates from breast cancer. Subsequent laboratory studies showed that atmospheric vinyl chloride triggers breast tumors in female rats, even at the lowest dosages; so does ingestion of PVC dust. Such an association is certainly biologically plausible, since vinyl chloride has an affinity for fat tissue.

Evidence for a link between vinyl chloride and breast cancer in women workers has broad implications for the rest of us. While vinyl chloride levels are very much lower outside the factory, significant exposures can occur among residents living near vinyl chloride and PVC facilities. The air currents that blow across hazardous waste sites also contain elevated levels of vinyl chloride. Vinyl chloride is a frequent contaminant of groundwater, where it can remain for months or years because there is no pathway to the atmosphere. The flesh of freshwater fish can also contain vinyl chloride.

According to the U.S. Agency for Toxic Substances and Disease Registry (ATSDR), each of these pathways exposes the general public to "neglible amounts" of this known carcinogen. However, no one knows what the cumulative lifetime risk from all of these neglible exposures is. The ATSDR also states that "exposure to vinyl chloride either in the prenatal period or during early childhood years may result in an increased risk of cancer" later in life. If vinyl chloride caused only a very rare form of liver cancer, perhaps these multiple routes of tiny exposures would be less cause for alarm. However, breast cancer is now the leading cause of death of American women aged thirty-five to fifty, and we are the first generation of women born after World

War II, when chlorinated chemicals such as vinyl chloride were first widely dispersed in the general environment.

In spite of all this preliminary evidence, no comprehensive study has ever been undertaken to examine vinyl chloride's contribution to breast cancer. In fact, the 1977 study of women PVC fabricators has never been followed up, even though cohorts of men exposed to vinyl chloride, and who demonstrate excesses of brain, liver, and lung cancer, have been periodically updated. This omission is especially frustrating to Peter Infante, the director of the Health Standards Program at the Occupational Safety and Health Administration, whose job it is to set limits on vinyl chloride levels in workplace air. Lack of interest in investigating a possible vinyl chloride–breast cancer link, says Infante, serves as an example of indifference to the plight of women in the workplace—indeed, to the plight of women everywhere.

The study of vinyl chloride in the workplace has one other lesson for us. Studies of male workers show that vinyl chloride exposure, even when it does not cause angiosarcoma, can cause noncancerous growths to form inside the liver, a result of scarring. These lesions can become malignant if the person is subsequently exposed to ethanol. In other words, drinking alcoholic beverages can cause a chemically induced, noncancerous growth to turn into liver cancer. In such a case, which kind of cancer is this: one caused by lifestyle or by the environment?

In response to the questions raised by the people of Normandale, two health studies were quickly conducted—one by the state health department and one by the county. Neither involved mapping disease patterns, identifying pollution sources, estimating actual exposures, locating those who had moved away, or, for those who had died, interviewing their next-of-kin. No blood, urine, or fat samples were collected to test for the presence of contaminants. In fact, the study design did not require public health officials to even set foot in Normandale.

In the first study, the Illinois Department of Public Health pulled up from its computerized cancer registry banks all the cancers diagnosed in Pekin's ZIP code area—as reported to the Illinois Cancer Registry between 1986 and 1989. From these data, researchers calculated an *actual* cancer rate for the whole town. Based on the statewide rates, researchers then generated an *expected* number of cancer cases for a hypothetical town the size of Pekin. Cancers were categorized by location in the body (colon, ovaries, breast, and so forth), the actual numbers were compared to the expected numbers, and . . . no statistically significant differences were found.

On December 19, 1991, the headline in the *Pekin Daily Times* read, STUDY: AREA CANCER RATES NORMAL.

❧

If cancer-causing chemicals in the environment play a significant role in actually causing cancer, then we should expect to find high rates of the disease in areas where carcinogens are highly concentrated. The industrial workplace, where such chemicals are manufactured or used, is one such area. Hazardous waste sites, where such chemicals are dumped, are another.

Although most studies are considered preliminary, links between the presence of hazardous waste and cancer in the surrounding community have indeed been documented. Corroborating the findings from a major study by the National Research Council, the ATSDR conceded in their 1991–92 report to Congress that "a variety of types of cancers are encountered more often than would be expected in human populations exposed to contaminants at waste sites." The ATSDR also reported that the blood, urine, and body tissues of many people living near such sites contain detectable residues of many of the chemicals found at the sites, including suspected carcinogens such as PCBs, chlordane, and other chemicals banned for use. However, the ATSDR concluded, "the extent of human exposure to these and other substances for all unplanned releases is, for the most part, not known."

Quite a few of us are included in these potentially exposed populations. As of 1990, the U.S. Environmental Protection Agency (EPA) had counted 32,645 sites of past chemical waste dumping in need of cleanup. Some of these are actual hazardous waste landfills, but many are former manufacturing sites where drums full of chemicals have simply been abandoned. The names of the most notorious appear on the EPA's National Priorities List. These are the so-called Superfund sites, and in 1996 there were 1,430 of them. In 1991, the National Research Council estimated that forty million people (about one of every six Americans) lives within four miles of one of them. (One in every sixty Americans lives within one mile.) Currently, Illinois is home to thirty-eight Superfund sites.

Most of these sites did not exist before the end of World War II, when most plastics, solvents, detergents, pesticides—and all the unwanted by-products of their manufacture—made their debut on the planet. Poor and dispossessed children have lived cheek by jowl with carcinogenic wastes ever since soot-encrusted chimney sweeps in eighteenth-century England were discovered to be at high risk for scrotal cancer. But those of us born after World War II are the first generation to grow up in such large numbers near such large amounts and diverse assortments of manufactured chemical refuse. Since the late 1950s, more than 750 million tons of toxic chemical wastes have been discarded. That their cancer-causing properties were not always known at the time they were dumped near our playgrounds, school yards, parks, and neighborhoods does not diminish our collective risk.

What of the federal programs designed to safeguard human health in the communities inhabited by these hazardous waste sites? Are they not effective? Charged with answering this question, the National Research Council recently gave up in frustration:

> Based on its review of the published literature on the subject, the committee finds that the question cannot be answered. Although billions of dollars have been spent in the U.S., an insignificant portion has been devoted to evaluate the attendant health risks. This has resulted in an inadequate amount of information about the connection between cause and effect.

In essence, conclusive evidence does not exist on the link between cancer and hazardous waste because money has never been appropriated to conduct the necessary studies. As long as the evidence remains inconclusive, the methods currently used to remediate risks to public health cannot be evaluated. With this evaluation, the notion that there is no proof environmental contamination causes cancer can continue to enjoy common currency.

Despite the miasma of uncertainty hanging over the whole subject, several large studies have detected elevated cancer rates around hazardous waste sites. One of them was conducted in New Jersey, a petite state with an astonishing 112 Superfund sites. Researchers asked whether cancer mortality was associated with environmental factors of various kinds, including the location of toxic waste dumps. Their results showed that communities near toxic waste sites had significantly elevated mortality from stomach and colon cancers. Additionally, in twenty-one different New Jersey counties, breast cancer mortality among white women rose as the distance from residence to dump site shrank. However, many of the clusters of excess cancer occurred in heavily industrialized counties so that air pollution from these sources confounded the results. Thus, a woman with breast cancer in northeastern New Jersey cannot know with certainty whether she is dying because of the air wafting down from the factory stacks or because of the water contaminated by the dump site.

In another large study, researchers scoured the United States for counties that met two criteria: first, their hazardous waste sites had contaminated the groundwater and, second, this groundwater served as the sole source of drinking water for the residents. Meeting these qualifications were 593 waste sites in 339 counties in forty-nine states. Next, researchers obtained for each of these 339 counties ten years' worth of cancer mortality data and compared them to cancer mortality data from counties without hazardous waste sites.

Here are the results: Men living in hazardous waste counties suffered significantly higher mortality from cancers of the lung, bladder, esophagus, colon, and stomach than did their contemporaries residing in counties without such sites. Women living in hazardous waste counties suffered significantly higher mortality from lung, breast, bladder, colon, and stomach cancers. Indeed, counties with hazardous waste sites were 6.5 times more likely to have elevated breast cancer rates than counties without such sites.

Researchers with the Public Data Access project corroborated these results. Looking specifically at breast cancer, they found that mortality rates at the county level were significantly correlated with Superfund sites. Counties with the highest breast cancer mortality had four times as many facilities that treated and stored toxic waste than the national average.

Once again, studies such as these two are considered preliminary rather than definitive because possible confounding factors could not be controlled. These include the possibility that residents living in counties with hazardous waste facilities are getting more cancers not because of the dumps but because they work for the companies that create the waste or because they smoke more and drink harder.

Among other things, the term *ecological fallacy* refers to the temptation of assuming that all associations are causative when one examines statistical patterns. My statistics professor was fond of telling the story of the boy and the department store escalator: The boy wondered what caused the escalator to move. After hours of observation, he concluded the escalator ran on the energy generated by the revolving door, because when the door ceased turning at the close of the day, the escalator stopped.

Ecological fallacy became a real issue for me when I started work as a field biologist. In Minnesota, I wanted to know why pine trees were failing to reproduce. Absence of new seedlings was correlated with high population levels of deer—but also with low frequency of forest fires and high populations of hazel shrubs. Which, if any, was the root cause of the problem and which were the confounders? Or, if fire, hazel, and deer all conspired to contribute to the demise of the pines, how exactly did they do so? Once I had established the pattern, I needed to design experiments that would uncover causal mechanisms. I found this work very exciting.

But as a woman with cancer who grew up in a county with fifteen hazardous waste sites, several carcinogen-emitting industries, and public water wells that, from time to time, show detectable levels of toxic chemicals, I am less concerned about whether the cancer in my community is more directly connected to the dump sites, the air emissions, the occupational exposures, or the drinking water. I am more concerned that the uncertainty over details is being used to call into doubt the fact that profound connections do exist between human health and the environment. I am more concerned that uncertainty is too often parlayed into an excuse to do nothing until more research can be conducted. "'We need more study' is the grandfather of all arguments for taking no action," says Peter Infante, who, in his daily struggle to set limits on workplace exposure to carcinogens, hears them all.

❧

By 1991, I am living a long way from Normandale. My sister still lives nearby.

"What's the latest?" I ask into the phone.

"People are worried about their dogs over there. They say there's a problem with cancer in pets. One man has a German shepherd with breast cancer."

I call an old high school teacher who has served a long term on the city council. The questions raised in Normandale have him thinking about other issues, such as emissions from the hospital incinerator and diesel exhaust from the produce trucks that rumble through town after the harvest. I ask him about the results of the Normandale investigation.

"The study found the cancer was due to chance."

"What do you think?"

"Perhaps it's more than chance."

❧

Epidemiologists investigate patterns of diseases in human populations. They look at the world through a wide-angle lens. While the focus of medicine is the treatment and prevention of diseases in individuals, epidemiology attempts to explain and prevent the occurrence of disease in large groups.

So far, I have presented the findings of what epidemiologists call ecological studies. In these, investigators compare the frequency of a given disease (e.g., cancer) in populations that differ in some factor of interest (e.g., the presence or absence of a leaking hazardous waste site). Statistics are then used to determine whether the frequency of disease is significantly different in the two types of communities. Researchers can often complete ecological studies without ever talking directly to any of the human subjects or assessing their exposure levels to the contaminants in question. The studies in Pekin and Normandale were ecological studies. As strange as it sounds to ecologists, the word *ecological* is used by epidemiologists simply to mean a descriptive, rather than an analytical, approach. Ecological studies, like circumstantial evidence, provide the weakest demonstration of proof.

Included in epidemiology's analytical category are two basic study designs. One is the case-control study. Here, a group of diseased people are identified (the cases) and compared to a group of people drawn from the larger population (the controls). The point of comparison is their exposure to possible disease-causing agents. Mary Wolff's study of DDT and breast cancer, discussed in Chapter One, is an example. Her cases were women with breast cancer; her controls were women without breast cancer (matched for age, menopausal status, and other variables of personal history); exposures were assessed by measuring blood levels of DDT and PCBs. Her results showed that women with breast cancer had significantly higher DDT levels than women without breast cancer.

Closely related to the case-control study is the cohort study, in which people are classified as exposed or unexposed and are followed through time until disease or death occurs. (This can often be done retrospectively.) In this way, we compare the rate of disease in people known to be exposed to a possible carcinogen to disease rates in unexposed persons. The difference between them is known as relative risk. Peter Infante's frustration with the lack of knowledge about vinyl chloride and breast cancer stems from a need for follow-up cohort studies. If a cohort of women exposed to vinyl chloride on the job had been followed for a period of years, their incidence of breast cancer could be compared to that of a cohort of nonexposed women (such as other female employees at the same plant with jobs not involving contact with vinyl chloride).

One needs to understand a bit about the inner workings of cancer epidemiology in order to understand why the topic of individual cancer clusters is such a vexing one. Determining from ecological studies that communities near hazardous waste sites tend to suffer from excessive rates of cancer is one kind of investigation. Determining that any one particular community has an elevated cancer rate due to any one particular waste site is a very different kind of project. The second kind is the one most people are interested in. We live in particular communities, not general ones, and our concerns are about the health of the particular people in our families and neighborhoods. Indeed, almost all cancer cluster studies are initiated by alert citizens contacting their health departments to request such investigations. Their phone calls and letters often tell of "cancer streets," along which the prevalence of cancer seems extraordinarily high, or of growing numbers of neighborhood children afflicted with disease. This is exactly what happened in Normandale.

In spite of public concern, many public health officials become dismissive—if not downright apoplectic—when the subject of community-level cancer clusters is raised. Some consider the investigation of alleged clusters a disparaged practice and lament the inability of common people to grasp the statistical concept of randomness. Too often, the message relayed back to those vigilant citizens seeking explanations is that their questions are misguided. Too rarely are they told that the tools of epidemiology are just too blunt to provide answers.

One problem with cancer cluster studies is that individual communities have limited power to identify existing problems. In this context, the word *power* refers to the ability to detect a significantly increased cancer rate if indeed the increase really exists. The word *significantly* also has a particular meaning. Significance is a statistical standard that limits a finding to only those increases in cancer rates we are reasonably sure did not occur by chance. *Reasonably sure* is traditionally defined as 95 percent sure, so 95 percent is accepted as the conventional cutoff for significance. If I roll a pair of dice six times and they always come up sixes, I can be more than 95 percent certain that this event is not due to chance. The finding is statistically significant. I conclude the dice are loaded. However, if I roll only one die one time and I get a six, the finding is not considered significant. By chance alone, the odds of this outcome are 16.7 percent. The dice may indeed be fraudulent, but my test does not have the power to say so.

Looking for a cancer cluster in a single, small community is like rolling the dice only once. Before the possibility that the cluster has occurred by chance can be ruled out, cancer rates in some small communities must reach extraordinarily high levels—sometimes as high as eight to twenty times higher than levels for the surrounding areas. Because of the small sample size, lesser increases will not attain sufficient power for the study to be conclusive.

The second problem with cancer cluster studies is that there often remain no unexposed populations to use as a comparison group. In cluster studies, epidemiologists look for an increase over and above some background level, but if the people in the background are also becoming increasingly contaminated, the researchers are paddling a boat in a moving stream. Differences are harder to see.

Suppose, for example, we want to know whether people living near a particular hazardous waste dump are getting cancer because of it. Suppose the chemicals wafting into the air and trickling into the groundwater include several pesticides, some vinyl chloride, and an industrial solvent called trichloroethylene (TCE), classified by the EPA as a probable human carcinogen. With such contents, this dump would be utterly typical: TCE is the most frequently reported substance at Superfund sites, vinyl chloride is not far behind, and half of all hazardous waste sites contain pesticides. We have already seen that almost all of us experience chronic, incremental exposures to vinyl chloride and pesticides from our air, food, and water.

Most of us are also exposed regularly to molecules of TCE. Used by industry to degrease metal parts, TCE is now estimated to be in 34 percent of the nation's drinking water. Most processed foods contain traces as well. TCE is also found in paint removers, spot removers, cosmetics, and rug cleaners. An estimated 3.5 million workers are exposed to TCE on the job. Not so long ago, TCE was also used as an obstetrical anesthetic, a fumigant for grain, an ingredient in typewriter correction fluid, and a coffee decaffeinater. These uses have been phased out, but there is still sufficient release of TCE into the general environment to ensure that traces of vaporized metal degreaser persist in the ambient air that we all breathe—including detectable amounts in the air above the Arctic Circle. Therefore, if we design a study that compares cancer rates between people living near this hypothetical dump site and a control group of people drawn from the general population, our results might reveal little about why either group is getting cancer.

At least two additional problems with cluster studies exist, and they both have to do with the nature of cancer. First, cancer usually requires a long period of time to develop after exposure occurs. This lag time makes exposure assessment very difficult. Researchers must rely on old, incomplete records—which may not exist at all—or people's memories, which may be imperfect. Second, the origins of cancer are multiple, often resulting from exposures to combinations of substances—vinyl chloride plus alcohol, for example. These two features of the disease dilute and confuse epidemiologists' finest efforts to understand its causes in any given community. In a case-control study, some of our cases may have contracted cancer from prenatal exposures, some from the dump site, some from their jobs, some from pesticide residues, and some from a combination of these. Furthermore, unexposed people migrate into the community, and exposed people move out. Epidemiologists cannot ask people living near carcinogens to stay put for ten years so that they may conduct a decent cohort study.

To appreciate the difficulty of elucidating the causes of cancer clusters, consider one of epidemiology's most dramatic successes in cracking a disease cluster not involving cancer: the Case of the Eleven Blue Men.

In 1953, New York City police reported to the health department that eleven homeless men in one neighborhood had all been discovered to be very ill and that all of them had turned sky blue. This particular skin color is the hallmark symptom of a disease called methemoglobinemia. Eleven cases in one neighborhood is thousands of times above background level. Knowing that this disease is associated with the ingestion of sodium nitrite, epidemiologists interviewed the men about their eating habits and discovered all had frequented a particular neighborhood diner and all had used the saltshaker there. The said shaker was impounded, laboratory tests were run, and the discovery was made that sodium nitrite had indeed been substituted for sodium chloride. The cook had made a mistake. Mystery solved.

Now imagine that cancer made people turn blue. And further imagine a skid row saltshaker containing a powerful chemical carcinogen that eleven customers unwittingly sprinkle over their food and that eventually causes them to develop cancer. In spite of their telltale hue, the reason for their disease would probably never be uncovered. Because of the delay between exposure and onset of disease, at least ten years would pass before any of the eleven turned blue, and some of them would undoubtedly move away during this time. Because cancer is a disease with multiple causes, other drifters with blue complexions, who contracted cancer for unrelated reasons, would move into the area. The saltshaker itself would be long gone. Thus, although a cluster of people did indeed contract cancer from a single, identifiable source, a study of all blue-faced people in the neighborhood would not likely be able to establish the fact.

❧

The kind of investigation conducted in Normandale was standard epidemiological fare. The statistical standards used in the analysis, however, were out of the ordinary.

Recall that statistical significance is traditionally defined as a less than 5 percent probability that any observed differences are attributable to chance. Curiously, state officials conducting this study chose 1 percent, rather than 5 percent, as their cutoff level for significance. This is an unusually strict measure, which, not surprisingly, causes differences to disappear. Two excesses in the Pekin area did in fact attain statistical significance at the usual 5 percent level: ovarian cancer and lymphoma.

No mention of the study's statistical methodology was made in the newspaper account headlined AREA CANCER RATES NORMAL. Nor was any discussion devoted to what *normal* meant in this context. Tazewell County's toxic emissions are high, but in comparison to those in the rest of the state, they are not off the chart. Indeed, Pekin is not even among the top ten. Thus, statistics aside, a discovery that Pekin's cancer rates are rising in tandem with the rest of the state's says nothing about whether our cancers, or anyone else's in Illinois, are or are not attributable to environmental exposures.

A rising tide raises all ships. Is this situation normal?

In spite of almost insurmountable difficulties, some communities have successfully documented cancer clusters and have begun tracing them to possible sources. Nearly all of these discoveries have come about through brilliant environmental sleuthing and heroic perseverance by ordinary citizens working together with sympathetic researchers.

One of these places is Long Island, New York. In 1994, the New York State Department of Health released the results of a case-control study of Long Island women that showed a significant association between residence near chemical plants and risk of contracting breast cancer. In other words, women with breast cancer were more likely to live near a chemical industry facility than women without the disease. Breast cancer risk rose with number of facilities: the more chemical plants in the community, the higher the incidence of breast cancer. Risk was also related to distance. The closer a woman lived to one of these plants, the greater her chance of developing breast cancer. These associations were most pronounced for women who had lived near these industries between 1965 and 1975, as compared with 1975 to 1985, when state air standards had become stricter. Although such links had already been established in animal studies, this study was the first to indicate that breast cancer in humans may be associated with air pollution.

The study's significance goes far beyond the statistical—and thereby hangs a tale. This particular investigation came on the heels of another study that dismissed environmental links to breast cancer in Long Island. It was begun in the 1980s after women residents became aware that their rates of breast cancer were 10 to 20 percent higher than the rates for the rest of New York State (which itself has rates higher than the national average). A group of cancer activists demanded an explanation.

After five years of study, the New York State Department of Health concluded that breast cancer incidence in Long Island is correlated with high socioeconomic status. No associations were found between breast cancer and proximity to toxic waste sites or closed, contaminated wells. The U.S. Centers for Disease Control reviewed these findings and, in 1992, recommended no further follow-up. Nevertheless, when women found various flaws in the study's design (such as the ways in which study subjects were selected) and also noticed that researchers had not addressed some of their main concerns (such as the extent of pollution in available drinking water), they insisted on more investigation. "After all, if they never looked for a link they could always say they hadn't found one," the journalist Joan Swirsky recalls.

At this point, spurred on by aggressive reporting, women took matters into their own hands. Some began creating their own maps and others started formulating their own questions for research. In the fall of 1993, a group of activists hosted their own scientific conference, the first in the nation to bring together scientists and women with cancer to design a program of study. It was within this atmosphere that the 1994 study linking breast cancer and proximity to chemical plants emerged.

In the meantime and after considerable pressure, the U.S. Congress directed two federal agencies, the National Cancer Institute and the National Institute for Environmental Health Studies, to begin a multi-million-dollar, multidisciplinary study: the Long Island Breast Cancer Study Project. This research is now in progress. Using a case-control approach, teams of researchers are investigating such environmental features as aircraft emissions, past pesticide practices, electromagnetic fields, and plumes of contaminants in groundwater and air currents. Exposures will also be measured directly. Blood and other tissues from Long Island women with and without breast cancer will be analyzed for pesticide residues and industrial chemicals. Household dust, indoor air, and carpet fibers will be scrutinized for carcinogens as well.

Ten years thus elapsed between the first call for a comprehensive environmental inquiry into breast cancer on Long Island and its actual inception. Ten years is a long time for people with cancer to wait.

Across Long Island Sound lie the shores of Connecticut and Rhode Island. Just beyond, past Buzzards Bay, Cape Cod emerges from the Massachusetts coastline like a girl's arm, curled and slender. As a midwestern adolescent, I became enchanted with Henry David Thoreau's account of walking the length of this narrowest of peninsulas as it unfurled into the Atlantic. Cape Cod seemed to me a place of danger, beauty, and wild escape. It still plays that role in the imaginations of many Bostonians, who line the highways on any given weekend to head for the Cape.

In the 1980s, year-round residents of the Upper Cape began agitating forcefully for an investigation into the relationship between environmental hazards and cancer rates. It seemed to them that cancer in their isolated communities was unusually common, and they were also aware of many sources of potential environmental hazard, including pesticide exposure from cranberry bogs and golf courses as well as groundwater and air contamination from the Massachusetts Military Reservation located there. Moreover, many could recall how the entire Cape had been drenched with DDT and other pesticides for several straight years during the early 1950s in a failed campaign to eradicate the gypsy moth.

Residents were correct about the cancer rates. Researchers found that mortality and incidence rates in the Upper Cape region were elevated above statewide averages. Records from the Massachusetts Cancer Registry revealed significant excesses in the incidence of prostate, colon, and lung cancers. Modest increases were also seen for cancers of the pancreas, kidney, and bladder. By 1993, the Massachusetts Department of Public Health had established that breast cancer in almost all the towns on Cape Cod exceeded the statewide average. Of the ten towns with the highest breast cancer incidence in Massachusetts, seven are located on the Cape, and nearly all towns throughout the Cape have higher than average rates of breast cancer. These elevated rates cannot be explained by differences in screening practices. U.S. census data show that the women of Cape Cod are similar to the rest of the state's women in ethnicity and income. If there is nothing unique about us, Cape women asked, could there be something unique about our environment?

Persistent citizen activism led to two major studies: an Upper Cape study completed in 1991 by two Boston University epidemiologists, Ann Aschengrau and David Ozonoff, and another, the ongoing Cape Cod Breast Cancer and the Environment Study, which began in 1994 with a $1.2 million appropriation from the state legislature. Bringing together scientists from many fields—epidemiologists, geographers, chemists, cell biologists, and physicians—this second study is modeled in part after the Long Island Breast Cancer Study Project. Founded by breast cancer activists and revealing its source of inspiration explicitly, the name of this organization is the Silent Spring Institute.

Of particular interest to the institute's investigators is the underground aquifer that supplies nearly all the area's drinking water. Covered with porous, sandy soil, the aquifer is vulnerable to all manner of contamination—from pesticides to septic tank effluent to jet fuel and solvents spilled at the military base. Ironically, environmental regulations that protect the Cape's coastal marine sanctuary mean that all waste water is discharged onto land, where it trickles through the sand and into the groundwater. Many of these chemicals contained in this waste are believed to play a role in breast cancer. Plumes of contaminated groundwater are currently being mapped and compared to maps of breast cancer incidence on the Cape.

While researchers in the ongoing study are focusing specifically on breast cancer, those in the 1991 Cape study, by contrast, were looking at nine different cancers. Organized in case-control fashion, the study's cases comprised Upper Cape residents diagnosed with cancer between 1983 and 1986, and the controls were a random sample drawn from the entire population of Upper Cape residents. Exposures were assessed through interviews. In this way, potential confounding factors such as smoking and other lifestyle habits could be uncovered and corrected for. The study was particularly thorough: when dealing with people who had already died from their disease, researchers matched them with nonliving controls—people who had died from other diseases and whose names were selected randomly from death certificate registries. To gather exposure information on cases and controls no longer alive, researchers interviewed their next-of-kin.

After three years of research, the study's chief investigators reached the following conclusion:

> In summary, this inquiry was begun because of concern about the generally increased cancer rates in the Upper Cape region along with the presence of known or suspected environmental hazards. After an extensive review of environmental factors it is clear that there was ample cause for concern.

While low statistical power prevented researchers from explaining all of the cancer increases, several interesting results emerged. The rates of both lung and breast cancer were elevated among residents living near the gun and mortar positions on the reservation. One possible explanation is airborne exposure to the military's chemical propellant, dinitrotoluene, which is used for firing artillery. Classified as a probable human carcinogen, dinitrotoluene has been shown to cause breast cancer in laboratory animals. The study also yielded evidence for an increase in brain cancer among people living close to cranberry bogs, and it revealed elevations in leukemia and bladder cancer among those whose homes were fed by a particular type of water-distribution pipe.

These water pipes had long been under suspicion. In the late 1960s, a new innovation in cement water pipes was introduced into New England: pipes with plastic liners that improved the water's taste. A large number were laid in the Upper Cape, which was then undergoing rapid development. In manufacturing these pipes, workers applied vinyl paste to the inside surface, using a solvent called tetrachloroethylene. For reasons known only to organic chemists, tetrachloroethylene is more commonly referred to as perchloroethylene, PCE, or simply perc. Like its chemical cousin trichloroethylene, perchloroethylene is classified by the International Agency for Research on Cancer as a probable human carcinogen.

The assumption by the manufacturers of these water pipes was that all the solvent would evaporate during the curing process. It did not. In fact, substantial quantities remained and slowly leached into the drinking water. Thus, the drinking water of the Upper Cape is contaminated not only by chemicals leaking from the land's surface into the sole-source aquifer from which public water supplies are drawn but also, in some areas, by the pipes carrying this water into individuals' residences. The knowledge that Upper Cape water pipes were shedding perc into the drinking water was not a new discovery. This phenomenon had been known since the 1970s, but perc was not a substance regulated in drinking water during that decade. In 1980, plastic-lined water pipes were finally banned for use.

The link between perchloroethylene and bladder cancer was also not a new discovery. Perc is a familiar substance to almost all of us. Since the 1930s, it has been the chemical of choice for dry-cleaning clothes. Compared to the general population, dry cleaners have twice the rate of esophageal cancer and twice the rate of bladder cancer. Thus, a discovery of a bladder cancer cluster among the folk of the Upper Cape should come as no surprise. Further studies of the Upper Cape's water pipes, published in 1993, showed that people's actual exposure to perc varied widely, depending on the length, shape, size, and age of the water pipe, the pattern of water flow, and the person's length of residence in that house. For those people with highest exposure, bladder cancer risk was four times higher and leukemia nearly twice as high when compared to people without such pipes.

The *Journal of the American Water Works Association* first reported on the problem of perchloroethylene leaching from drinking water pipes in 1983. The following words, written by scientists studying the Upper Cape, were published exactly ten years later:

> In conclusion, we have found evidence for an association between PCE-contaminated public drinking water and leukemia and bladder cancer. In some EPA surveys, 14–26 percent of groundwater and 38 percent of surface water sources have some degree of PCE contamination. Thus, its carcinogenic potential is a matter of significant public health concern.

Ten years is a long time for people with cancer to wait.

The second study in Normandale supported the first one. Because it could not provide cancer incidence data on any scale smaller than ZIP code level, the state health department turned the remainder of the investigation over to the county. County officials promised to conduct a door-to-door survey with the goal of determining whether "the cancer cases in Normandale are out of sync with the rest of the ZIP code." They did not. Instead, questionnaires, which recipients were asked to fill out and mail back, were sent to 184 Normandale residences. Sixty-seven completed forms came back—a 37.5 percent response rate—and among these, eight cases of cancer were described.

The headline on March 6, 1992 announced, STUDY: NO CANCER CLUSTER, and the accompanying story read, in part:

> The Tazewell County Health Department found no significant cancer problem [in Normandale], officials said Thursday in announcing results of the department's cancer survey of the 40-acre subdivision. . . . The findings put an end to five months of investigation by state and county health officials into some residents' fears that they were living amidst a cancer cluster.

One does not need to be an epidemiologist to see the glaring problems with this study. First, the numbers are too small to draw conclusions one way or the other. Second, there is no way of knowing whether respondents represent a random sample of the community. Perhaps responding households were, on average, healthier or better educated than nonresponding households. Perhaps households providing care to a family member with cancer were more likely to misplace the mailing or were more likely to be too grief-stricken or too overwhelmed to sit down and fill out answers to lengthy questions. Perhaps they were too busy fighting with insurance companies or planning funerals to write up a detailed family history and remember to mail it. Perhaps families with cancer were more likely to be out of town. Perhaps illiteracy prevented some from responding. Perhaps those angriest about the county's broken promises chose to boycott the questionnaire. In short, without direct human contact, no one can know why nonresponding households, the majority, remained silent.

Furthermore, anyone who had lived alone and died from cancer had no chance to be counted at all. The local newspaper obtained county death certificates showing that at least five cancer deaths in the community were never reported to the county's survey. These included one case of liver cancer, two cases of breast cancer, one case of leukemia, and one of ovarian cancer.

How can silence be statistically evaluated? How can such a flawed, limited response to a questionnaire lead to an assertion that there is no problem?

These questions were not lost on the people of Normandale, many of whom expressed doubts about the study's validity. Still, the inhabitants of Normandale are not the residents of Cape Cod nor the women of Long Island. They are not positioned to reject the results of a county investigation and insist on a multi-million-dollar federal study. They have no friends in Congress. They are not armed with fax machines and university connections. They are unlikely to invite world-renowned scientists to convene proceedings in the parking lot of the A & W root-beer stand.

The citizens of Cape Cod and Long Island have struggled mightily to bring scientific attention to the link between cancers and environmental contamination in their communities. Still, the resources they command are starkly different from those among Normandale's residents. My meetings with the breast cancer activists of Long Island have taken place on college campuses and convention hotels. I have spoken with the cancer activists of Cape Cod in a beachfront conference center. When I met with a community leader in my own hometown, we held our discussion in the back room of an auto repair shop and towing company.

The Massachusetts report concerning the alleged cancer clusters on Cape Cod is more than five hundred pages long. It is considered preliminary. A ten-year, million-dollar, state-of-the-art investigation is in the works. The two reports detailing the state and county investigations into cancer rates in Pekin and the Normandale subdivision together total eight pages.

Said a man from Normandale who lost his wife to ovarian cancer, "I think the state has a way of putting things to the side or overlooking what's the real truth."

FIVE

war

When my father, at age sixty-nine, wrote his memoirs on a manual typewriter and sent copies to all surviving members of his family, he did so to commemorate the fiftieth anniversary of the Allies' victory in the Mediterranean theater. The significance of this event is emphasized throughout the text. It was his defining moment.

I have often imagined my father as a soldier in Italy. His two desires: to stay alive and to avenge the capture of his brother, my Uncle LeRoy, held as a prisoner of war in Germany. His one fear, which the Allied victory in Europe very nearly realized, was to be sent to the other theater—the blood-soaked Pacific.

My father firmly believes his life was saved by excellent typing skills. This was not a lesson to be lost on his daughters. The ninth child of a poor Chicago family, he moved a dozen times before finishing school and enlisting. How exactly he learned to type a hundred words per minute *with no errors* I do not know. It is part of my father's mystique. Throughout my childhood, the sounds of rapid, flawless typing filled my parents' bedroom. According to legend, his remarkable talent with the typewriter saved him for two reasons: first, because he was selected to work in correspondence at a U.S. Army office safely away from the front and, second, because he was therefore privy to orders about upcoming troop deployments. Thus forewarned, he deftly reenlisted in the right unit at the right moment and kept himself out of harm's way. His skills as a tank destroyer (motto: Seek, Strike, and Destroy), for which he was trained, would go untested.

With these stories, I was encouraged to spend time practicing penmanship, dictation, and typing at my father's big desk. Like him, I am near-sighted and left-handed. Neither were allowable excuses for sloppy work. But if I became more attracted to the sounds of the words than to the speed with which I could produce them, it was both for my plain lack of clerical talent and for the irrelevance introduced into the whole endeavor by electric, self-correcting machines—and, later, computers. Still, until I read his error-free autobiography while sitting at my own big desk, I did not realize how deeply my father's stories had influenced me or how much I am like that nineteen-year-old army clerk furiously typing up casualty reports. My own work as a writer is a legacy of a war ended years before I was born.

World War II is mentioned throughout the chapters of *Silent Spring*. Carson's references are casual, and they seem designed to remind already-aware readers that the technologies developed for wartime purposes had changed chemistry and physics forever. The atomic bomb was only the most arresting example. More intimate aspects of the human economy were also changed. The multitude of new synthetic products made available after the war altered how food was grown and packaged, homes constructed and furnished, bathrooms disinfected, children deloused, and pets de-flea'd. Carson described this transformation almost offhandedly, as though the connection between lawn-care practices and warfare was perfectly obvious.

Carson made at least two other points about World War II.

First, because many of these new chemicals were developed under emergency conditions and within the secretive atmosphere of wartime, they had not been fully tested for safety. After the war, private markets were quickly developed for these products, and yet their long-term effects on humans or the environment were not known. Second, because wartime attitudes accompanied these products onto the market, the goals of conquest and annihilation were transferred from the battlefield to our kitchens, gardens, forests, and farm fields. The Seek, Strike, and Destroy maxim of my father's antitank unit was brought home and turned against the natural world. This attitude, Carson believed, would be our undoing. All life was caught in the crossfire.

When *Silent Spring* was published, the victory days of the Second World War had not yet reached their twentieth anniversary. Compared to Carson's generation, those of us born after World War II are not as aware of the domestic changes wrought by this war. We have inherited its many inventions—as well as the waste produced in their manufacture—but we do not have a keen sense of their origins. In seeking explanations for the unprecedented cancer rates among our ranks, we need to examine them.

Taped above my desk are graphs showing the U.S. annual production of synthetic chemicals. I keep them here to make visible a phenomenon I was born in the midst of but am too young to recall firsthand. The first consists of several lines, each representing the manufacture of a single substance. One line is benzene, the human carcinogen known to cause leukemia and suspected of playing a role in multiple myeloma and non-Hodgkin's lymphoma. Another is perchloroethylene, the probable human carcinogen used to dry-clean clothes. A third represents production of vinyl chloride, a known cause of angiosarcoma and a possible breast carcinogen. They all look like ski slopes. After 1940, the lines begin to rise significantly and then shoot upward after 1960.

A second graph shows the annual production of all synthetic organic chemicals combined. It resembles a child's drawing of a cliff face. The line extending from 1920 to 1940 is essentially horizontal, hovering at a few billon pounds per year. After 1940, however, the line rockets skyward, becoming almost vertical after 1960. This kind of increase is exponential, and in the case of synthetic organic chemical production, the doubling time is every seven to eight years. By the end of the 1980s, total production had exceeded two hundred billion pounds per year. In other words, production of synthetic organic chemicals increased 100-fold between the time my mother was born and the year I finished graduate school. Two human generations.

The terms *organic* and *synthetic* are slippery ones and require explanation. *Organic* has two definitions that very nearly contradict each other. In popular usage, *organic* describes that which is simple, healthful, and close to nature. Similarly, in the language of agriculture, *organic* refers to food grown only with the aid of substances derived from plant and animal matter. Food certified as organic is supposed to be free from manufactured pesticides, antibiotics, hormones, and other additives—that is, fruits, vegetables, meat, eggs, and milk produced without the use of artificial, *synthetic* chemicals.

In the parlance of chemistry, however, *organic* simply refers to any chemical with carbon in it. The study of organic chemistry is the study of carbon compounds. The word *synthetic* means essentially the same as it does in everyday conversation: a synthetic chemical is one that has been formulated in a chemical laboratory, usually by combining smaller substances into larger ones. Most often, these substances contain carbon. Indeed, many organic chemicals now in daily use are synthetic—they do not exist in nature.

Of course, not all organic substances are synthetic. Wood, leather, crude oil, sugar, blood, coal—these are all carbon-based, organic substances found in the natural world. But, insofar as they have carbon atoms in their structures somewhere, the vast majority of synthesized chemicals are also organic. Plastic, detergent, nylon, trichloroethylene, DDT, PCBs, and CFCs are all synthetic organic compounds. The close alignment between organic and synthetic leads to the absurd but truthful concept that organic farmers are those who shun the use of (synthetic) organic chemicals.

Most synthetic organic compounds are derived from either petroleum or coal. Recognizing this fact brings the widely divergent definitions of the word *organic* together. To a biologist, organic substances are those that come from organisms—living or dead. Long chains of carbon atoms compose the chemical infrastructure of all life forms, including the liquefied organisms and the petrified organisms who lived on the planet eons ago and who have since been extracted from their burial grounds. Nothing manufactured from these so-called fossil fuels is really "unnatural." A molecule of DDT is made up of rearranged carbon atoms distilled from some creature's once-living body.

And here lies the problem. Many synthetic molecules are chemically similar enough to substances naturally found in the bodies of living organisms that, as a group, they tend to be biologically active. Our blood, lungs, liver, kidneys, colon—with the help of an elaborate enzyme system—are all designed to shuttle around, break apart, recycle, and reconstruct carbon-containing molecules. Thus, synthetic organics easily interact with the various naturally occurring biochemicals that constitute our anatomy and participate in the various physiological processes that keep us alive. By design, petroleum-derived pesticides have the power to kill because they chemically interfere with one or another of these processes. DDT, for example, interferes with the conduction of nerve impulses. The weed killer atrazine hinders the process of photosynthesis. The phenoxy herbicides bring about death by mimicking the effect of plant growth hormones.

Recall from Chapter Three that chlorofluorocarbons (CFCs), the famous ozone depleters, were exceptional because they did not share this property of biological activity. And because they are so chemically stable, CFC molecules can be swept into the stratosphere in their still intact state. Only when hit by a beam of ultraviolet light do they finally fall apart, releasing the chlorine atom that begins the destructive chain reaction culminating in the loss of ozone. CFCs were invented in 1928 but came into large-scale production only after World War II. Since the 1950s, the total amount of chlorine in the stratosphere has increased by a factor of ten.

Plenty of other synthetic organics are similarly inert in their finished forms. Indeed, this is why they are not biodegradable: their molecules are so large or otherwise so complex that they do not decay. They are thus exempt from the global carbon cycle that is constantly building up and breaking down organic molecules. And, of course, this exemption is what you want in a roof gutter, a water pipe, or a window frame.

For several reasons, however, this unreactiveness is misleading.

First, many of these compounds are themselves synthesized from synthetic chemicals that are highly reactive. By accident or on purpose, these industrial feedstocks are routinely released, dumped, or spilled in the general environment. While PVC plastic is, biochemically speaking, quite lethargic, the vinyl chloride from which it is manufactured exerts striking effects on the human liver. Second, inactive synthetic substances can shed or off-gas the smaller, more reactive molecules from which they are made. Third, new reactive chemicals can be created if these substances are subsequently burned—as when perfectly benign piles of vinyl siding are shoveled into a garbage incinerator, and poisonous dioxin rises from the stack. The incinerator itself, in this case, acts as a de facto chemical laboratory synthesizing new organic compounds from feedstocks of discarded consumer products.

Through all of these routes, we find ourselves facing a rising tide of biologically active, synthetic organic chemicals. Some interfere with our hormones, some attach themselves to our chromosomes, some cripple the immune system, and some overstimulate the activity of certain enzymes. If we could metabolize these chemicals into completely benign breakdown products and excrete them, they would pose less of a hazard. Instead, a good many of them accumulate. In essence, synthetic organic chemicals confront us with the worst of both worlds. They are similar enough to naturally occurring chemicals to react with us but different enough to not go away easily.

A number of these chemicals are soluble in fat and so collect in tissues high in fat content. Synthetic organic solvents, such as perchloroethylene and trichloroethylene, are an example. They are specifically designed to dissolve other oil- and fat-soluble chemicals. In paint, they work well to carry oil-based pigments. As degreasing agents, they work well to clean lubricated machine parts. As drycleaning fluids, they excel at dissolving human body oils and greasy fabric stains. They also all work splendidly to dissolve human body oils still on our skin and can thus easily enter our bodies upon touch. In addition, they are readily absorbed across the membranes of our lungs. Once inside, they take up residence in fat-containing tissues.

Many such tissues exist. Breasts are famous for their high fat content and often serve as repositories for synthetic organic chemicals circulating within the female body. But organs less renowned for fat content also collect these chemicals. The liver, for example, is surprisingly high in fat. So is bone marrow, the target organ for benzene. And, amazingly enough, because nerve cells are swathed in a fatty coating, so are our brains. Consider that many solvents have been used as anesthetic gases due to their ability to affect brain functioning. Chloroform is one.

Its medical uses long since discontinued, chloroform continues to be used as a solvent, fumigant, and ingredient in the manufacture of refrigerants, pesticides, and synthetic dyes. U.S. annual production of chloroform is currently about 600 million pounds, and it is found in nearly half of the hazardous waste sites on the Superfund National Priorities List. As we shall see in Chapter Nine, trace amounts are also formed when drinking water is chlorinated. Chloroform is classified as a probable human carcinogen. Its residence time in the body is actually quite brief. DDT, for example, has a half-life of at least seven years, while that of chloroform is a mere eight hours. (Half-life is the time required to convert half the body's burden of a given substance into excretable by-products.) The problem, then, with chloroform is not so much biological persistence but the fact that we are continuously exposed through multiple routes. All human beings, according to the U.S. Agency for Toxic Substances and Disease Registry, receive at least low levels through water, food, and inhalation.

As noted earlier, in the last half of the twentieth century, cancers of the brain, liver, breast, and bone marrow (multiple myeloma) have been on the rise. These are all human organs with high fat content. In the last half of the twentieth century, the production of fat-soluble, synthetic chemicals has also been on the rise. Many are classified as known, probable, or possible carcinogens. We need to ask what connections might exist between these two time trends.

First synthesized in 1874, DDT languished without purpose until drafted into World War II, and it proved its mettle by halting a typhus epidemic in Naples. My father arrived in this occupied city not long after. According to his wartime account, Naples lay in ruins, its people hungry, dirty, and in great despair. Little wonder they were also vulnerable to typhus. DDT's ability to annihilate the insect carriers of this disease—fleas, lice, and mites—must have seemed miraculous. Shortly thereafter, DDT was loaded onto American bombers and sprayed over the Pacific Islands to control mosquitoes. War production of DDT soon exceeded military requirements, and by 1945, the U.S. government allowed the surplus to be released for general civilian use.

As documented by the historians Thomas Dunlap and Edmund Russell, this decision marked a profound change in purpose. It is one thing to fumigate war refugees falling ill from insect-borne epidemics and quite another to douse the food supply of an entire nation not at risk for such diseases. It is one thing to rain insecticide over war zones ravaged by malaria and quite another to drench suburban Long Island. The skillful advertising that accompanied this transformation advocated a whole new approach to the insect world. Various insect species—some, mere nuisances—were recast in the public's imagination as deadly fiends to be rooted out at all cost. Cohabitation was no longer acceptable. In demonizing the home front's new enemy, one cartoon ad even went so far as to place Adolf Hitler's head on the body of a beetle.

Synthetic pesticide use thus began in the United States in the 1940s. Two other chemicals participated in this debut: parathion and the phenoxy herbicides 2,4-D and 2,4,5-T. Parathion—and its sibling malathion—belong to a group of synthetic chemicals called organophosphates, which are created by surrounding phosphate molecules with various carbon chains and rings. Like the chlorinated pesticides, they attack an insect's nervous system, but they do so by interfering with the chemical receptor molecules between the nerve cells rather than by affecting the conduction of electricity, which is DDT's mode of action. Like the chlorinated pesticides, organophosphate poisons played a starring role during the war—but as villain rather than hero. Developed by a German company as a nerve gas, members of the first generation of organophosphate poisons were tested on prisoners in the concentration camps of Auschwitz.

By contrast, the phenoxy herbicides were an Allied weapon. As we have already seen in Chapter Three, they were mobilized in the

1940s with the goal of destroying enemy crops. Another American invention—the atomic bomb—ended that war before field testing could yield to full-scale chemical warfare. Twenty more years would pass before 2,4-D and 2,4,5-T would reenter combat—this time in Vietnam's rainforests under the nom de guerre Agent Orange. In the meantime, they were introduced into U.S. agriculture for weed control and into forestry for shrub control. By 1960, 2,4-D accounted for half of all U.S. herbicide production. The hoe was fast on its way to becoming obsolete.

The graphical picture of pesticide use in the United States closely resembles the graphs of synthetic chemical production: a long, gentle rise between 1850 and 1945 and then, like the side of a mesa rising from the desert, the lines shoot up. Insecticide use begins ascending first; herbicide use closely follows. The line for fungicide use rises more gradually. All together, within ten years of their introduction in 1945, synthetic organic chemicals captured 90 percent of the agricultural pest-control market and had almost completely routed the pest-control methods of the prewar years. In 1939, there were 32 pesticidal active ingredients registered with the federal government. At present, 860 active ingredients are so registered and are formulated into 20,000 different pesticidal products. Current U.S. annual use is estimated at 2.23 billion pounds.

While agriculture consumes the lion's share of this total, with only about 5 percent used by private households, family pesticide use is emerging as an important source of exposure for those of us not living on farms. According to the EPA's National Home and Garden Pesticide Survey, 82 percent of U.S. households use pesticides of some kind. In a survey of families in Missouri, nearly 98 percent said they use pesticides at least once a year, and almost two-thirds said they use them five or more times. Yard and garden weed killers are used by about 50 percent of U.S. families, as are insecticidal flea collars, sprays, dusts, shampoos, and dips for household pets. These kinds of uses place us in intimate contact with pesticide residues, which can easily find their way into bedding, clothing, carpets, and food. Pesticidal residues persist much longer indoors than outdoors, where sunlight, flowing water, and soil microbes help break them down or carry them away. Yard chemicals tracked indoors on the bottoms of shoes can remain impregnated in carpet fibers for years. Some researchers now believe that infants and toddlers experience significant exposure to pesticides by crawling on carpets and ingesting house dust—perhaps even more so than by ingesting pesticide residues on food.

Several studies have linked childhood cancer to home pesticide use. Childhood cancer in Los Angeles was found to be associated with parental exposure to pesticides during pregnancy or nursing. In a 1995 study in Denver, children whose yards were treated with pesticides were four times more likely to have soft tissue cancers than children living in households that did not use yard chemicals. In another case-control study, researchers found statistically significant associations between the incidence of brain tumors in children and the use of several household pesticidal products: pest-repelling strips, lindane-containing lice shampoos, flea collars on pets, and weed killers on the lawn. All together, these findings may represent the beginning of an explanation as to why brain cancer in children under age fourteen has risen sharply during the past twenty years.

Of course, the postwar boom in synthetic organics was not limited to pesticides. Industrial products manufactured from fossil fuels also exploded onto the scene. In this case, World War II simply accelerated a process set in motion years earlier.

Historians of chemistry date the twentieth-century rise of the petrochemical industry back to the near extermination of whales in the nineteenth century: lack of whale oil for lamps created a market for kerosene, one of the lighter fractions of petroleum. Another petroleum derivative, gasoline, found purpose with the advent of the automobile. With the blockades against imported materials during World War I, the chemical industries of all warring nations were stimulated to invent new products. Germany, for example, developed artificial fertilizers when its supplies of Chilean saltpeter were cut. The same manufacturing process proved quite useful for producing explosives—as the fertilizer-derived bomb that destroyed the Oklahoma City federal building in 1995 illustrates.

With a large supply on hand for making dyes, Germany turned to chlorine gas to serve as a wretched weapon of chemical warfare in the trenches of France. Chlorinated solvents were also introduced during this time. After the war ended, new chemical products in the United States were protected by high tariffs, the war's losing parties surrendered their chemical secrets to the victors, and considerable wealth and prestige accrued to the chemical industry. By the 1930s, petroleum began to outpace coal as the source of carbon for new chemical inventions.

The cliff face of exponential growth in synthetic organic chemicals, however, did not begin until the 1940s. The all-out assaults of World War II created instant demands for explosives, synthetic rubber, aviation fuel, metal parts, synthetic oils, solvents, and pharmaceuticals. The innovations in chemical processing developed in the wake of World War I—such as the cracking of large, heavy petroleum molecules to produce many lighter and smaller molecules—were perfected and tested in large-scale production. When the war ended, the resulting economic boom, housing boom, and baby boom created unprecedented consumer demands as wartime chemicals, aided by skillful advertising, were transferred to civilian posts. Fearing a return to economic depression, national leaders encouraged the conversion of military products to civilian use. "In the United States," the historian Aaron Ihde has wryly noted, "peace did not prove catastrophic to an industry grown to monstrous proportions in response to the needs of war."

From an ecological point of view, World War II was a catalyst for the transformation from a carbohydrate-based economy—as it has been called by some analysts—to a petrochemical-based economy. For those of us born in the last fifty years, a review of petroleum's displaced, replaced, and discarded natural chemical predecessors is a fascinating exercise. I found myself amazed at how many products now derived from a barrel of oil were once manufactured from vegetation.

You may be excited to learn, as I was, that plastic existed before it was synthesized from petroleum. It was derived from plants, invented in the 1870s, and called celluloid. Clear plastic film derived from wood pulp with adhesive on one side was introduced in the 1920s as cellophane tape. Plant-derived substances were once used to make steering wheels, instrument panels, and spray paint for cars.

Thus, while the carcinogen vinyl chloride was actually first synthesized in 1913, its production did not begin to skyrocket until after World War II when research on the industrial uses of plant matter was replaced by an emphasis on petrochemistry. Automobile interiors would no longer come from cotton fibers or wood pulp, but from oil.

Guess, if you can, what formaldehyde and soybeans have in common. Imported from Asia in the nineteenth century, soybeans are a low-growing legume that produces round, yellow seeds inside of fuzzy pods. One of the oldest and simplest synthetic organic chemicals, formaldehyde consists of a single atom each of carbon and oxygen, plus two hydrogen atoms. These two substances could hardly be more different. Both happen to be deeply familiar to me, since soybeans cover the Illinois prairies and formaldehyde is the standard preservative of biological specimens destined for dissection. (Anyone who has ever confronted a pickled frog in a biology class would instantly recognize its distinctive odor.) Classified as a possible chemical carcinogen, formaldehyde is consistently ranked among the top fifty chemicals with the highest annual production volumes in the United States. In 1990 alone, 6.4 billion gallons were produced. Formaldehyde serves as an embalming fluid in funeral homes. It is also sprayed on fabric to create permanent press. In the 1970s, formaldehyde-based foams became popular for thermal insulation of houses. But nearly half of formaldehyde's annual production is used for synthetic resins to hold pieces of wood together as plywood and particle board. The subsequent evaporation of formaldehyde vapors from construction materials and furniture makes this chemical a significant contributor to indoor air pollution. As with chloroform, the problem with formaldehyde is not that it accumulates in our tissues but that we are exposed to small amounts of it almost continuously and from so many sources—from our subflooring to our wrinkle-free sheets.

Now the answer to the riddle: What formaldehyde shares with the soybean is an ability to act as an adhesive. Before formaldehyde was synthesized in such gargantuan quantities, soybean resins were used to hold particle board and plywood together. Soybean oil was also used in fire-suppressant foam and wallpaper glue, and as a base for paints, varnishes, and lacquers.

Other plant-based oils also played leading roles in industry before the war. Oils extracted from corn, olives, rice, grape seeds, and other plant parts were used to make paint, inks, soaps, emulsifiers, and even floor covering. The word *linoleum* echoes the name of its original key ingredient: linseed oil. Castor oil, from the tropical castor bean tree, was used to lubricate machine parts.

Countless examples of synthetic substitutions have occurred in the last half century and have provided us with new exposures to known or suspected carcinogens. In the 1950s, for instance, synthetic cutting oils were introduced into machine shops. Used for cooling metal parts during both cutting and grinding, cutting fluids come into close association with machinists through both touch and inhalation. Synthetic degreasers, such as perc, are then often used to clean the parts once they are cut. These have become a common contaminant of hazardous waste sites and therefore of drinking water. Researchers have recently discovered that synthetic cutting fluids can expose workers to N-nitrosamines, a contaminant formed during their manufacture. By the 1970s, cancer among machine operators and its possible relationship to synthetic cutting fluids began receiving attention. In one study the researchers concluded:

> Until now, N-nitrosamines have not been directly associated with human cancers because no population groups had been identified that were inadvertently exposed. Cutting fluid users have the dubious honor of being the first such population group to be identified.

❧

The rapid birthrate of new synthetic products that began in 1945 far surpassed the ability of government to regulate their use and disposal. Between 45,000 and 100,000 chemicals are now in common commercial use; 75,000 is the most frequently cited estimate. Of these, only about 1.5 to 3 percent (1,200 to 1,500 chemicals) have been tested for carcinogenicity. The vast majority of commercially used chemicals were brought to market before 1979, when the federal Toxics Substances Control Act (TSCA) mandated the review of new chemicals. Thus, many carcinogenic environmental contaminants likely remain unidentified, unmonitored, and unregulated. Too often, this lack of basic information is paraphrased as "there is lack of evidence of harm," which in turn is translated as "the chemical is harmless."

Pesticides are regulated by twin laws: the Federal Food, Drug, and Cosmetic Act (FFDCA) and the Federal Insecticide, Fungicide, and Rodenticide Act (FIFRA). FFDCA governs pesticide tolerances on agricultural commodities—that is, it sets legal limits for pesticide residues allowed in foodstuffs ranging from raw vegetables to animal feed. FIFRA, on the other hand, requires companies manufacturing pesticides to test their products for toxicity and submit the results to the federal government. Amendments to FIFRA require reevaluation of old, untested pesticides approved before the current requirements for scientific testing were put into place. Initially scheduled to be completed in 1976, this reregistration process is still under way, has been repeatedly delayed, and is now scheduled for completion in the year 2010. Until then, the old, untested pesticides can be sold and used. As one critic has noted, it is as if the bureau of motor vehicles issued everyone a driver's license but did not get around to giving us a road test until decades later. According to the National Research Council, only 10 percent of pesticides in common use have been adequately assessed for hazards; for 38 percent, nothing useful is known; the remaining 52 percent fall somewhere in between.

In the 1970s and 1980s, various right-to-know laws began springing up as a response to this ever-expanding mosh pit of toxic chemicals. The first group of laws established employees' right to know about hazardous substances in their workplaces. A second group sanctioned citizens' right to know about the presence of toxic chemicals in their communities and, finally, about the routine release of some of these chemicals into the environment. For nearly four decades after the widespread introduction of such chemicals into our environment, these rights were not ours. The identity of chemicals released by industry was considered privileged information—trade secrets. Those of us born during this time—the 1940s until the mid-1980s—will never know with certainty what we were exposed to as children and what carcinogenic risks we have assumed from such exposures. We can, however, obtain partial information about our current exposures.

Significantly, neither set of laws came about because legislators and manufacturers calmly agreed that citizens should be made aware of their chemical exposures. Rather, workplace right-to-know laws are rooted in a long history of labor struggle, and the community-based laws—codified as the Emergency Planning and Community Right-to-Know Act (EPCRA)—passed the U.S. Congress in 1986 over intense industry opposition. This legislation was a response to citizen activism at the state and local levels, as well as a direct reaction to the 1984 chemical disaster in Bhopal, India, which occurred when a feedstock for pesticide manufacture escaped from a Union Carbide plant and killed many thousands of sleeping residents in their homes. Emergency medical efforts were frustrated by the fact that no one knew what the chemical was. A similar chemical release occurred at a sister plant in West Virginia. Shortly thereafter, Congress voted EPCRA into law. Key parts of this legislation passed by a one-vote margin.

The linchpin of EPCRA is the Toxics Release Inventory (TRI). As the SEER Program registry is to cancer incidence, TRI is to carcinogens and other toxins. It requires that certain manufacturers report to the government the total amount of each of some 654 toxic chemicals released each year into air, water, and land. The government then makes these data public information. As a pollution disclosure program, TRI has many deficiencies. Its main shortcoming is that it relies completely on self-reporting and lacks adequate procedures for checking data quality. In addition, it does not address the presence of carcinogens in consumer products; small companies are exempt from reporting; the compliance rate among industries that are required to file is only about 66 percent; and 654 is a small fraction of the total chemicals they use.

Furthermore, loopholes in reporting requirements allow industries to play an elaborate shell game with their wastes. Some analysts believe the substantial decline in emissions from 1987 to the present, for example, partly consists of phantom reductions—such as changes in accounting methods or the contracting of highly polluting processes to other facilities. Researchers tracking the flow of toxic chemicals through the economy point out that declines in toxic waste *releases* have not always been accompanied by parallel declines in toxic waste *production*: the generation of toxic waste by TRI-reporting facilities remains high. Where, then, is the waste going? Without thorough materials accounting, which is not currently required, no one is exactly sure.

Nevertheless, under EPCRA, for the first time in history, any citizen can request from the Environmental Protection Agency (EPA) a list of the reported toxic releases in his or her home county. Access to this information is now acknowledged by our government as a fundamental public right.

In some communities, the TRI has served as a powerful tool for pressuring factories to reduce pollution. Its most important function may be the implicit recognition that a so-called private industry is engaging in a very public act when it releases toxic chemicals into a community's air, water, and soil. Conceptually, we all know the industries in our communities pollute the environment. We may even be able to see and smell the results. But very often, the picture does not come into focus for us until we actually stare at the list of specifics, as when the names and the numbers are printed in our local newspapers: how many pounds of which known or suspected carcinogens were released by which companies into the air we breathe or into the rivers we fish and from which we draw our drinking water?

The TRI's first report, released by the EPA in 1989, had just such an effect. It revealed that *billions* of pounds of toxic chemicals were being routinely emitted each year into the nation's air, water, and land. Nearly all who read the report were amazed. This was the first attempt to gather together routine toxic releases, and the sum was an unquestionably staggering amount. Said a representative from the Chemical Industry Council of New Jersey: "I'll be honest with you. [Our reduction in emissions] probably would not have occurred if that data had not become public information. It was something that caught everyone's attention, including the corporate leaders." A Monsanto spokesman was even more blunt: "The law is having an incredible effect. . . . There's not a chief executive officer around who wants to be the biggest polluter in Iowa."

In the first year of reporting, only about 5 percent of toxic releases in the country were reported under TRI, and yet the effect of ending the silence about toxic releases was huge. Some companies who found themselves on the list of the worst toxic offenders immediately entered into voluntary programs to reduce their emissions. Several communities began using their local data to force more recalcitrant industries to follow suit. Concerned citizens who also happened to be computer wizards came together to provide technical support to communities wishing to access their local TRI data electronically (a task that can now be accomplished on a home computer or at almost any local library). Public Data Access, Inc., mapped the information and, by bringing TRI data together with death certificate data, correlated areas of severe environmental contamination with areas of elevated cancer mortality. The results are the black, white, and gray maps described in Chapter Four.

According to the most recent TRI, which is about the size of an average telephone book, 2.26 billion pounds of toxic chemicals were released into the environment in 1994. Of these, 177 million pounds were known or suspected carcinogens.

In a favorite photograph of myself as a child, I am hanging determinedly onto a tricycle, wearing a goofy expression and my father's army hat. The determination came from trying to salute my father, the photographer, while simultaneously pedaling. It is 1962. The setting is the concrete patio on the south side of our house. A construction worker before the GI Bill returned him to the typewriter as a college student, my father poured this patio himself and laid the brick walkway leading out into his 1.5 acres of former cow pasture.

After the war, my father married a farmer's daughter with a degree in biology and another in chemistry. He built his house on Pekin's east bluff and planted lines of silver maple and white pine in the sod. Before the trees grew up to form a wall around the borders of his property, the patio offered a spectacular view. To the east, cows grazed. Although afraid of them, I liked to stand at the fence and watch them eat—their purply tongues and black plumes of flies, the ripping sounds of the grass.

Just beyond, the bluff's pastures unrolled into what was once—and I am guessing here—hill prairie. Here lay vast fields of corn and soybeans. I liked the corn—each stalk a green man waving his arms. In September, the soybeans turned brilliant yellow and then deepened into an orange-brown far richer than my burnt sienna crayon.

"What color would you call soybeans?" I inquired of Aunt Ann, who farmed two counties east from us.

She didn't miss a beat. "At six dollars a bushel, I would call them gold."

My father drove west to work every morning. Looking through the patio screen from my tricycle, I could see the smokestacks, cooling towers, and distillation chambers of the river valley's three dozen industries. I liked the steam clouds, trails of smoke, and mysterious shimmering vapors. I was especially fond of the pink-and-white-striped towers, which reminded me of giant candy canes. These stacks belonged to the ethanol distillery and the coal-burning power plant just upriver. At night, they became lighthouses—great blinking columns warning planes away. To my sister and me, my father referred to this scene as "progress, girls, progress."

Tazewell County, Illinois, is home to two distinct cultures, one emblemized by the lone figure on his tractor and the other by picket lines of striking plant workers. Our house was situated in the transitional zone between the two.

Among farmers, Tazewell is known as the birthplace of Reid's Yellow Dent, a famous strain of field corn that became the ancestor of many hybrid seed lines. Among industrialists, Tazewell County is known for the 127-acre Pekin Energy Company, one of the nation's largest producers of ethanol, and as the manufacturing site and proving grounds for Caterpillar tractors, backhoes, and bulldozers. Caterpillar's management offices are headquartered across the Illinois River in Peoria. A hydrologist's description of the area from 1950 is as good as any: "The Peoria-Pekin area is a highly industrialized district requiring an enormous volume of water. The industrial areas are surrounded by the fertile agricultural prairie lands of the corn belt."

Settled before the prairie was sod-busted, Pekin began as a military fort. War and manufacturing have frequently danced together here. Distilling and brewing began as a means of transforming grain into a nonperishable cash commodity that could easily be shipped east. Wartime needs for industrial alcohol then provided a huge new market and inspired new production technologies. One of Pekin's distilleries was founded in 1941 expressly to provide the U.S. military with ethanol. In 1916, the U.S. Army ordered the first Caterpillar tractors, which it used to drag cannons, ammunitions, and supplies to the front. During World War II, Caterpillar machines were used to bulldoze airstrips, grade roads, clear bomb wreckage, and topple palm trees. By 1945, 85 percent of Caterpillar's production was shipped overseas for military work. Collected photographs of "Cats" in action during both world wars and the Korean War are still a hot item in the local bookstore.

Other photographs from the turn of the century show child laborers posing in the sugar-beet fields. The black smoke of the sugar factory forms a dramatic backdrop against the little white faces. The sugar works later become Corn Products, which produced Argo cornstarch and Karo syrup. In 1924, a starch explosion incinerated forty-two workers. In 1980, Corn Products became Pekin Energy.

In September 1994, I drove along the Illinois River banks to pass by where those old beet fields must have been—less than two miles from the house I grew up in. The floodplains are now a landscape of docks, stacks, rail yards, conveyers, elevators, hopper bins, pits, lagoons, coal piles, tailings ponds, settling tanks, power lines, and scrap heaps—all that I had seen at a distance as a child. A union billboard announced, "You Are Now Entering a War Zone," a comment not on the environment but on labor's latest showdown with management at the nearby Caterpillar plant.

There are some places in this world that prompt one to ask, "Where did all this come from?" The fish, vegetable, and flower markets of New York City always bring me to this question. Tazewell County is another kind of place. Spend some time on the Pekin docks. Watch the barges of coal, grain, steel, chemicals, and petroleum products. "Where is all this going?"

There are partial answers. The grain elevators and the mills ship corn and pelletized animal feed south to New Orleans and from there to Asia and Europe. The coal-fired power plant called Powerton sends electricity 165 miles north to Chicago via high-power lines. In

1943, its smoke prevented landings at an airport forty-one miles away. In 1974—the year I turned fifteen—this plant was named the worst polluter in the state of Illinois. Trace Chemicals formulates pesticides. I do not know where they end up. The brass foundry makes huge cylinders, called bushings, for draglines, drills, and crushers used in strip mines. Caterpillars end up everywhere. In 1986, I looked out the window of a bus heading south along the Nile River in eastern Sudan and found myself face to face with Caterpillar's familiar logo—a giant capital *C*—painted on a billboard near a military installation.

About the other industries lining the river valley I know less. I do not know what goes on at Airco Industrial Gases, the Sherex Chemical Company, the Agrico Chemical Company, or the aluminum foundry. I know that Keystone Steel and Wire makes nails and barbed wire out of scrap metal. In 1993, the company faced charges for polluting the sand aquifer below its facilities with TCE and another synthetic degreaser, 1,1,1-trichloroethane, a suspected carcinogen. The promises it made to clean up and switch to less toxic chemical technologies have, so far, kept Keystone off the Superfund's National Priorities List.

Like a film of gasoline on a pond's surface, an emotional blankness coats my words here. There are, of course, many ways of expressing the relevance of the historical past to the personal present. Surely there is one that could describe the private thoughts of an East Bluff girl returning home from Boston and passing by the hospital where, years before, she was diagnosed with a type of cancer known to be caused by exposure to environmental carcinogens. Surely there is a language able to explain why such a woman would now drive along Distillery Road, breathing the acrid air, searching for nineteenth-century sugar-beet fields and twentieth-century hazardous waste sites.

A silence spreads out. I cannot make her speak.

It is not the silence of resignation or paralysis. It is the fear that speaking intimately about this landscape—or myself as a native of this place—would make too exceptional what is common and ordinary. I feel protective of my hometown. Its citizens are not unusually ignorant or evil or shortsighted. And, away from the river, the city itself is lovely. Between the fields and the factories are nice, old neighborhoods, beautiful parks, the county fairgrounds, and reasonably good schools. There is nothing unique or even unusual about Tazewell County, Illinois. As true everywhere else, its agricultural and industrial practices—from weed control to degreasing parts—were transformed by chemical technologies introduced after World War II. As true everywhere else, these chemicals, many of them carcinogens, have found their way into the general environment. As true almost everywhere else, no systematic investigation has been conducted to determine whether any connection exists between the release of these chemicals and the rates of cancer here.

"We know the emissions are present, and the cancer, but we don't know if the two are related," said a state toxicologist quoted in the local paper in March 1995. This article concluded:

> The impact of tons of toxic emissions on thc health of industrial workers and the public never has been systematically studied and may be impossible to determine. . . . Health statistics in Peoria and Tazewell counties are troubling, but the connection between emissions and health problems is not clear.

There is nothing special or unusual about the toxic release inventories for Tazewell and Peoria Counties. Of seventy-eight regions in Illinois, the Pekin-Peoria area ranks only thirteenth in TRI emissions. Nonetheless, I cried when I first read through these inventories. Hundreds of pages of computer print itemize the toxic emissions for area industries during the years since 1987, when this information was first compiled. In 1991, for example, large manufacturers in Peoria and Tazewell Counties legally released 11.1 million pounds of toxic chemicals into the air, water, and land. Among the known and suspected carcinogens released were benzene, chromium, formaldehyde, nickel, ethylene, acrylonitrile, butyraldehyde, lindane, and captan. Captan is a carcinogenic fungicide prohibited for many domestic uses in 1989. In 1987, according to the TRI, 250 pounds of captan ended up in the Pekin sewer system. In 1992, 321 pounds were released into the air.

Tips of all kinds of icebergs are revealed in other right-to-know documents. For example, I have a partial record of pre-TRI toxic releases in Tazewell County dating back to 1972. The carcinogens catch my eye first—PCBs, vinyl chloride, benzene—but the list also includes other frightening and curious items: printing ink, jet fuel, asphalt sealer, dynamite, scrubber sludge, fuel oil, antifreeze, fly ash, coal dust, herbicides, furnace oil, and "explosive vapors."

In addition, I possess a twenty-four-page list of facilities in Tazewell County with permits to discharge wastes into particular rivers and streams ("local receiving waters: Farm Creek . . . local receiving waters: Illinois River," etc.). I have also obtained a thirty-four-page list of each and every facility—from the local crematorium to the auto body shop—permitted to deal in any way with hazardous materials. Right-to-know legislation has given me access to a hefty off-site transfer report, a document particularly revealing because it shows the flow of toxic wastes coming into Tazewell County. I know, for example, that the Sun Chemical Corporation of Newark, New Jersey, sent 250 pounds of friable asbestos to the Pekin Metro landfill for disposal in 1987. Tazewell doubled the amount of hazardous waste it generated and shipped off-site between 1989 and 1992, but, as one of the state's top receiving counties, it still received four times more waste than it produced.

The spill report for Tazewell County details chemical accidents. Here is the first entry as it appears on the list:

DATE: 6/11/1988

STREET: RTE 24

MATERIAL SPILLED: METHYL CHLORIDE

AMOUNT SPILLED: 2,000 LBS

WATERWAY/OTHER: AIR RELEASE

EVENT DESCRIPTION: WEIGH TANK/WHILE PREPARING FOR INSPECTORS, VALVE INADVERTENTLY OPENED/EXACT CAUSE UNDER INVESTIGATION

ACTION TAKEN: TEMPORARILY EVACUATED AFFECTED BUILDING FOR TWO HOURS . . . SHUT VALVE TO STOP RELEASE

Route 24 is an old highway. To the west, it follows the Illinois River valley for some miles before shooting across the plains to the

Mississippi River town of Quincy. To the east, it connects Pekin to the Indiana border, passing four miles south of my grandparents' farm in Forrest. I can tell you about every small town between here and there, describe every moraine, name every creek.

Methyl chloride is classified as a probable human carcinogen. It causes mutations in bacteria and kidney cancers in mice. It also causes birth defects and degeneration of the sperm-carrying tubules in rat testicles. Used in the manufacture of silicone products, fuel additives, and herbicides, methyl chloride is synthesized by attaching a chlorine atom onto a molecule of wood alcohol. By 1981, annual production reached 362 million pounds per year. Domestic consumption expands approximately 6.5 percent per year. Methyl chloride's long-term effects on human health have never been studied directly.

Amid a flooded sea of information, an absence of knowledge. Amid a thousand computer-generated words, a silence spreads out.

Seek. Strike. Destroy. Of all the unexpected consequences of World War II, perhaps the most ironic is the discovery that a remarkable number of the new chemicals it ushered in are estrogenic—that is, at low levels inside the human body, they mimic the female hormone estrogen. Many of the hypermasculine weapons of conquest and progress, are, biologically speaking, emasculating.

This effect occurs through a variety of biochemical mechanisms. Some chemicals imitate the hormone directly, while others interfere with the various systems that regulate the body's production and metabolism of natural estrogens. Still others seem to work by blocking the receptor sites for male hormones, which are collectively called androgens. In 1995, fifty years after its triumphant return from the war and entry into civilian life, DDT again made headlines when new animal studies showed that DDT's main metabolic breakdown product, DDE, is an androgen-blocker.

Our enzymes quickly convert DDT into DDE. But because the next step is much slower (recall DDT's seven-year half-life), we accumulate DDE as we age—much as a fine stream of sand grains gradually forms a heap at the bottom of an hourglass. DDE molecules can cross the human placenta and can also accumulate in breast milk. Thus, those of us too young to have been sprayed by DDT directly nevertheless have accumulated DDE in our bodies through at least two routes: from our mothers (both before and after birth) and our consumption of milk, meat, eggs, and fish. Animals, like the humans who eat them, lack the biochemical hardware needed for efficient conversion of DDE to something excretable.

For boys and men, the consequences may include physical deformities such as undescended testicles, lowered sperm counts, and testicular cancer. No one knows what effect DDE exposure has on the reproductive development of girls or women; no research has been done. The only thing we know for a fact is that DDE is biochemically different enough from anything else in the human body—male or female—that it is not completely metabolized as are our own natural sex hormones. This is one reason why, more than two decades after DDT's forced retirement in the United States, we still have DDE molecules floating around in our tissues.

Much of the concern about hormone-disrupting chemicals has been focused on their possible role in contributing to birth defects, reproductive failures in wildlife, and infertility in humans. At times, these discussions seem nearly to eclipse the quieter, but longer-running conversations about the possible contributions of estrogen-mimicking contaminants to cancer. Certain breast cancers, for example, are notorious for growing faster in the presence of estrogen, which is why prescribing antiestrogenic drugs is standard chemotherapeutic protocol. Many other cancers—those of the ovary, uterus, testicle, and prostate, for example—are also known to be, or suspected to be, hormonally mediated. Thus, identifying pollutants that interfere with hormones is important to public dialogue about human cancers of all kinds.

The relevance of endocrine disruption for cancer is not a new subject. Rachel Carson mentioned it explicitly in *Silent Spring*. Nevertheless, a mysterious event in a Tufts University laboratory a few years ago brought renewed attention to the topic.

The cell biologists Ana Soto and Carlos Sonnenschein were working out the details of estrogen's relationship to breast cancer when something puzzling happened in their laboratory. Breast cancer cells growing in plastic dishes containing no estrogen started dividing rapidly, as though they were being hormonally stimulated. "This indicated that some type of contamination had occurred," Soto remembers. "We made an accidental discovery."

Soto and Sonnenschein traced the contamination to the plastic tubes they were using to store blood serum. Together, they purified the contaminant and identified it as nonylphenol, a synthetic organic chemical added during the manufacture of plastic to prevent it from cracking. Molecules of nonylphenol were being shed from the tubes into the serum.

In a series of follow-up experiments, the two researchers demonstrated that nonylphenol is estrogenic. It activates estrogen receptors within cells so equipped, which in turn alters the activity of certain genes and changes the rate at which these cells divide. Nonylphenol makes breast cancer cells—at least those growing in petri dishes—grow faster. Soto and Sonnenschein began testing other chemical aliens—certain common pesticides, detergents, and other types of plastics—and discovered estrogenic activity in a whole variety of petrochemically derived substances. Other researchers were inspired to do the same. Approximately forty such chemicals have so far been identified as capable of mimicking estrogen.

This flurry of attention has shed light on the biological activities of two ubiquitous but almost totally unknown groups of synthetic compounds: plasticizers and surfactants. Plasticizers are chemicals that are mixed with plastics to give them more strength and flexibility. Surfactants are added to, for example, detergents, herbicides, and paints to help the active ingredient stick to the surface of its target—dirt particles, weeds, or the wall of a house. Alkylphenol polyethoxylates (APEOs) are surfactants widely used in household detergents. Since their introduction in the 1940s, they have become widely disseminated in rivers, lakes, and streams via sewage systems. APEOs have been detected in drinking water in New Jersey. In 1994, in the wake of Soto and Sonnenschein's discovery, a team of researchers in England reported that APEOs can, in trace amounts, stimulate the growth of breast cancer cells and feminize male fish exposed to contaminated sewage. Fish collected from many U.S. rivers also display hormonal abnormalities consistent with exposure to estrogenic substances in river-borne sewage. However, it is not at all clear at this point—either in England or the United States—whether the feminization of fish downstream of sewage outfalls can be totally explained by exposure to chemicals such as APEO surfactants. New evidence suggests that at least some of the problem may stem from exposure to natural and synthetic estrogens found in women's urine—and so researchers investigating the gender-bending potential of sewage are now turning their attention from washing machines to toilets.

Phthalates, the plasticizers with the nearly impossible name, turn out to be the most abundant industrial contaminant in the environment. At least two have now been identified as estrogenic, and traces of both have been found in food. One is used in plastic food wrap and the other in papers and cardboard designed for contact with liquid, dry, and fatty foods.

Some phthalates are known to be overtly carcinogenic. For example, DEHP—which stands for the even more impossible di(2ethylhexyl)phthalate—gives PVC plastic its flexibility. It is also classified as a probable human carcinogen and because of this, its use in baby pacifiers, plastic food wrap, and toys has been discontinued. Residues of DEHP have been found in food items, especially those with high fat content, such as eggs, milk, cheese, margarine, and seafood. Because DEHP, like nonylphenol, can leach from plastic containers holding bodily fluids, it has also been found in blood used for transfusions. In 1993, the yearly production of DEHP was 270 million pounds. According to TRI data, in 1991 alone 3.76 million pounds of DEHP were released into the environment or transfered off-site for disposal.

About half of the synthetic materials known to function as endocrine disrupters belong to a chemical group called organochlorines. Not all estrogenic materials are organochlorines, and not all organochlorines are estrogenic, but the overlap is impressive. Moreover, organochlorines are such a large group—around eleven thousand exist—and they tend to be so persistent in the environment, so reactive within human tissues, and so frequently associated with cancer that they merit special consideration.

Many of the chemicals we have already discussed belong to this group. Lindane, DDT, heptachlor, chlordane, PCBs, CFCs, TCE, perc, 2,4-D, methyl chloride, vinyl chloride, polyvinyl chloride, dioxin, and chloroform are all organochlorines. Benzene, formaldehyde, nonylphenol, and phthalates are not.

Organochlorines, which involve a chemical marriage between chlorine and carbon atoms, are not strictly a human invention. A few are formed during volcanic eruptions and forest fires and some by living organisms such as marine algae. For the most part, however, chlorine and carbon move in separate spheres in the natural world—and in the bodies of humans and other mammals. To force the two together, elemental chlorine gas is required.

Although it holds a rightful place in the periodic table of elements, pure chlorine *is* a human invention. It can be produced by passing electricity through salt water in a procedure that was first undertaken on an industrial scale in 1893. A powerful poison, chlorine gas became known to the world during World War I, but its manufacture grew slowly until World War II, then rose exponentially. About 1 percent of this production is used for disinfecting water and about 10 percent for bleaching paper, and the majority is combined with various carbon compounds, usually derived from petroleum, to make organochlorines.

In its elemental form, chlorine (but not the ion chloride) is highly reactive with carbon, which is why so many different combinations are possible. Like houses of different architectural styles, some organochlorines are very small and plain, and others huge and ornate. One of the simplest is chloroform, which consists of a single carbon atom with one hydrogen and three chlorine atoms attached to it like four spokes on a hub. Consisting of one chlorine, two carbon, and three hydrogen atoms, vinyl chloride is not much more complicated. The dry-cleaning solvent perchloroethylene is two carbon and four chlorine atoms, while the industrial degreaser trichloroethylene consists of two carbon and three chlorine atoms.

On the more elaborate side are chlorinated phenols. These consist of a hexagonal ring of six carbons with various chlorinated groups hanging off the corners. The pesticide lindane, for example, consists of a carbon hexagon with six chlorine atoms attached all around. The herbicide 2,4-D is a hexagon with chlorines attached to the second and fourth carbon atoms and a carbon chain waving like a flag from the first carbon atom. DDT is more complicated yet. It consists of two hexagonal rings, each with one chlorine atom attached, yoked together by a single carbon atom from which dangles a chlorinated carbon tail.

And then there are the PCBs. PCBs are the elders of the group, and they are referred to in the plural for a reason. As their name implies, polychlorinated biphenyls comprise two rings of carbon atoms welded directly together, around which are attached any number of chlorine atoms. In fact, there are 209 possible combinations and therefore 209 different PCBs. Some of these chemical combinations are estrogenic and some appear not to be, but no one has worked this out definitively.

As a group, organochlorines tend to be persistent in air and water. When they evaporate and are swept into the wind currents, some fall back to the earth close to their origins, while others can circulate for thousands of miles before being redeposited into water, vegetation, and soil. From there, they enter the food chain. Diet is thus believed to be a major route of exposure for us.

Not all organochlorines are deliberately constructed. Whenever elemental chlorine is present, the natural environment will synthesize additional, unwanted organochlorine molecules. These reactions can take place when water containing organic matter, such as decayed leaves, is chlorinated. It can happen in pulp and paper mills during the process of bleaching or when chlorinated plastics are burned. It can happen during the manufacture of other organochlorines. The production of 2,4,5-T, the burning of plastic, and certain methods of bleaching paper all contribute to the birth of dioxin. A chemical of no known usefulness and never manufactured on purpose, dioxin has been linked to a variety of cancers and is now believed to inhabit the body tissues of every person living in the United States. Dioxin is a beautifully symmetrical molecule, consisting of two chlorinated carbon rings held together by a double bridge of oxygen atoms.

The development of industrial chemistry in this century has been driven by the exigencies of war. Out of this crucible came new chemicals of all sorts. Some, such as organophosphate nerve gas, seem to have been born from truly evil intentions; others, from admirable ones. But few were invented solely for the purposes to which they were turned after the war's end. And few were adequately tested for long-term health effects.

As the daughter of a World War II veteran, I am grateful that my father did not die in a typhus epidemic in Naples. But as a survivor of cancer, as a native of Tazewell County, and as a member of the most poisoned generation to come of adult age, I am sorry that cooler heads did not prevail in the calm prosperity of peacetime, when careful consideration and a longer view on public health were once again permissible and necessary. I am sorry that no one asked, "Is this the industrial path we want to continue along? Is this the most reasonable way to rid our dogs of fleas and our trees of gypsy moths? Is this the safest material for a baby's pacifier or for a tub of margarine?" Or that those who did ask such questions were not heard.

These questions are finally beginning to receive a hearing. In 1993, the American Public Health Association issued a resolution calling for the gradual phaseout of most organochlorine compounds and for the pursuit of safe alternatives. In doing so, it followed another august agency, the International Joint Commission on the Great Lakes. Citing rising rates of breast cancer within the Great Lakes basin, the commission recommended scrapping the current practice of regulating persistent toxic chemicals after they have been produced, used, and released. Taking its place would be a preventive strategy recognizing that all such substances are "deleterious to the human condition" and must no longer be tolerated in the ecosystem, "whether or not unassailable scientific proof of acute or chronic damage is universally accepted."

In the fall of 1994, the esteemed epidemiologist David Ozonoff addressed a group of five hundred breast cancer activists in Boston and expressed his support for these concepts:

> The ability to make these chemicals [organochlorines] in high volume did not even exist prior to World War II. . . . They are not a legacy from the industrial revolution of the 19th century, but of the rise of the chemical industry of the 20th. They are

not woven into the warp and woof of our national fabric, but on the contrary, are recent and unwelcome newcomers.

I do not contend that all synthetic organic chemicals should be banned. Neither do I advocate a return to the days of celluloid and castor oil. From what I understand, celluloid was flammable and brittle, and I'm sure castor oil had its own problems. However, I am convinced that human inventiveness is not restricted to acts of war. The path that chemistry has taken in the last half of this century is only one path—and not even a particularly imaginative one.

Some solutions may indeed be found through the rescue of chemical processes abandoned years before—as in the quiet decision of many daily newspapers to switch to soy-based inks—while others may be sought through altogether new applications of knowledge. Chlorine-free methods of bleaching paper are possible and are already in small-scale commercial use both here and in Europe. Citrus-based solvents, ultrasonics, and old-fashioned soap and water can often replace chlorinated solvents used for degreasing operations and precision cleaning of electronic parts. New methods of embalming and different attitudes about the role of funeral services can reduce the use of formaldehyde in mortuaries.

Sweeping changes are immediately possible in the dry-cleaning industry. Most clothing tagged as "dry-clean only" can in fact be professionally cleaned with the use of water, special soaps, and reengineered washing machines that allow computerized control over humidity, agitation, and heat. (Pressurized carbon dioxide also holds promise as a nontoxic solvent for cleaning textiles.) The Boston area, for example, is home to one such wet-cleaning operation, a pilot project of the Toxics Use Reduction Institute. I recently delivered to this shop a down coat, a silk dress, a badly stained antique kimono, and a pile of my best wool, cashmere, and rayon suits. All came back clean, beautifully pressed, and odor-free. The white streak across the sleeve of the green blazer—the result of an encounter with a freshly painted doorframe—was gone. Best of all, the proprietor, who appeared about eight months pregnant, expressed to me her relief at not having to be exposed to perc.

Most of the perchloroethylene manufactured in the United

States is used by the textile and dry-cleaning industry. In 1992 alone, 12.3 million pounds of this organochlorine and suspected carcinogen was released into air, ten thousand pounds to rivers and streams, and nine thousand pounds to land. Thirteen thousand pounds were directly injected into underground wells. The recycling of perchloroethylene produces contaminated sludge and filters, which are subsequently deposited in landfills where they poison soil. Traces of perchloroethylene have been found in breast milk, cow's milk, meat, oil, fruit, fish, shellfish, and algae. Perc has been detected in rainwater, seawater, river water, groundwater, and tap water. More than 650,000 workers are thought to be exposed to perc on the job, and an estimated 99,000 New York City dwellers are exposed to elevated levels just from breathing—many because their office or apartment shares the same building with a dry-cleaner. A 1993 survey found that 83 percent of New York City apartments located above a dry-cleaning establishment had ambient perc levels in excess of state health guidelines.

It is time to start pursuing alternative paths. From the right to know and the duty to inquire flows the obligation to act.

SIX

animals

Bathed in a brilliant yellow-green light, they look like bats floating in a perfectly round pond. I have seen many micrographs of cancerous tissue—reproduced neatly in atlases of human tumor cell lines or on the shiny pages of medical journals—but never before have I stared at living cancer cells. Alive, they look to me like bats.

"Now compare that one to this one."

The first petri dish is removed and replaced by another, and I look again through the microscope. In this second watery landscape, they look more like fallen leaves—some drift together in large masses, others in smaller clusters.

"Okay, here's dish number three."

Now they are everywhere. A mosaic of islands and jutting peninsulas. Pieces of a crazy quilt tossed into a lake. A raft of vines tangled with shards of crockery. There is no one way to describe them. Collectively or alone, cancer cells are more chaotically arranged than the shy, scurrying animals from which the disease—as well as the zodiac constellation—derives its name. Cancer, carcinogen, carcinoma, from the Greek *karkinos*, "the crab."

The three petri dishes I have been asked to compare contain estrogen-sensitive breast cancer cells derived from a human cell line called MCF-7. The first dish is the control. Its culture medium, the broth that nourishes the growing cells, contains no estrogen. The third dish is a control of the opposite sort. Its medium was innoculated with the most potent known form of human estrogen, which is called estradiol. It's also the dish with the most luxuriant growth. By definition, estrogen-sensitive breast tumors grow faster in the presence of estrogen, and MCF-7 cells are well-known exemplifiers of this principle.

It is the second dish, the one with the intermediate growth rate, that reveals the significant finding. Its culture medium has been laced with trace amounts of endosulfan, an organochlorine pesticide. These three dishes are part of a series of experiments showing that endosulfan—introduced in 1954 and now widely used on salad crops—is estrogenic. Like the hormone it mimics, endosulfan stimulates breast cancer cells to divide and multiply.

In this ability, endosulfan is much less effective than a woman's own estradiol. However, studies similar to this one have shown that endosulfan can act in concert with other xenoestrogens, that is, chemicals foreign to the body that, directly or indirectly, act like estrogens. For example, when ten different synthetic chemicals, all estrogen mimics, are added to the culture medium at one-tenth the minimal dose required for proliferation of MCF-7 cells, proliferation ensues. Like raindrops eroding a boulder, quantities of weakly estrogenic chemicals too small to exert observable effects on their own have a significant impact when combined. Furthermore, some xenoestrogens may have the ability to interact with naturally occurring estrogens and amplify their effect. If confirmed, such results imply that "safe" levels of exposure to individual estrogen-mimicking chemicals may not exist. (The actual cellular pathways followed by xenoestrogens are described in Chapter Eleven.)

The discovery that xenoestrogens can work additively was made by the cell biologists Ana Soto and Carlos Sonnenschein, whose laboratory in downtown Boston I am visiting. Since their 1991 discovery that nonylphenol stimulates the growth of MCF-7 cells, they have continued to probe the phenomenon of estrogen mimicry and its implications for breast cancer. In addition to plastic additives, Soto and Sonnenschein have identified estrogenic activity in a variety of pesticides. Some, like endosulfan, are still in use. Others, such as dieldrin and toxaphene, are now banned.

That toxaphene—fat soluble and stubbornly persistent—should prove estrogenic is particularly frightening. Identified as an animal carcinogen in 1979 and banned in 1982, toxaphene was not so long ago the most heavily used insecticide in the United States. It was the chemical weapon of choice against boll weevils in cotton fields, where it was used in extraordinary quantities. In 1950, northern Alabama cotton fields received an average of sixty-three pounds per acre. Rachel Carson herself denounced toxaphene as an indiscriminate killer of fish, and in *Silent Spring* she described in detail the die-offs of crappies, bass, and sunfish in southern streams and farm ponds. Ironically, it rose to even greater popularity after pesticides like DDT fell into disfavor.

Toxaphene's continuing effects on wildlife are what led Soto and Sonnenschein to become concerned about its possible relationship to breast cancer. When field researchers linked toxaphene to reproductive damage in seals and documented its ongoing accumulation in the muscle fat of Arctic and Baltic salmon, these two laboratory researchers decided to test its effects on breast cancer cells. Not only does toxaphene cause MCF-7 cells to proliferate, the pair discovered, but it does so at levels well within the range of concentrations now found in the flesh of some salmon.

Soto and Sonnenschein's work thus depends on a collaboration between cell biology, which peers through magnifying lenses at the smallest units of life, and wildlife biology, which monitors the world's animals. In this way, changes in the growth rate of breast cancer cells in a Boston laboratory help elucidate the reasons for reproductive failures among sea mammals living thousands of miles away—and vice versa. The evidence from animals, in turn, provides reasons for rising cancer rates among humans, as well as our routes of exposure to cancer-promoting agents.

❧

But let's go back for a moment to the microscope and look once more at the cells named MCF-7. Whose breasts did they come from, and what was her fate?

Finding answers to such questions isn't easy. Medical researchers maintain a comfortable distance between themselves and the cancer patients who provide the human tissues used in their experiments. The results of research involving MCF-7 cells are reported in numerous published articles. Even as the cells' various properties are described in depth, these papers mention almost nothing about their human origins.

Here is what I do know. All successfully established cancer cell lines, including MCF-7, are immortal, meaning that they will reproduce endlessly in covered dishes so long as they are provided with the proper nutrients. Under such conditions, most human cells—even most cancer cells—tend to die out after a finite number of cell divisions. No one knows why some cancer cells can attain immortality while others cannot. Because they can be shipped all over the world, immortal cell lines allow many laboratories to conduct research on cells from the same tumor over long periods of time. Immortal cells are to cancer researchers what sourdough starter is to bread bakers.

BT-20, VHB-1, MDA-MB-241, CAL-18B, T47D: these are the names of other famous breast cancer cell lines. MCF-7 is among the oldest and is also considered the most reliable—the coin of the realm, according to one researcher. Its name reveals a few interesting clues. *MCF* stands for Michigan Cancer Foundation, the Detroit institution that makes this cell line available to laboratories around the world. The trailing seven refers to the number of attempts that were required to establish a self-perpetuating stock of cells from the body of the particular woman patient who consented to this effort. Immortality was finally achieved on the seventh try.

"Does this mean cancerous cells were withdrawn multiple times?" I ask into the phone, trying to imagine the procedure, wondering if it was painful, wondering how many attempts she was willing to submit to.

"Yes, that's right," says Joe Michaels of the Michigan Cancer Foundation.

I learn that her birth name was Frances Mallon. At the time of her diagnosis, she was a nun—Sister Catherine Frances—at the Immaculate Heart of Mary Convent in Monroe, Michigan, a small town midway between Detroit and Toledo on the west bank of Lake Erie. Strangely enough, I have been there. The Immaculate Heart of Mary, which has a long history of involvement with social issues, was the setting for a conference I attended in 1992 concerning organochlorine contamination of the Great Lakes. So, not only have I looked at the cells of her breasts, but I have walked through the corridors of her home and eaten in her dining room.

Sister Catherine Frances died of her disease in 1970. An old newspaper clipping reports that "she was a slightly built woman of medium height, with auburn hair, gray eyes and hands that were remarkable for their delicate beauty." Before entering Immaculate Heart in 1945, she had worked for twenty-five years as a stenographer at the Mueller Brass Company in Port Huron. Both her mother and sister had died of cancer before her. Her father had died of tuberculosis. The cancer cells that ultimately begat the MCF-7 line were extracted from fluid trapped in her chest cavity. This is all I know.

In 1995, at a national breast cancer meeting, I am introduced to a well-known researcher whose work I admire. Over dinner we discuss his current experiments, and I ask which cell line he uses.

"MCF-7. It's a very well-described line."

"Did you know that she was a nun?"

There is a long pause. I watch him grope toward this unexpected bit of information. He blinks several times and takes a few swallows from his glass of ice water.

"Then, MCF is her name, her initials?" His voice is low and gentle.

"Actually, no . . ."

Now, as I'm writing, I propose a rechristening of MCF-7. Let them be called IBFM-7: the Immortal Breasts of Frances Mallon, attempt number seven. Let them be known as a sacrament: *This is my body, which is broken for you. This do in remembrance of me.*

In science, an assay is an evaluation of a biological or chemical substance. Estrogens, for example, are defined as substances that stimulate proliferation of uterine and vaginal cells. Thus, the traditional assay for estrogenicity involves injecting the substance to be evaluated into female rats or mice, letting a period of time go by, killing the animals, and then noting whether or not their genital tracts have gained weight, in comparison to the tracts of a control group.

These assays are complex, messy, and expensive. For these and other reasons, screening of environmental chemicals for possible hormone-mimicking effects is not routinely done. The question is whether human breast cancer cells growing in petri dishes can serve as an alternative to rodents for an assay. So far, concordance between animal assays and breast cancer cell line assays has been high. The pesticide endosulfan, for example, not only makes breast cancer cells proliferate but also lowers testosterone levels in male rats and causes their testicles to shrivel. Together, these results tell a consistent story.

In the attempt to identify environmental carcinogens, human studies and animal assays remain the standard yardstick. The strongest evidence for associations between particular chemicals and particular cancers comes from epidemiology, but accurate information about exposure is often hard to come by in these studies. Animal assays have a few distinct advantages over epidemiological studies. Most important, confounding factors can be controlled more easily. Laboratory rats do not smoke cigarettes or move out of state or change jobs. They can be made to have identical diets, exercise habits, and reproductive practices. Their exposures to the substance in question can also be made identical. Also, rodents have much shorter life spans. Cohorts of rats and mice can readily be followed from birth to death. In human studies, twenty to thirty years are often required between exposure and onset of cancer. Furthermore, animal assays can be conducted before a substance is marketed. Epidemiological studies, in contrast, are initiated only after evidence for harm has accumulated. Epidemiology relies on body counts.

For these reasons, evidence for carcinogenicity in laboratory animals often precedes evidence from human studies. About one-third of the agents now classified as human carcinogens were first discovered in animals. Obviously, if animal assays worked perfectly—and if human exposures to known animal carcinogens were adequately prevented—this proportion would be much higher. No human being would have to die to prove that certain chemicals cause cancer.

The history of carcinogenicity testing in animals is intimately linked to the history of organized labor. In 1918, two Japanese scientists reported that coal tar, suspected of causing cancer in workers, induced skin tumors when applied to rabbits' ears. By the 1930s, researchers working with mice were able to determine which specific chemicals within coal tar—a mixture of many ingredients—were to blame.

In 1938, in a series of now-classic experiments, exposure to synthetic dyes derived from coal and belonging to a class of chemicals called aromatic amines were shown to cause bladder cancer in dogs. These results helped explain why bladder cancers had become so prevalent among dyestuffs workers. With the invention of mauve in 1854, synthetic dyes began replacing natural plant-based dyes in the coloring of cloth and leather. By the beginning of the twentieth century, bladder cancer rates among this group of workers had skyrocketed, and the dog experiments helped unravel this mystery. The International Labor Organization did not wait for the results of these animal tests, however, and in 1921 declared certain aromatic amines to be human carcinogens. Decades later, these dogs provided a lead in understanding why tire-industry workers, as well as machinists and metal workers, also began falling victim to bladder cancer: aromatic amines had been added to rubbers and cutting oils to serve as accelerants and antirust agents.

The researcher who carried out the original dog-dye studies was none other than Wilhelm Hueper, whose work formed the basis of Rachel Carson's chapter on cancer in *Silent Spring*. All of Hueper's papers—including his typewritten autobiography—are housed in the chambers of the National Library of Medicine in Bethesda, Maryland. I once spent a bright spring day poring through them. Now that animal testing has become associated with cruelty, reading about the various attempts to squelch the results of his work is a lesson in shifting cultural perceptions. While employed as an industry scientist, Hueper endured stonewalling, harassment, threats of lawsuits, defunding, firing, and gag orders. Whatever we may now think about the ethics of exposing dogs to carcinogens, animal studies such as these were highly threatening to industries whose manufacturing processes had become dependent on certain chemicals and who feared disclosure of trade secrets.

Routine screening of chemicals for carcinogenicity in laboratory animals began in earnest in the early 1970s. As of 1993, the International Agency for Research on Cancer (IARC) had assayed about a thousand chemicals—a small fraction of the total number used in commerce—and had identified 110 definite or very probable human carcinogens. IARC is quite clear on the relevance of animal experiments for human cancers: "In the absence of adequate data on humans, it is biologically plausible and prudent to regard agents and mixtures for which there is sufficient evidence of carcinogenicity in experimental animals as if they present a carcinogenic risk to humans."

Here in the United States, the Environmental Protection Agency (EPA) combines animal evidence with the results of epidemiological studies in order to classify substances into one of five categories. These evaluations generally match those of IARC. Group A includes the *known* human carcinogens. To be so ranked, evidence from epidemiological studies alone must be strong enough to make the case. Group B are the *probable* human carcinogens. Members of this group include chemicals for which there is sufficient evidence from animal studies to regard it as a human carcinogen as well as limited evidence from human studies. A Group B ranking often means the needed human studies have never been conducted. Group C, the *possible* human carcinogens, are all those chemicals for which some evidence exists for carcinogenicity from animal studies. Group D includes chemicals not classifiable because there are simply no data on which to base a decision. Group E are noncarcinogens, chemicals that show no indication of causing cancer in any species.

Known. Probable. Possible. The fact that numerous chemicals with long-standing membership in Groups A through C are still allowed to be manufactured, sold, released, dumped, imported, exported, or otherwise used comes as a surprise to many knowledgeable people. I include myself here. It is comfortable to assume such substances are automatically expelled from human society as soon as their cancer-causing potential is demonstrated. This is not the case, but the ethical implications of any other alternative seem too much to bear.

As an antidote to innocence, I recommend a document produced every two years by the National Toxicology Program of the U.S. Department of Health and Human Services: the *Biennial Report on Carcinogens* (formerly the *Annual Report on Carcinogens*). The report exists because the National Toxicology Program is charged by law with publishing "a list of all substances (i) which either are known to be carcinogens or may reasonably be anticipated to be carcinogens; and (ii) to which a significant number of persons residing in the United States are exposed." The edition that stands next to the Toxics Release Inventory on my bookshelf is 473 pages long and features nearly 200 entries.

Some of these listings describe chemicals with large production volumes, such as benzene. When lead was outlawed as an antiknock additive in gasoline, benzene replaced it. We are therefore exposed to benzene every time we fill our cars with gasoline. Benzene is classified as a known human carcinogen.

Other listings in the National Toxicology Program's carcinogen report describe old chemical chestnuts no longer manufactured but still present, such as the PCBs. About one-third of the world's total production of PCBs is believed to have escaped into the general environment. Like thousands of tiny bombs exploding in slow motion, pieces of discarded equipment containing the oily fluid—electrical transformers, television sets, old french friers—leak their contents drop by drop into soil and water. From here, PCB molecules rise into the atmosphere, circulate with the wind, and are redeposited all over the globe. They then enter the food chain. The fatty tissues of nearly all Americans are believed to contain PCB molecules. We have accumulated most of them by eating food derived from animals: eggs, meat, milk, fish, and shellfish. In rodent assays, PCBs cause liver cancer, pituitary tumors, leukemia, lymphoma, and intestinal cancers. Manufactured from 1929 until 1977, PCBs are classified as a probable human carcinogen, as noted earlier. As more PCBs escape from obsolete equipment, human exposure is expected to continue.

As the report clearly explains, carcinogens are regulated differently than noncarcinogens, and special rules intended to monitor their each and every move through human society exist. The appearance of a chemical on the government's official roster of carcinogens is only the first step toward a program of intense surveillance and assessment. Nevertheless, the very existence of this list means that trading in cancer-causing chemicals is still a perfectly legal activity.

We all know this, of course. We may even read the signs on the gas pumps that warn against inhaling the shimmery vapors that rise from the end of the nozzle. But we also all know that enforcement of any rule is imperfect, that accidents happen even when rules are followed, and that many chemicals were released into our environment before these regulations went into effect. In 1995, for example, the reopening of some public schools in Cape Cod, Massachusetts, was postponed after high levels of the banned pesticide dieldrin were found in the soil surrounding one of the grade schools. While attempting to establish the extent of this problem, investigators discovered this school was also contaminated with PCBs at levels two thousand times greater than acceptable under state regulations. Four of the district's schools are located on the Massachusetts Military Reservation, a heavily regulated place.

Perhaps most amazing is the fact that aromatic amines—the first officially designated group of workplace carcinogens—are still among us. Benzidine dyes, for example, remained in commerce for nearly forty years after Hueper's dog studies. From a 1980 review of benzidine-based dyes conducted by the National Institute for Occupational Safety and Health:

> Benzidene-based dyes are chiefly used in the leather, textile and paper industries, but they are also used by beauticians, craft workers, and the general public. The common starting material for the manufacture of these dyes, benzidine, is acknowledged by both industry and government to cause bladder cancer. This is based on considerable evidence from studies with humans as well as with animals. . . . Both brief and prolonged exposures to

> benzidene have been associated with the development of bladder cancer in workers.

From a 1994 report:

> Benzidine is no longer manufactured for commercial sale in the United States. All benzidine production is for captive consumption and it must be maintained in closed systems under stringent workplace controls. . . . Prior to 1977, U.S. production of benzidine amounted to many millions of pounds per year.

From a 1996 report:

> Some benzidine-based dyes (or products dyed with them) may still be imported. Benzidine has been found in waste sites and becomes part of the bottom sediment in water. It exists in the air as very small particles, which may be brought back to the earth's surface by rain or gravity.

Perhaps it's the very matter-of-fact tone of these reports that contributes to the sense of unreality. Perhaps it is the absence of words like *pain*, *surgery*, *chemotherapy*, *support group*, *recurrence*, *hospice care*, *palliative treatment*, and all the other terms that those with cancer and those who love them learn to speak. When I read these reports, I see a urologist's waiting room full of patients and a pencil-thick, telescopic tube called a cystoscope lying on its stainless steel tray in the examination room that awaits each of them. I remember the cystoscope sliding up my urethra during an inspection for bladder tumors. And I remember the nervous, red-haired woman whose cystoscopic checkup was scheduled before mine. When I walked out of outpatient surgery, she was sobbing into the pay phone. I never found out what happened to her.

❧

Typically, about eight hundred animals are required for a carcinogenicity assay. The first step is to assign animals of two different species—usually rats and mice—and both sexes to one of four groups. Groups one through three represent high-, medium-, and low-dose exposures to whatever substance is being tested. The fourth serves as the unexposed control. Each group thus contains approximately fifty animals of each sex and species. Next, the test substance is administered by inhalation, ingestion, or skin application on a regular basis throughout the animals' life spans. At the end of the experiment, researchers compare tumor patterns in exposed and unexposed animals and determine whether differences exist among the four groups.

Using animal assays to ascertain potential human carcinogens requires two leaps of faith. One is the supposition that what causes cancer in one species will cause cancer in another. This is called transpecies extrapolation, and there is good evidence that we are on firm footing when we make this leap. Almost all substances discovered to cause cancer in humans also cause cancer in at least one species of laboratory animal. And they very often cause cancer in the same organ or tissue. The inference we make is that the reverse is also true. Significantly, five of the top ten sites for chemically induced tumor formation in laboratory rodents correspond to five of the top ten sites of cancer in the U.S. human population: lung, breast, bladder, uterus, and the blood-producing cells of the bone marrow.

Consider once again the female breast, that gland that defines all mammals. While rodent breasts and human breasts would seem to have little in common—after all, mice have ten of them and we have only two—the similarities in development and anatomy are remarkable. Early in life, both female rodents and female humans possess immature breast tissue that consists of a bundle of slender tubes. These structures—the mammary ducts—open into the nipple, which is essentially a sieve. At the other end, giving each duct the shape of a canoe paddle, are the terminal buds. Under the direction of female hormones, the ducts begin to branch. Meanwhile, the paddle-shaped buds metamorphose into clusters of lobules designed to produce milk. A cushion of fat surrounds both the ducts and their lobules, and the whole apparatus comes to resemble an orchard of fruiting trees. This process unfolds in all female mammals according to the same basic blueprint.

When women get breast cancer, the tumors most often form in the interior lining of the ducts. This is also where breast cancer in rodents is usually located. Nevertheless, some differences between species do exist. In most strains of mice, breast cancer is largely estrogen independent, while in most rats, it is estrogen dependent. Human females are vulnerable to both kinds. These sorts of variations are the reason for using more than one species in cancer assays.

Agreement between species is high—75 percent of chemicals that cause cancer in rats also do so in mice—but it is not absolute. Moreover, some anatomical differences between species do create obvious complications. Mouth breathing, for example, is impossible for a mouse. All rodents must breathe through their noses. Humans, on the other hand, often inhale through their mouths, especially when in the presence of a bad smell. Animal assays that rely on exposure to airborne contaminants become tricky. Rodents will filter the chemical through a highly developed defense system—and possibly place their sinuses and pharynx at greater risk for cancer—while humans may inspire the same chemical directly into their lungs, sparing their sinuses but possibly raising the exposure level of other tissues.

The overall concept of using one species to predict cancer in another is, nonetheless, receiving renewed support from molecular biology. This field of study focuses on the intricate mechanical workings inside the body's cells. The basic truism emerging from this line of inquiry is that nature does not often reinvent the wheel. Life is conservative. Certain essential processes—such as the regulation of cell division or the extraction of energy from a molecule of glucose through its stepwise disassembly—are held in common by all members of the animal kingdom, including humans. These activities are accomplished by nearly identical enzyme systems and are governed by nearly identical genes. Because chemical carcinogens very often strike directly at these minute structures and interfere with processes basic to all life, they tend to inflict the same kind of cellular sabotage across species lines. Animals and people exposed to the same environmental contaminant frequently possess identical genetic mutations. Evidently, what is bad for the goose is bad for the gander—and for the rat, the mouse, the fish, and the bank accountant.

The second leap of faith is the assumption that the animal assays are relevant to human cancers because the results of laboratory studies conducted with high exposures to potential carcinogens can be extrapolated to human situations involving low exposures. The rationale for this extrapolating is simple. High doses must be used in the laboratory because the number of animals exposed is small compared to the human population at risk. A substance with the power to cause cancer at low doses in 1 percent of the American population, for example, could potentially kill more than two million people if all of us were so exposed. One percent of fifty rats, however, is less than one individual. Therefore, exposures are raised in excess of that to which humans are subjected in order to raise the chances of detecting any possible problems.

Animal assays are a yes-or-no instrument designed simply to pinpoint potential human carcinogens. They are not intended to provide quantitative information on responses to a wide range of doses. Too often, the necessary follow-up studies that would elucidate this relationship are never conducted. Overwhelming the protocol are the tens of thousands of chemicals in commercial use still in need of basic testing.

Without these tests, we can only guess at the number of chemical carcinogens in our midst. As of 1995, the National Toxicology Program had completed animal assays on 400-odd chemicals. Based on these results, researchers have estimated that of the 75,000 chemicals now in commercial use, somewhat fewer than 5 to 10 percent of these might reasonably be considered carcinogenic in humans. Five to 10 percent means 3,750 to 7,500 different chemicals. The number of substances we have identified and regulate as carcinogens is, at present, less than 200.

In the summer between my sophomore and junior years in college, I was diagnosed with bladder cancer of a type called transitional cell carcinoma. It is something I have in common with Hueper's dogs, as well as with at least one beluga whale in the St. Lawrence River.

Proceeding northeast from Lake Ontario, the St. Lawrence River slants through the Canadian province of Quebec and flares open like a trumpet as it pours itself into the North Atlantic. Nova Scotia stands to the south, Newfoundland to the north. Where the river's current meets the ocean's tide, in the neck of the Gulf of St. Lawrence, is one of the world's deepest, longest estuaries. About five hundred beluga whales, a remnant of the thousands that once lived here, inhabit this transition zone. This estuary also receives tributarial waters that have traversed some of the most industrialized landscapes of southern Canada and the northeastern United States.

Belugas are small, toothed whales. Their skin is pure white.

Transitional cell carcinoma among the belugas was first discovered during an autopsy of a carcass that had washed ashore in 1985. It was a particularly provocative finding because workers in nearby aluminum smelters, which release their wastes into the St. Lawrence, had also been found to have an elevated incidence of this type of bladder cancer.

Gross hematuria, or noticeable blood in the urine, is the usual way bladder cancer presents itself. I do not know how a whale would experience this—perhaps through sense of smell. As for myself, gross hematuria arrived as I was finishing up a morning shift at a truck-stop diner. After making my final rounds with the ketchup bottles and syrup dispensers, I stopped in the restroom. Turning to flush, I froze. My urine looked like cherry Kool-Aid. I stood there a long time.

And then I remembered the beets—sliced red beets, which the cook had prepared for the lunch special and which I had eaten in great quantity during my break. Could beets make urine turn pink? Asparagus was certainly famous for its ability to transmit pungent odors to urine. What other explanation could there be? I felt fine.

I swore off beets. Three weeks later, I returned home from a night shift at a pancake house, tore off my waitress uniform, went to the bathroom, turned to flush, and . . . the toilet was full of blood. Brilliant and thick. I drove to the emergency room.

I was wrong about the beets.

Bladder cancer is one of several cancers striking the beluga population of the St. Lawrence. In 1988, a team of veterinarians found tumors in the bodies of four dead whales from a group of thirteen that had washed up over a period of ten months along a polluted stretch of the river. In addition, the immature breast ducts of one young female showed abnormal proliferation. Called ductal hyperplasia, this condition is considered a strong risk factor for breast cancer in women. (Whales are mammals and so have breasts; in belugas, they are located on either side of the vagina, with only the nipples visible and the mammary glands themselves hidden beneath a layer of blubber.)

Autopsy reports on twenty-four other stranded carcasses were published in 1994. Twenty-one tumors were found in twelve carcasses. Among these tumors, six were malignant. The researchers concluded, "Such a high prevalence of tumors would suggest an influence of contaminants through a direct carcinogenic effect and/or a decreased resistance to the development of tumors." Both possible mechanisms are currently receiving close attention.

To date, cancers identified in the beluga include bladder, stomach, intestinal, salivary gland, breast, and ovarian. The prevalence of intestinal cancer is especially high. Of seventy-three stranded whales autopsied since 1983, fifteen had cancerous tumors somewhere in their bodies, and one-third of these were intestinal tumors. No cases of cancer have been reported in belugas inhabiting the less contaminated Arctic Ocean.

The beluga whales of the St. Lawrence estuary have more wrong with them than cancer. They also have trouble reproducing. Even though belugas have been protected from hunting since the 1970s, their numbers have failed to rebound. When chemical analyses of their blubber were conducted to illuminate possible causes of both problems, PCBs, DDT, chlordane, and toxaphene—at some of the highest levels ever recorded in a living organism—were all found dissolved in the whales' fat. All four chemicals are endocrine disrupters, as well as probable carcinogens. All were banned decades ago. All are chemically very persistent.

Unlike PCBs and DDT, chlordane and toxaphene do not have a history of use in the St. Lawrence basin. And yet these two chemicals are found in the waters and sediments of the estuary, presumably because they are carried into the seaway by winds blowing up from the southern United States, where both were once used heavily. The St. Lawrence basin drains a 500,000-square-mile area; any contaminant that rains down within its vast perimeter is, sooner or later, flushed into the estuary.

There is another route of exposure. Beluga whales love to eat eels, which run through the icy, deep Lawrentian channel on their autumn migration from Lake Ontario to the warm waters of the Sargasso Sea. The eels may explain why the belugas are contaminated with Mirex, an organochlorine pesticide, now banned, that was once used against fire ants. There is no Mirex in the water of the lower St. Lawrence nor in its sediments and hardly any in the bodies of other marine mammals living in the estuary. But there is Mirex in the flesh of St. Lawrence eels. And there are two sources of Mirex in the Lake Ontario basin where the eels originate: a pesticide-manufacturing plant near Niagara and a river called the Oswego, where Mirex was once accidentally spilled. The eels are the apparent courier between these contaminated sites and the beluga whales living six hundred miles away.

Eels are very strange. Like salmon, they migrate thousands of miles to spawn, crossing between fresh and salt water to do so. However, eels make their journey in reverse: they spend twelve to twenty-four years living in lakes and rivers, and then they head out to the ocean to lay their eggs. Baby eels, each the size and shape of a willow leaf, spend their first year of life trying to swim back.

Less is known about what toxins eels might bring back from their birthplace, which is also a contaminated site. Elliptical and still, the Sargasso Sea lies within the clockwise current of the Gulf Stream. The islands of Bermuda rise from its center. Eels from freshwater rivers in North America, Europe, and Africa all converge here to spawn. The Sargasso sits at the center of a whirling gyre of currents, and so accumulates seaweed and debris from all over the Atlantic—but especially from the U.S. and Caribbean coasts. Along with the ocean's other detritus, chemical pollutants—such as DDT and balls of tar—also slowly drift in and accumulate here to join what the poet Ezra Pound once called "this sea-hoard of deciduous things."

I tried to be kind to my hospital roommate. No one else was. We were both recovering from surgery, but her situation was more typical of what happened to girls in Pekin: A fast car. Drunk boys. She was the only one pulled out alive, and the story had made the front page. When the nurses refused to tell her what had happened, I read aloud to her from the newspaper account. Mostly, she slept and watched TV. I spent a lot of time staring at her.

Outside this room, our lives were on two different tracks—in my view, at least—and I was trying to figure out how I had ended up here with her. I was the clean-living winner of the local Elks Club scholarship who viewed drink, drugs, TV, and junk food as tickets to nowhere, who was only back in this town for the summer, and now college had resumed and I was still here. Some malevolent current had deposited us together in this hospital. But unlike my partner in the next bed, no one had any explanations for my situation. The newspaper said she was expected to survive. Was I?

I examined the outline of my legs under the thin blanket, the shadow my hand cast on the sheet. Between the sheet and the blanket snaked the wretched catheter tube. I felt flattened down, like an animal wounded by something cruel and meaningless. My roommate looked over at me and touched her hands to her discolored face. Her boyfriend and brother were both dead.

"I think I'm going to stop partying for a while."

It was the kind of moment where laughing and crying were synonymous. What happened to her was pathetic. What was happening to me was pathetic. We started laughing.

"I think I'm going to start."

The belugas have local problems as well as global ones. Aluminum smelters and other industries lining the river basin have contaminated their waters with benzo[a]pyrene, a potent and well-known carcinogen.

Benzo[a]pyrene is seldom manufactured on purpose. With molecules consisting of twenty carbon atoms arranged as two hexagonal rings nestled on top of three hexagonal rings, it is created during the combustion of all kinds of organic materials from wood to gasoline to tobacco. It also occurs in coal tar, which is distilled to make a couple of familiar products. One of these is creosote, used to preserve wood (think of the smell of telephone poles on a hot summer day), and another is pitch, used in roofing and in aluminum smelting. Coal tar itself is classified as a known carcinogen, but because humans are almost always exposed to its constituent ingredients in mixtures, the data from human studies are inadequate to so classify benzo[a]pyrene individually. Animal assays, however, are unequivocal. Benzo[a]pyrene is thus listed as a probable carcinogen.

Benzo[a]pyrene causes cancer in a simple, direct way. Nearly all living things have in common a group of cellular enzymes responsible for detoxifying and metabolizing possibly harmful chemical invaders. When this enzyme group encounters benzo[a]pyrene, it inserts oxygen into the foreign molecule, the first step toward breaking it down. However, in a strange twist of fate, this addition *activates* benzo[a]pyrene rather than detoxifies it. The altered molecule now has the ability to bond tightly to a strand of DNA—that is, to one of the cell's chromosomes along which lie the organism's genes. A chemical invader so attached is called a DNA adduct, and it has the power to alter the structure of the DNA strand and produce a genetic mutation. If uncorrected, this type of damage can become a crucial step leading to the formation of cancer.

The number of adducts attached to an organism's DNA is considered a useful measure of benzo[a]pyrene exposure. DNA from the brain tissue of stranded St. Lawrence belugas bore impressively high numbers of adducts. They approached values found in laboratory animals exposed to levels of benzo[a]pyrene sufficient to cause a response in cancer bioassays. In contrast, DNA adducts were not detectable in beluga whales inhabiting Canada's more pristine estuaries.

Now we have reached the moment when, discharged at last from the hospital, I opened the door to my dormitory room and saw the bare mattress. I became secretive and territorial. I staked out a favorite stall in the women's bathroom. In my return every third month to the hospital for cystoscopic checkups, cytologies, and other forms of medical surveillance, I told no one where I was going. The interval between checkups approximated one semester's worth of time, one season. These seasons went by. I waited tables and pursued perfect grades. I finished college and began graduate school. I stopped studying grasses and started studying trees. I endured chronic bladder infections. I married.

Like breast cancer, bladder cancer can recur at any time, lying quiescent for years—sometimes decades—and then reappearing inexplicably. "Once a year for life" is the National Cancer Institute's guideline regarding cystoscopic examination of the bladder in patients, beginning five years after diagnosis. Most of my doctors seemed to agree with this. There has been more dispute about the risks and benefits of annual IVPs—intravenous pyelograms—which require X-ray imaging of the entire urinary tract, including the kidneys, to check for secondary tumors that might have seeded themselves upstream. IVPs involve considerable radiation. I navigated this controversy on my own. Transitional cell carcinomas are categorized according to aggressiveness, from stage 0 to stage 4. One pathology report describes my tumor as a stage 1; the other classifies it as stage 2.

After five years, my checkups became annual, and I was no longer tethered so tightly to the medical system. This change was almost unnerving—as though it were normal to think of the interior landscape of one's body as a study site that required constant data collection.

I immediately accepted a fellowship in Costa Rica, where I became involved in a field study of ghost crabs—delicate creatures that occupy burrows along the Pacific beaches at the edge of the rainforest. At the study's conclusion, the night before we were to fly out, I had a vivid dream: I am walking by the ocean and discover a pale orange crab, big as a whale, washed up on the beach. It is dying. I lie down next to it, and slowly it wraps a great, clawed arm around me. Reaching my arm over its carapaced body, I return the embrace. I am not afraid. As if in the final frame of a movie scene, giant letters appear in the sky above us, spelling out a single word—G-R-A-C-E.

Among those of us who had spent days out in the tropical sun trying to monitor the movements of these reclusive, lightning-fast animals, the dream was hugely funny. ("Sleep with any arthropods last night, Steingraber?") Not until I returned home did I connect the dream to the end of five intense years of monitoring the possible movements of cancer. Ghost crabs.

I mean to say two things here. First, even if cancer never comes back, one's life is utterly changed. Second, in all the years I have been under medical scrutiny, no one has ever asked me about the environmental conditions where I grew up, even though bladder cancer in young women is highly unusual. I was once asked if I had ever worked with dyes or had ever been employed in the rubber industry. (No and no.) Other than these two questions, no doctor, nurse, or technician has ever shown interest in probing the possible causes of my disease—even when I have introduced the topic. From my conversations with other cancer patients, I gather that such lack of curiosity in the medical community is usual.

Steadily increasing in incidence, bladder cancer is associated with a few lifestyle habits—especially cigarette smoking—as well as with more than a few occupations. Besides dyestuffs workers and tire manufacturers, these include janitors, mechanics, miners, printers, hairdressers, painters, truck drivers, drill press operators, and machinists. In one small factory in England, all fifteen workers distilling one particular aromatic amine, naphthylamine, developed bladder cancer. In 1984, bladder cancer was found in excess among inhabitants of Clinton County, Pennsylvania, where a forty-six-acre toxic waste site is contaminated with aromatic amines and benzene.

Other environmental patterns exist. Among U.S. males, bladder cancer is significantly higher in counties with chemical-manufacturing plants. Bladder cancer and exposure to perchloroethylene from drinking-water pipes were linked in Massachusetts, as we have seen in Chapter Four. In Taiwan, an investigation of bladder cancer deaths among children and adolescents found that almost all those afflicted lived within a few miles of three large petroleum and petrochemical plants. In a case-control study of bladder cancer in young women, risk of developing the disease was significantly associated with having undergone a thyroid procedure that involved the use of radioactive iodine.

Household dogs have a few interesting bladder cancer trends of their own. Transitional cell carcinomas of the bladder among pet dogs are significantly associated with direct exposure to insecticidal flea and tick dips, especially if dogs were obese or lived near another potential source of pesticides. A study of more than eight thousand dogs showed that bladder cancer in these animals was significantly associated with residence in industrialized counties, a pattern that mirrors the geographic distribution of bladder cancer among humans.

In 1990, at the International Forum for the Future of the Beluga, the conservationist Leone Pippard of the Canadian Ecology Advocates asked the following questions:

> Tell me, does the St. Lawrence beluga drink too much alcohol and does the St. Lawrence beluga smoke too much and does the St. Lawrence beluga have a bad diet . . . is that why the beluga whales are ill? . . . Do you think you are somehow immune and that it is only the beluga whale that is being affected?

❧

In Washington, D.C., the National Museum of Natural History, part of the Smithsonian Institution, can be found on Constitution Avenue midway between Capitol Hill and the Washington Monument. On any given day, its exhibit halls are busy with tourists and troops of schoolchildren. Central among the objects of their interest are the animals posed in freeze-frame action inside decorated dioramas, each an exquisite hybrid of taxidermy and landscape painting. Many contain the spoils of Teddy Roosevelt–era safaris. In a reconstructed African savanna, a lion forever slinks toward his unsuspecting prey: zebras foolishly grazing with their backs to the hunter. Will they look up in time? In the glass boxes all around, other dramas are unfolding. Animal mannequins ready the attack, startle at a sound, confront an intruder, or display their antlers. In this world of showcased trophies, nature is still a place of virility.

In an upstairs wing away from all this activity is a suite of rooms where nature appears very differently. These are the offices of the Registry of Tumors in Lower Animals, a project jointly sponsored by the Smithsonian Institution and the National Cancer Institute. (Since my visit, the registry has moved to the campus of George Washington University.) Here, too, one can look through glass at preserved animals. Whole specimens float in jars; their tumors are sliced and mounted on slides. Fish with liver cancer. Salamanders with skin cancer. Clams with genital cancers. Up here, animals are not so much triumphant warriors as casualties of toxic encroachment.

In 1964, the pathologist and physician Clyde Dawe discovered white suckers with liver cancer in Deep Creek Lake, Maryland. This was the first time this disease had been found in a wild population of fish, and Dawe was worried. While isolated incidents of fish tumors had been previously described, liver cancer in large numbers of individuals outside of hatcheries had never been seen before. What might be going on with other populations of fish or with other species?

The following year, at Dawe's instigation, the registry was founded to facilitate the study of tumors in cold-blooded vertebrates (fish, amphibians, and reptiles), as well as various invertebrates (such creatures as corals, crabs, clams, snails, and oysters). Field biologists who discover animals with tumors can send them live, frozen, or preserved to the registry. So can anyone else. Indeed, a number of the registry's accessions—and there are now over sixty-four hundred, representing nearly a thousand species—are submitted by ordinary concerned citizens.

After more than three decades of data collection and experimental research, several important patterns have emerged. All indicate that cancer in at least some species of lower animals, but especially liver cancer in fish, is intimately linked to environmental contamination. In these patterns is writ an urgent message to us higher animals.

First, the preponderance of cold-blooded animals with cancer are aquatic bottom feeders. And the dark beds of rivers, lakes, and marine estuaries are precisely where the highest concentrations of contaminants are found. Each year industry releases millions of tons of suspected carcinogens directly into surface water. Adding to this burden are motorboat exhaust and toxic runoff from the surrounding landscape, as well as deposition of airborne pollutants. Adhering to fine particles of sediment and pulled to the bottom by gravity, these chemicals slowly accumulate. Scavenger fish and detritus-grazing mollusks are thus highly exposed. They are also the animals most severely afflicted by tumors. The more contaminated the sediments, researchers have found, the higher the prevalence of certain tumors among those who haunt the cool, murky bottoms of rivers, lakes, bays, and estuaries. What's more, when extracts of these sediments are painted onto healthy fish, injected into their eggs, or added to clean aquarium tanks in the laboratory, these fish contract cancer in significant numbers.

Two trends among the fish echo human trends: frequency of cancer has increased substantially during the past thirty years, and the distribution of these cancers tends to be geographically clustered around areas of environmental contamination. Obviously, no one fills out a death certificate when a fish dies. We cannot, therefore, ascertain rates of cancer incidence and mortality per every 100,000 lake trout. Instead, cancers in animals such as fish are gauged by frequency of epizootics. Like an epidemic among people, an epizootic refers to large numbers of individual animals in the same area having the same disease at the same time.

According to the registry director, John Harshbarger, epizootics of liver tumors in fish are rising and coincide with the big increase in production of synthetic organic chemicals since 1940. In North America, there are now liver tumor epizootics in sixteen species of fish in at least twenty-five different fresh- and saltwater sites. Each of these sites is considered polluted. In contrast, liver cancer among members of the same species who inhabit nonpolluted waters is virtually nonexistent. Some skin cancers among North American fish also appear linked to chemical carcinogens. Other fish epizootics under investigation include testicular cancer among yellow perch in the Great Lakes and various connective tissue cancers in Great Lakes walleye, northern pike, and muskellunge.

The registry's survey of epizootic liver cancer in marine fish from around the world identified twelve affected species on three continents. In all cases, disease was correlated with the presence of chemical pollutants, as compared to a virtually zero background rate in nonpolluted waters. William Hawkins of the Gulf Coast Research Laboratory in Mississippi collaborates with registry staff in his laboratory studies of fish cancers. "We should listen to what the fish have to tell us about our environment," says Hawkins. "When fish get certain kinds of cancers, it's almost always the result of human activities."

The new field of ecotoxicology is uncovering additional patterns. Among other projects, this line of study looks for "genetic signatures" along the organism's chromosomes that identify exposure to particular toxins that have been released into particular ecological systems. It is as though fugitive chemicals carve their initials into the genetic code, leaving their marks for us to read. DNA adducts, as found in the chromosomes of St. Lawrence belugas, are one such signature. These can be quantified, in which case, the genes themselves serve as an instrument for measuring the dose of a carcinogen received by the organism. So far, this system has been worked out for only a small number of chemicals. But for those in which it has, increased exposure correlates with increased rates of cancer. This relationship seems especially strong for liver cancer in fish.

These findings have direct implications for us. The remarkable conservatism of genetic systems spans even the four hundred million years that have passed since fish diverged from the rest of the vertebrates. "The same genes that seem to go awry in rodents and humans are also going awry in trout when they are exposed to cancer-causing agents," says the biochemist George Bailey at Oregon State University. "Cancer is cancer, is what we're finding out at the molecular level." Aquarium studies in the laboratory show that the same carcinogens known to cause cancer in humans and rodents also cause cancer in fish and mollusks—and they are often metabolized in the same way. Concordance is not perfect, however, and plenty of exceptions exist. Lobsters, for example, do not get cancer; they seem able to sequester carcinogens in their tissues in a way that prevents damage to their chromosomes.

The appearance of cancer in wild animals—especially, it seems, liver cancer among fish—may announce the presence of carcinogenic agents in the environment. Unlike laboratory animals but very much like ourselves, wild animals living in contaminated habitats are exposed to low levels of ever-changing combinations of chemicals throughout their lifetimes. And like the proverbial canaries in the coal mine, whose sudden collapse warned miners of poisonous gas, wildlife with tumors are environmental sentinels. Not all species suitably fill this role. In some animals, infectious viruses can also cause tumors. The question of which animals are the most accurate indicator species for which environmental carcinogens is receiving considerable attention among wildlife biologists.

Turtles seem to serve as good sentinels for the presence of hormone-disrupting chemicals in the environment. As adults, they are not any more sensitive than other vertebrates, but as embryos, turtles possess a peculiar characteristic that dramatizes the presence of such chemicals in the environment: sex determination lability. This means that slight changes in the environment during fetal development can change the animal's sex. Temperature can have this effect, and so can certain chemicals. When painted onto their eggs, PCBs turn red-eared sliders from male to female. Moreover, this happens at very low concentrations—levels comparable to the average level of PCBs now found in the breast milk of women living in industrialized countries. Might the turtles have something to tell us about hormone-sensitive breast cancers?

Deer Island, Massachusetts, is situated in Boston Harbor. It is actually not an island, although it once was. When a hurricane filled a narrow channel with sand in 1938, Deer Island became a swelling at the end of a peninsula. The former site of a military base and a prison, Deer Island is the current site of one of the world's largest sewage treatment plants, recently expanded and upgraded. The access road along the tongue of land leading out to the gate lies just below the flight path for jets landing at Boston's Logan Airport. It is not a tranquil place. Nonetheless, I like to come out here every now and then. The tidy, modest homes lining this road and the flocks of children swooping around on bicycles remind me a bit of Normandale. The island itself was granted to Boston in 1634. Deer once walked out here across the ice.

If you park your car and stand facing the harbor—as though to watch planes land—you can look out over Deer Island Flats, a shallow area where winter flounder congregate. They are down on the bottom, working hard at imitating mud and sand.

Flounder would certainly win my vote for most bizarre looking fish. A flounder starts out life normally enough, but then its left eye begins to migrate across its face until it crosses sides and bumps up against the right one. The top half of its face now appears to be at a ninety-degree angle to the bottom half. The flounder then lies down on its eyeless left side and remains in the sediment, unmoving as a saucer. Unlike some bottom dwellers, it relies on the ocean sediments only for camouflage, not for food. A lie-in-wait predator, the flounder periodically shoots up from its concealed position to chase down unwary marine worms and other prey.

None of this activity is visible from the surface. Nor is the fact that the winter flounder of Deer Island Flats suffer from high rates of liver cancer. The cancer epizootic among this population—and its connection to the harbor's chemical contamination—is one of the most well-described case studies on file at the Registry of Tumors in Lower Animals. Populations of winter flounder in cleaner waters away from the harbor do not have tumors.

Two elegant experiments involving contaminated sediments from a harbor in Bridgeport, Connecticut, further support an environmental explanation for the tumors in the Boston Harbor winter flounder. In the first study, researchers exposed uncontaminated adult oysters to sediments from Bridgeport's Black Rock Harbor in Long Island Sound. This procedure was done in both the laboratory and the field (by suspending caged oysters in the harbor waters). The sediments of Black Rock Harbor are known to be contaminated with a variety of toxins, from PCBs to pesticides. Both sets of oysters accumulated these contaminants in their tissues, and both sets developed tumors. Oysters exposed to uncontaminated sediments did not acquire tumors.

In the second experiment, blue mussels were exposed to Black Rock Harbor sediments. They similarly absorbed the contaminants into their bodies. They were then fed to winter flounder. The flounder developed tumors in their kidneys and pancreases, as well as precancerous lesions in their livers. No tumors developed in flounders fed on mussels that were exposed to sediments from uncontaminated sites.

I wish the flounder beds were as visible to us as the bright interior of a Smithsonian diorama. It is easy to dismiss the problems of species we seldom see—especially bottom-dwelling, cold-blooded ones. So I wish we could watch them down there—the clownish Picasso-esque faces twisted over top of the blind, white undersides that lie in direct and constant contact with the harbor's contaminated sediments.

Out of this desire, I have developed an idea for a pilgrimage that involves people with cancer traveling to various bodies of water known to be inhabited by animals with cancer. It involves an assembly on the banks and shores of these waters and a collective consideration of our intertwined lives. We could start with the question posed by Leone Pippard at the beluga conference: Do you think you are somehow immune?

I have worked out a possible itinerary, too. Beginning on the edge of Deer Island, we would travel north to Cobscook Bay in northern Maine. Here, more than 30 percent of softshell clams have tumors in their reproductive organs. In this otherwise-pristine habitat, clams and other marine animals have been exposed to the herbicides 2,4-D and 2,4,5-T both as runoff from nearby blueberry bogs and as drift from commercial forests where they were sprayed aerially. Rebecca Van Beneden, a zoologist and molecular biologist from the University of Maine, studies the patterns of tumors in clams from these estuaries and is looking for genetic changes that might reveal the sequence of molecular events leading to carcinogenesis. Her research is made urgent by the fact that rates of ovarian and breast cancers among the women living in this county are higher than the national average. "Maybe there's a correlation; maybe there's not," she says, referring to clams, women, and herbicides. "At the least, we need to pursue it."

Next, we would travel across the continent to Puget Sound where it reaches into the mouth of the Duwarmish River. The story of liver cancer in the English sole inhabiting these waters greatly resembles that of the winter flounder. Flounder and sole, both flatfish, are mirror images of each other: one lies off Boston's shore, the other off Seattle's. The molecular epidemiologist Donald Malins and his colleagues have captured sole in the lower reaches of the Duwarmish, which is contaminated with PCBs and benzo[a]pyrene, and analyzed their tissues for DNA damage. It was diverse and extensive. Indeed, these fish had hundreds of times more DNA alterations than sole captured in Washington State's more pristine rivers. Happily, in other studies, when fish from polluted streams were moved into cleaner water, their DNA began to repair itself, and their risk of developing cancer presumably declined.

Our journey would then take us to the south branch of the Elizabeth River in Virginia, which is home to a small, common fish called the mummichog. The males are handsome—dark green or steel blue with white and yellow spots and narrow silvery bars. The females are olive green. Unlike winter flounder or English sole—all famous travelers—the mummichog is a homebody and spends its life scavenging for food within the same spot. We would gather on both banks of the river branch. On one side is a wood treatment facility. Here, residues from coal-tar creosote are found in very high concentrations in the river's sediments. About one-third of the mummichogs on this side of the river have liver cancer. On the other side of the river, where sediment contamination is much lower, no mummichogs have liver cancer.

From Virginia, we would travel north to the steel mills where the Black River of Ohio empties into Lake Erie. In the early 1980s, researchers discovered that Black River catfish living near the discharge pipes of coking facilities, where coal is turned into fuel for the mills, had very high rates of liver cancer. The facilities closed in 1983. By 1987, sediment contamination had declined dramatically, liver cancer among brown bullhead catfish had declined by 74 percent, and the percentage of fish with normal livers had doubled. In the conclusion of their study, the researchers expressed amazement at "the efficacy of natural, unassisted remediation once the source of pollution is removed." This is a phrase we would want to think about while walking the length of this river.

Finally, I would bring us to the Fox River, which flows south from Wisconsin and joins the Illinois River near the river town of Ottawa, about seventy-five miles upstream from my hometown. Tumors from an assemblage of Fox River fish—walleyes, pickerel, bullheads, carp, and hog suckers—were among some of the first identified. But I would not take us to the old industrial sites thought to be responsible. Instead, I would bring us up to Buffalo Rock, a ninety-foot-tall bluff that abuts the Illinois River just a mile or two downstream from its confluence with the Fox. From here we could see the entire river valley spread out before us.

The ecology of Buffalo Rock, which is about the size of Deer Island, was ravaged in the 1930s by a strip mine that tossed highly toxic shale and pyrite onto the topsoil. All life forms, animal and plant, were killed. For decades, it remained a landscape of jagged escarpments that funneled acidic runoff into the river.

Then in 1983, the artist Michael Heizer was commissioned to

help reclaim Buffalo Rock. Inspired by ancient Native American earth mounds, Heizer used bulldozers to sculpt the thirty-foot furrows into the shapes of five river animals: water strider, frog, catfish, snapping turtle, and snake. Each of these earth sculptures is hundreds of feet long. Rye grass now grows on top of them. You can climb the catfish's whiskers, walk along the strider's legs, lie down on the snake's head, touch the grass, breathe the air.

I would choose a September afternoon with a sky as blue as cornflowers. Cancer patients and ex–cancer patients would gather on the backs of these monumental animals and, in this place of damage and reclamation, talk about all that we have seen.

SEVEN

earth

All flesh is grass.
—ISAIAH 40:6

To hear my mother tell it, butchering day was a festive occasion. Preparation started weeks in advance when all five girls got new outfits, each hand sewn by my grandmother. Butchering dresses, they were called. These were not work clothes, my mother stresses here, but real dress-up dresses. They honored the company that would be arriving to assist with the slaughter and share in its spoils.

To hear my Aunt Lucy tell it, butchering day was a time of somber anticipation. Before dawn, the yard was filled with fire and cauldrons of boiling water, the kitchen with knives and large strangers. She recalls tiptoeing through the house, listening for and finally hearing . . . the gunshot. Aunt Lucy firmly disputes the existence of butchering dresses.

Girls do not appear in my uncles' accounts. In their rendition, butchering day was a ceremony of manhood. Sons accompanied their fathers. They shot pigs. They shared secret knowledge. And they partook of hog testicles.

The story of Hog Butchering on the Farm—featuring gleefully indelicate food items, special costumes, and gunfire—easily withstood repeated retellings to the next generation of children, which included me and my various cousins. I always preferred my mother's version because she could provide the most vivid details of sausage making. She starts by describing the scraping of bristles from the skin of a scalded hog, spends considerable time on the part where the women clean out the intestines (soon to become casings), and then ends with a vision of the sausage press as manned by Great-uncle Sander. Better than anyone else, Sander knew just when and how to twist a pork-engorged casing to form a chain of perfect links. Coiled into five-gallon crocks and covered with melted lard, these sausages could safely pass the winter in the crypt of a farmhouse cellar.

Somehow, all these images explain why great jars of beans, flour, grain, rice, cereal, and pasta line the shelves of my tiny, urban apartment. Something in my kitchen is always boiling or soaking, as though an enormous family dined here. I like the purchase of commodities in bulk quantity, the laying up of food, the heft and volume of it all. I like fruit in the fruit bowl, greens in the crisper, a sack of onions in the cupboard, potatoes under the sink, a wreath of garlic bulbs nailed to the wall. The more dirt on the carrots the better. I like to visualize my food growing in the field. And although I store no crocks of sausage, I like to imagine old Uncle Sander bent over the extruding press, with a lengthening garland of pork looped about his shoulder like a string of Christmas lights.

In many respects, American agriculture has changed dramatically since *Silent Spring* was published. For one thing, the number of farms has declined sharply. Illinois now has half the number of farms it had in 1960, and at last count, my home county, a leading hog producer, could lay claim to only 1,008 of them. Secondly, land ownership has become separate from farming: over half the farmland in the United States is now farmed by persons who do not own it. My cousin John, for example, leases most of the thirteen hundred acres he farms near Saybrook. The agricultural landscape has also become more uniform. Farm animals have vanished from the barnyards as the raising of livestock has become an enterprise separate from the growing of crops. And the diversity of crops grown on any one farm has itself declined. Over the past three decades in Illinois, harvests of fruits, vegetables, hay, wheat, and oats have all fallen off. Orchards, pastures, vegetable plots, and woodlots have been plowed into ever-larger fields of corn and soybeans, which are now running neck and neck in number of acres planted.

All of these changes conspire to make farming an increasingly remote activity. With meat, milk, eggs, and produce trucked to our supermarkets from distant and unknown locations, we know less and less about how and where our food is grown and by whom. And the number of folks around who can explain such things to us is diminishing. The disappearance of farm talk from our communities is not included in compilations of agricultural statistics, but it is a very real change.

I feel lucky to have grown up around such talk—not only because my mother descends from a line of farmers but because I waitressed in roadside restaurants. Rainy mornings brought farmers in from the fields to sit and visit for a while. One had only to pour coffee and listen. Another group came in very early before heading out to work. Indeed, a magical hour occurred between 4:00 and 5:00 A.M. when the last of the shift workers still occupied the booths and the first wave of farmers began lining up at the counter. Talk of union contracts mingled with discussions of weather and grain prices. Outside the windows, corn and bean fields slowly took shape as the darkness faded into gray light.

Two decades later, I take my nephews to the cafés where I once carried platefuls of pancakes and chili-mac. We don't see many farmers there. Among the idle talk, there is a silence.

By other measures, farming has changed remarkably little since Rachel Carson described what she saw as a dangerous trend toward an increasing reliance on pesticides. Indeed, many of the changes outlined thus far have come about, directly or indirectly, because agriculture has traveled even farther down the path of chemical dependency depicted in *Silent Spring*.

Synthetic chemicals introduced into agriculture after World War II reduced the need for labor. At about the same time, profits per acre plummeted. Both these changes pressed farmers into managing more acres to earn a living for their families. Average farm size increased. (Or as my Uncle Roy says, "These days, you have to plow half the state or forget it.") Federal farm programs that encouraged farmers to specialize in a single crop further increased the need for chemicals to control pests. And the use of these chemicals themselves set the stage for additional ecological changes that only more chemicals could offset.

The increasing use of herbicides to control weeds, for example, has discouraged the practice of crop rotation, a kind of repertory theater in which a sequence of crops is cycled through a field—corn, oats, hay, corn—each one altering the chemistry and physics of the soil in slightly different ways. Because their residues can overwinter in the soil, herbicides prevent the sowing of chemically sensitive crops the following season. Herbicide-insensitive corn, therefore, may end up planted in the same field year after year. Lack of crop rotation, in turn, encourages insect pests—whose reproductive cycles are no longer disrupted by changing vegetational patterns—and these outbreaks invite the use of insecticides. Alfalfa has natural weed-suppressing effects when rotated in with corn and other grain crops, but without farm animals, there is little local market for it. Simple plowing helps control many kinds of weeds and pests, but the enormous size of most fields locks farmers in to low-till and no-till practices in order to prevent topsoil from blowing away.

Some entomologists routinely refer to insecticides as ecological narcotics because, as with any chemical dependency, routine use invariably makes worse the very problem it was supposed to fix. Drowning one's sorrows in alcohol begets more sorrow. Eradicating insects with pesticides incites more severe pest outbreaks.

The astonishing ability of insects to repel chemical attacks is accomplished in several ways. The development of genetic resistance, a simple evolutionary phenomenon, is one. The few individual pests who happen to possess an inborn ability to detoxify particular insecticides (or whose behavior or anatomies in some way protect them from being poisoned) become the progenitors of the next generation as their more chemically sensitive compatriots are killed off. With repeated exposure and over many generations, this minority eventually becomes the majority until entire populations of pests carry genes that confer some degree of pesticide resistance. Higher dosages or different, more potent chemical weapons are then required to maintain control.

In 1950, fewer than 20 species of insects showed signs of pesticide resistance. By 1960, Rachel Carson had documented an alarming 137 species resistant to at least one pesticide and urged that we should hear in this statistic the early rumblings of an avalanche. She was right. By 1990, the number of pesticide-resistant insect and mite species stood at 504.

In creating pests impervious to the arsenal of chemical weapons directed at them, the story of herbicides reiterates the story of insecticides. Herbicide-resistant weeds are not mentioned in *Silent Spring*, as they did not yet exist. Today, weed scientists have identified 273 such species. In tracing the explosion of herbicide resistance among weed species that began in the late 1980s, researchers conducting a recent study were forced to conclude that the "short-term triumphs of new pest control technologies have carried with them the seeds of long-term failure."

There are two other means by which campaigns to eradicate unwanted life forms with pesticides become self-defeating. Both are the result of ecological, rather than evolutionary, forces. The first is called resurgence, and it describes the uncanny ability of some pests to come roaring back in even greater numbers soon after their ranks have been decimated by chemical spraying—far sooner than can be accounted for by the development of resistance. The second is the appearance of altogether new miscreants—so-called secondary pests—usually involving an insect that had never before presented a serious agricultural threat. Both phenomena are attributable to the death by poisoning of insect predators and parasites.

The explanation, then, for these two apparent mysteries is contained in a childhood riddle: the enemy of my enemy is a friend of mine. Unfortunately, under the prevailing methodology, chemical poisons are laid down over the entire ecological system, of which the natural enemies of pests, too, are a part. In this way, we sabotage, rather than assist, our potential allies. These include spiders, beetles, praying mantises, dragonflies, lacewings, and other insects familiar to most casual gardeners, as well as a myriad of species known only to the entomologists. Many kinds of tiny, inconspicuous wasps, for instance, play important roles in natural pest control. Their habits involve drilling holes into the bodies of plant-consuming insects, injecting clusters of eggs, and transforming their hosts from speedy eating machines into mummified incubators. Upon hatching, the young parasites chew their way out, mate, and continue the cycle.

All of these beneficial species are poisoned along with the creatures on which they feed. Even more unfortunately, many natural enemies are more sensitive to the toxic effects of pesticides than are the pests, and their ranks do not rebound as quickly. In essence, pesticides punch into the food web gaping holes that pest species are free to pour through, their population growth unchecked. Other formerly harmless organisms, likewise released from predatory and parasitic pressures and on their way to becoming secondary pests, quickly follow. Now only more pesticides can avert complete disaster, and conditions have been created for chemical dependency.

All this is well described in *Silent Spring*. What has been documented since then is just how thoroughly we undermine our own efforts at pest control by trying to solve an ecological problem with a chemical tool.

The agroecologist David Pimentel of Cornell University has devoted his career to understanding the network of tiny, unseen interactions surrounding crop plants and has described dozens of unforeseen consequences of chemical pest control. Many insect predators, for example, rely on elaborate search-and-attack strategies to capture their prey. Trace exposures to insecticides can alter these behaviors: natural enemies are spared from death but rendered incapable of locating the pests. Thus, even at sublethal levels, chemical pesticides can impair effective biological control mechanisms. Pimentel has observed that fungicides, too, can trigger insect outbreaks—as when poisons sprayed to control plant molds end up serving as healthful tonics for crop-eating insects troubled by fungal diseases. Healed from their infections by the fungicide, the pests enjoy enhanced survival. Soybean fields so treated, for example, were shown to harbor greater numbers of both cabbage loopers and velvet bean caterpillars, both serious pests that can significantly reduce yields. In this way, the use of one poison (fungicides) can create the need for another (insecticides).

The disturbance of ecological interactions by chemical poisons has created secondary pests out of weed species as well. Since herbicides have been in widespread use, a number of shade-intolerant grasses, formerly weeds of minor status, have emerged as major problems. In this case, absence of competition, rather than absence of predation, drives the process. The hard, tight ground and full sunlight created by weed killers also create ideal conditions for the sun-loving grasses, whose growth is otherwise kept in check by the shade thrown down from broad-leafed weeds. In the absence of these shadows, grass seedlings soon become established, reach maturity, set seed, and ensure the need for additional chemical weed killers.

What I am writing here is hardly an exposé. The concepts of resistance, resurgence, and secondary pests have been taught in introductory science classes for decades. Indeed, some college textbooks now provide examples of pesticide resistance to illustrate Darwin's theory of evolution. In the ways in which it reveals, through destruction, the invisible workings of the ecological world, pesticide-dependent agriculture has become pedagogically useful. These textbook principles bear repeating here because so little has changed. In spite of a growing awareness of its various irrationalities and in spite of various declarations of reform, the use of agricultural pesticides continues to rise. And so do crop losses caused by the pests.

Since synthetic pesticides were introduced into agriculture at the end of World War II, total crop loss due to insect damage has doubled—from 7 percent in the 1940s, when all agriculture was essentially organic, to 13 percent by the end of the 1980s. Higher yields have more than offset this forfeiture, and reliance on pesticides is still economical in the short run. But the pests clearly have not been bludgeoned into submission. When grown in rotation, corn had few insect pests, for example. In 1945, almost none of the U.S. corn acreage was treated with insecticides. Crop losses due to insects, according to the U.S. Department of Agriculture (USDA) averaged 3.5 percent. Now less than half of corn is grown in rotation, and corn is the largest user of insecticides. Despite the more than 1,000-fold increase in the use of insecticides on corn, losses to insects now average 12 percent.

In one recent study, researchers compiled a variety of agricultural censuses and surveys to investigate changing pesticide use in specific U.S. regions. In all three regions they examined—Sun Belt, Wheat Belt, and Corn Belt—reliance on pesticides in general, but especially herbicides and fungicides, has grown substantially since the 1960s, both in intensity (amount of chemicals used per acre) and extensiveness (percentage of cropland treated). In many cases, these increases continued into the early 1990s—contrary to the commonly held belief that pesticide use had become more judicious. The *rate* of increase, however, began to stabilize in the 1980s, and the use of insecticides has actually begun to decline—in part because of the banning of toxaphene in 1982, in part because newer formulations of pesticides are more acutely toxic at smaller doses, and in part because some farmers are moving toward more ecologically based methods of insect control.

The vegetable fields and fruit orchards of the Sun Belt, according to this study, receive the highest rates of application—especially of insecticides and fungicides. Dates, peaches, grapes, and tomatoes are sprayed with more fungicide (as measured in pounds of active ingredient per acre) than any other crops, for example. Pears, on the other hand, top the list of crops most heavily sprayed with insecticides. Corn ranks only eighteenth on the herbicide list. But because we grow so much of it, this crop alone consumes 53 percent of the total herbicides used in this country. Together, corn and soybeans are responsible for nearly three-quarters of all herbicides used, which in turn account for the majority of all pesticides used. In short, Corn Belt weeds have become the number one target of agrichemical warfare.

A cornfield and a soybean field are very different places.

Bean fields are humble; they start out that way and stay that way. For reasons I can't explain, they are also a little bit sad. Walking through a soybean field, I feel like myself, only sadder. A soybean is a delicate plant. Like all other legumes—clover, peas, alfalfa—the soybean plant has a softness in its leaves. Fully grown, it is mostly shaped like a little bush that never extends much above the thighs, but, late in the season, an inconspicuous twining reveals its origin as an Asian vine. In spite of their modesty, the high-yielding varieties of soybeans are given brawny names—Jack, Burlison, Pharoah—that sound like brands of condoms.

Illinois soybeans bloom in midsummer and produce somewhere between sixty and eighty pods per plant. The pods are fuzzy—almost bristly—and bulgy. Each holds three (on average) round, pale-yellow beans. At the end of the season, soybeans seem to lose all sense of individuality. Turning from deep green to brilliant yellow to an indescribable shade of red-brown, they all blur into each other and sink back toward the earth, forming an upholstered surface that invites one to walk in and lie down.

Not so with cornfields. There is something animate about corn, which starts out as a merged, green expanse and then sharpens and stiffens into lines of individual, human-shaped forms. Corn is proud; it seems to stand in judgment somehow. It has conspicuous body parts—ears, tassels, silks, stalks—and it has the power to alter the landscape. By summer's end, county roads that had once offered unbroken views of the horizon become chutes between solid walls of corn. But this view offers only a one-dimensional perspective. To really understand how corn occupies space, you have to push past this screen and step into the deep interior of the field.

Walking through a field of corn can feel very sheltering. It can also make you have crazy thoughts or bring on panic, as when swimming in a large lake and suddenly realizing you are too far from shore. Much of corn's power to soothe and disorient surely has to do with the fact that it becomes taller than we are. By the end of July, you cannot see over it or walk through it without being touched along the length of your body by long, fibrous leaves. Moreover, even when green, corn makes a continual noise. It is a kind of hiss, not unlike the sound of snow falling. *Listening to the corn grow* is a phrase of respect, not an expression of boredom.

Corn is a grass and, as such, owes its existence to the wind. The silk-shrouded ear is the female flower of the corn, and each stalk has between one and three ears. The entire harvest depends on the ability of pollen grains to blow loose from the reddish-gold tassels at the top of the stalks, float as a cloud over the field, and alight on the ends of the silks as they wave from the ears. Each strand leads back to a single kernel. Like following a thread through a maze, a grain of pollen sends its chromosomes down the entire length of a silk to reach one kernel and fertilize it. As each ear holds about six hundred kernels and each acre about thirty thousand individual cornstalks, this process must happen independently and successfully innumerable times.

Most high-yielding corn varieties are hybrids. Therefore, fields of corn destined for seed consist of two different varieties planted together in alternating stripes. The so-called male rows, the pollen donors, are allowed to tassel out. To ensure cross-pollination, the stalks of the other variety are castrated. Their tassels yanked out before they mature, they are thereafter known as the female rows. This task is accomplished by sending into the fields crews of teenaged detasselers who either walk the rows or ride above them in little baskets pulled by machines. I can attest that both methods are equally tedious. Still, good money could be made from a few long, hot weeks of removing reproductive organs from cornstalks and flinging them to the dirt.

More than soybeans, corn in central Illinois defines certain experiences. "Corn-growing weather" means hot, sunny, and humid. Such conditions are probably also good bean-growing weather, but no one ever calls it that. Soybeans, introduced from China in the late 1800s, are relative newcomers to Illinois farming. Today, Illinois is responsible for fully 10 percent of the world's total production of soybeans, but we have not yet completely incorporated them into our myths about ourselves.

Corn parables, on the other hand, abound. My mother is an especially skillful narrator. Consider, for example, the Story of the Scarred Knee.

As a young girl, my mother was running through a newly harvested cornfield when she tripped and tore open her knee on the sharp end of a broken-off stalk. My grandmother, while tending to the wound, noticed a bit of stringy debris stuck to the flesh. She gently pulled on it. Out from inside my mother's leg came a wedge of cornstalk *as long as a pinky finger.*

As a means of encouraging caution in wild children, this proved a successful tale. My sister and I could be stopped in our tracks by the mere mention of it. In fact, I am still spooked at the thought of running through a cornfield.

Corn and soybeans are harvested at about the same time with modified versions of the same, basic, all-in-one machine called a combine. Again, however, the two activities have a very different feel to them. Combining corn seems almost violent—even more so from the cab of the combine. As the long tines of the combine's corn head advance down the rows, the stalks first begin to tremble and wave their leaves wildly, then jerk, snap back, and disappear. Somewhere out of sight, beneath the driver's feet, a series of augers, chains, cylinders, rasp bars, sieves, screens, and fans dissassemble the ears, strip the corn from the cobs, and shoot the ground-up trash back into field. In the meantime, out the rear window, the hopper fills with a thick stream of gold kernels.

From a distance, harvesting beans looks peaceful. Delicate spinning sickle bars sweep the pods, stems, and leaves into the hidden chambers of the combine (where different settings allow the same machinery to dissociate beans from pods as kernels from cobs). Showers of red, glittering chaff fall gently back to the ground in its wake. Beneath this apparent serenity, however, lies high-stakes anxiety. Soybeans nestle close to the ground. One good-sized rock pulled into the grain table can completely disable a twenty-five-foot combine. Moreover, bean pods cannot withstand repeated rounds of soaking and drying out once they are ready for harvest. As my cousin puts it, "A little rainfall on the wrong day in October and you're done."

A bushel of corn weighs about 56 pounds, and a bushel of soybeans about 60. An average acre of corn in Illinois yields 149 bushels, and an average acre of soybeans around 43. In the fall of 1996, corn sold for $2.69 a bushel, soybeans for $6.80.

These are the names of the enemies: velvetleaf, foxtail, cocklebur, pigweed, smartweed, ragweed, morning glory, lambs-quarters, jimsonweed, dogbane, milkweed, nightshade, fall panicum, shattercane, nutsedge, Canada thistle. Each weed has its own manner of surviving and reproducing in Illinois corn and bean fields. Canada thistle is a perennial with creeping underground stems that give rise to new shoots. Cockleburs promulgate the usual way but hedge their reproductive bets: of the two seeds housed inside each prickly bur, one germinates in the spring, while the other waits patiently in the soil and sprouts a year later. Lambs-quarters, foxtail, and pigweed are now-or-never annuals that pelt the landscape with a yearly seed rain, as weed scientists refer to their prodigious reproductive output. The seeds of these three species are among those most commonly encountered in Corn Belt fields, where the soil contains between 600 and 160,000 living weed seeds per square meter.

Cultivation by plowing, hoeing, or disking was once the dominant strategy for coping with weeds of all kinds in corn and bean fields. This was done by hooking various implements—the spring tooth harrow was a popular one—to the back of a tractor and dragging them through the soil. For perennial weeds, cultivation worked to repeatedly disturb the above-ground parts and eventually starve the roots; the aim for annual weeds was to cut them apart before their seeds formed. Since the early 1970s, however, virtually no agricultural research has been conducted on mechanical weed control, which, for all the reasons described earlier, fell out of favor. Instead, almost all weed control research has been conducted on herbicides—most of it directed at formulating chemicals poisonous to weed species but relatively harmless to crop plants. Most recently, geneticists have been working on developing crops that are herbicide resistant, which would allow even more intensive applications of chemical weed killers.

By 1993, according to the Illinois Agricultural Statistics Service, herbicides were applied to 99 percent of corn and bean acreage. Lest anyone assume that the habitudes of warfare are no longer part of chemical pest control, here are the trade names of some the ones in common use: Arsenal, Assault, Assert, Bicep, Bladex, Bullet, Chopper, Conquest, Contain, Dagger, Lasso, Marksman, Prowl, Rambo, Squadron, Stomp, and Storm.

These herbicides kill by a variety of different poisoning mechanisms. Some interfere with plant hormones. The original synthetic herbicide 2,4-D, for example, causes its target to grow faster than it can obtain nutrients. A weed exposed to 2,4-D has a peculiar, twisted posture—its stems are swollen and bent. As tissues burst, disease-causing agents invade and deliver the coup de grâce. In 1993, about 13 percent of Illinois soybean fields and 14 percent of cornfields were sprayed with 2,4-D. Other herbicides halt the production of amino acids, from which proteins are made. Still others strike directly at the process by which plants use sunlight to transform water and carbon dioxide into sugar and oxygen. These are the triazine herbicides, and they are used most often in cornfields but also in orchards and lawns, as well as cotton, sugarcane, and sorghum fields. One member of the group, atrazine, is currently one of the top two most widely used pesticides in United States agriculture.

The triazines have been in use since the 1950s, but the actual mechanism behind their ability to kill plants was not discovered until years later. We now know these herbicides poison a chain reaction that takes place inside the chloroplasts. Scattered across the leaf's surface like tiny Quonset huts, chloroplasts provide housing for the pigment chlorophyll, as well as the rest of the cellular machinery that runs the light-driven reactions of photosynthesis.

At its heart, photosynthesis depends on the handing off of electrons from one acceptor molecule to another, like a bucket of water passed down a fire brigade. These electrons are cleaved from molecules of water, and their liberation is intimately linked to the creation of oxygen. For photosynthesis to work, the electrons must reach a central reaction center. This is the link poisoned by triazine herbicides. By binding to a protein in the reaction center, triazines effectively block the bucket brigade's action. Without the transfer of these crucial electrons, the entire interlocking process grinds to a halt. "Excessive radiative excitement" builds up in the plants' green pigments, along with the toxic products of oxidation. The chloroplast swells and eventually ruptures.

The wisdom of broadcasting over the landscape chemicals that extinguish the miraculous fact of photosynthesis—which, after all, furnishes us our sole supply of oxygen—is, in and of itself, questionable. Applied directly to the soil, triazines are absorbed by the roots of plants and transported to the leaves. They poison from within. Triazines are thus water soluble. And because of their solubility, they tend to migrate to many other places.

Their capacity to inhibit photosynthesis does not stop once they leave the farm. Traces of triazine herbicides are now found in groundwater, as well as about 98 percent of all midwestern surface waters. They have demonstrated a remarkable capacity to poison plankton, algae, aquatic plants, and other chloroplast-bearing organisms that form the basis of the freshwater food chain. According to the Environmental Protection Agency: "Field monitoring studies . . . indicate that the actual toxic concentration of the triazines and their degradages in the environment trigger deleterious effects at much lower concentrations than predicted by the laboratory guideline studies."

Triazines have also been detected in raindrops in twenty-three states in the upper Midwest and the Northeast, including pristine areas such as Isle Royale National Park in northern Minnesota. The possible effects of "triazine rain" on the forests and fields over which it is deposited are not yet known. It is known that triazines inhibit the growth of native prairie species: like corn, mature plants are tolerant of the herbicides, but seedlings are quite susceptible.

If photosynthetic inhibition were the only problem, the widespread use of triazine herbicides in the Corn Belt would be worrisome enough. But triazines also have effects inside our own bodies, where they gain entry as contaminants of drinking water and as residues on food. Three of the triazines—cyanazine, simazine, and atrazine—are classified as possible human carcinogens. Atrazine, a known endocrine disrupter, is restricted for use in Germany, the Netherlands, and several Nordic countries. However, in the United States, atrazine is used on about two-thirds of all cornfields (81 percent in Illinois). Simazine is used on lawns and on fruit crops, including oranges, apples, plums, olives, cherries, peaches, cranberries, blueberries, strawberries, grapes, and pears. Until 1994, it was also employed to kill algae in swimming pools and hot tubs, but these uses have since been disallowed after the EPA determined they posed "unacceptable cancer and non-cancer health risks to children and adults." In 1995, the manufacturer of cyanazine voluntarily agreed to phase out production over a four-year period in response to concerns about the herbicide's carcinogenic potential as a contaminant of food and drinking water. In the meantime, cyanazine continues to be used to control grasses and broadleaf weeds in cotton, sorghum, and, most of all, cornfields. In 1993, it was sprayed on about 21 percent of Illinois corn.

The triazine herbicides are under particular scrutiny for their possible role in breast and ovarian cancers. Evidence comes from both animal and human studies. Atrazine, for example, causes breast cancer and significant changes in menstrual cycling in one strain of laboratory rats—but not in another. Scientists are currently trying to determine which rat strain is the more accurate model for breast cancer formation in humans. Atrazine has also been demonstrated to cause chromosomal breakages in the tissues of hamster ovaries growing in culture. This damage occurs even at trace concentrations—that is, at concentrations comparable to those found in Illinois's public drinking waters. However, researchers do not yet know if the levels found in drinking water represent the doses received by human tissues. "The results provide evidence for further investigations as to potential health risk of consuming water contaminated with atrazine," concludes the author of this study, the cytogeneticist A. Lane Rayburn at the University of Illinois. Animal studies show that atrazine inhibits the metabolism of male hormones even as it induces certain pituitary hormones that play a role in ovulation. Meanwhile, across the Atlantic Ocean in northern Italy, case-control studies have shown a correlation between exposure to triazine herbicides and ovarian cancer among women farmers.

In response to these and other studies, the EPA has begun a special review of these three triazine herbicides. A sixty-three-page position document outlines the main areas of concern, one of which is residues of triazines found in foodstuffs. This is a complicated issue. We are exposed to these herbicides not only through consumption of produce—oranges and apples appear to be major sources of dietary exposure to simazine, for example—but also through meat, milk, poultry, and eggs, resulting from the ubiquitous use of corn in animal feed. Crop plants that are tolerant of triazines, such as corn, will metabolize the herbicide molecules—as will the animals who feed on these crops—but it is not at all clear how complete this process is nor how carcinogenic the metabolites themselves might be. The EPA's description of what is known about the movement of triazines through the human food chain is full of qualifiers, disclaimers, and expressions of frustration:

> These data introduce uncertainty. . . . The percentage of the total estimated risk accounted for by these data is not known, but will always lead to an underestimate of risk when detectable residues are present. . . . Registrants have been unable to develop an analytical methodology which measures total triazine ring residues in non-radiolabel food trials.

In other words, more than three decades after the introduction of chloroplast-destroying chemicals into American agriculture, the cancer risks we have assumed from eating food grown in fields sprayed with them—and from drinking water that has percolated through these fields—have yet to be determined.

When Rachel Carson wrote *Silent Spring*, maximum permissible levels of pesticide residues in food were called tolerances. These limits were set by the federal government and regulated on an individual basis: each food item was assigned a separate tolerance level for each and every pesticide used in its production. Very often, these assignments were based on inadequate knowledge of the chemicals used, most of which were quite new to agriculture at the time. Further research sometimes revealed that the health risks from eating foods legally contaminated with these chemicals were in fact more serious than initially presumed. These results led to stepwise reductions of certain tolerances—and sometimes to their revocation altogether.

Carson argued that this system was flawed from the start. After-the-fact adjustments in tolerances exposed people for months or years to levels of pesticides that were later admitted to be unsafe. Furthermore, the whole concept of setting "safe" limits for any one pesticide in any one food item was meaningless, she asserted, because it did not take into account our total exposure to multiple chemicals in multiple food items. Finally, enforcement was woefully inadequate and underfunded: the federal government had jurisdiction only over food moved across state lines and then tested only a tiny fraction of these shipments for illegal residues. She labeled as intolerable a regulatory methodology that involved "deliberately poisoning our food, then policing the result." Nevertheless, this was the food safety system in place when those of us born in the first decades after World War II were infants and children.

It is essentially still the same system. As of 1994, there were 9,341 tolerances. The vast majority of these govern residues on raw commodities; the remainder apply to residues known to concentrate in processed foods. The EPA is now the federal agency responsible for establishing these limits, while the Food and Drug Administration (FDA) continues to be charged with enforcement. (The enforcement responsibility for pesticide residue in meat and poultry, however, rests with the USDA.)

In 1993, the National Research Council concluded that the current regulatory arrangement permits pesticide levels in food that are too high for children and infants. Tolerances are insufficiently protective, according to the council's report, for two basic reasons. First, they are not based solely or even primarily on health considerations. The actual values chosen as legal limits reflect the results of field trials designed to measure the highest residue concentrations likely under normal agricultural practice.

Second, the safety margins supposedly ensured by tolerances assume adult eating habits. However, children eat far fewer types of food in proportionally greater quantities. A nonnursing infant consumes fifteen times more pears than the average adult, for example. And pears, as we have seen, are one of the most heavily sprayed fruits on the market. Children also differ sharply in their ability to activate, detoxify, and excrete contaminants. Finally, childhood exposures to pesticides may lead to greater risks of cancer and immune dysfunction than exposures later in life. For all these reasons, the National Research Council in 1993 called for a new health-based approach to establishing tolerances that would take into consideration the unique biology of children, as well as their other nondietary exposures to pesticides—including air, dirt, carpets, lawns, and pets.

While baby food itself appears to have levels of pesticides well below tolerance levels, detectable residues are nevertheless found in all major brands sold in the United States. In one recent study, researchers found traces of sixteen pesticides in eight different baby foods purchased in U.S. grocery stores. Five of these were possible human carcinogens.

Carson's concern about deficient enforcement also still reverberates. In 1994, the General Accounting Office—the investigative arm of Congress—reported that the federal system of food policing is fragmented, inconsistent, and fails to protect us adequately against illegal residues of pesticides. There are several ways that pesticide residues can be in violation of the law. They can exist in quantities above the tolerance level, they can exist on foods for which they are not registered for use, or they can exist on foods after their tolerance has been revoked. Detectable pesticide residues are found in about 35 percent of food consumed in the United States, but the proportion of this food that contains illegal residues is a contested figure.

As an example, consider produce. The FDA reports that 3.1 percent of the fruits and vegetables consumed by the public contains pesticide residues above the legal tolerance level. One research institution, the Environmental Working Group, places this estimate higher. In their review of FDA monitoring data, EWG staff discovered that many more violations were detected by FDA chemists than were reported by FDA enforcement personnel. The actual violation rate, according to their reanalysis, is 5.6 percent—nearly double the FDA's offical claim—and involves sixty-six different pesticides, including many banned or restricted for use. Green peas showed a violation rate of nearly 25 percent, pears 15.7 percent, apple juice 12.5 percent, blackberries 12.4 percent, and green onions 11.7 percent. In response, the FDA asserted that many of the additional violations uncovered by the group were technical in nature and "of no regulatory significance." In any case, these numbers reflected only the violations caught by FDA inspectors, who still examine a tiny fraction—less than 1 percent—of food shipments sent through interstate commerce. How representative these samples are of the entire food supply is anyone's guess. Confounding the uncertainty is that the methods used for evaluating compliance are only sensitive enough to detect about half of the pesticides known to be used on food.

In 1991, the USDA itself began collecting data on pesticide residues in certain fresh fruits and vegetables. According to its latest report, the USDA's Pesticide Data Program, using more sensitive tests, found 10,329 detectable residues on 7,328 samples. Actual violations were discovered in 1.3 percent of domestically grown commodities and 2.4 percent of imported fruits and vegetables. "These data reinforce the fact that the Nation's food supply is one of the safest in the world," according to the report's opening paragraph.

These percentages do look reassuring. But fruits and vegetables are not lobsters or licorice or eggnog. With current dietary guidelines advising five to nine daily servings of fruits and vegetables, a few percentage points of illegality add up quickly. Anyone following official dietary recommendations is consuming from one to four servings of illegal pesticide residues every twenty days—or somewhere between eighteen and seventy-two servings a year. This tally, of course, does not include illegal residues also received from meat, dairy, eggs, fish, or grains.

Food items containing *legal* residues of *illegal* pesticides also find their way into our grocery carts. Some of these foods have been imported. Some food items contain residues of long-banned pesticides that persist in the environment and therefore continue to trickle into food and forage crops. Some foods contain recently banned pesticides whose remaining stocks have not yet been depleted. Typically, the EPA has allowed about two years for leftover supplies of canceled pesticides to be used up and for products treated before the ban took effect to move through commerce. On average, the government takes more than six years to revoke tolerances on pesticides that have lost their registrations. One EPA official has estimated that 100 or more illegal pesticides may still have legal tolerances.

In response to public concern over pesticides in food, the Food Quality Protection Act was signed into law during the summer of 1996. In the attempt to revise the amounts and types of legally allowable residue, this act struck a Faustian bargain.

On the positive side, the act accepts the National Research Council's 1993 recommendations to consider the special vulnerability of infants and children when setting tolerances. These are now law. It also requires the government to screen pesticides for endocrine disruption. It further mandates that all tolerances be reassessed within ten years and that these assessments take into account multiple routes of exposure.

As with any deal with the devil, these gains exacted a high price. The new law overturns an old law that strictly forbade the presence of cancer-causing pesticides and food additives in processed foods. Enacted in 1958 as the Delaney Clause, the old law had been openly flouted for years. To comply with it, the EPA was faced with disallowing dozens of different pesticide uses. Instead, it effectively removed pesticides from Delaney restrictions by transfering regulation of processed food to another section of the FFDCA.

In the place of an absolute, but widely violated, prohibition on cancer-causing pesticides in processed food, we now have a law that legalizes carcinogens in food but limits them to levels determined to be "safe."

In discussions about pesticide residues, fruits and vegetables often dominate. Perhaps this is because, no matter how remote our relationship to farming, we can hold a peach, a cucumber, or a cluster of grapes and can easily picture the spray gun, the nozzle, the crop duster. When we lean over the fish counter, however, we do not necessarily picture pesticide mists from orchards, fields, and vineyards settling onto the creek surfaces; creeks running into lakes; pesticide-tainted plankton consumed by protozoa; protozoa eaten by insect larvae; or larvae eaten by minnows, and so forth up the food chain to salmon, walleye, and trout. And yet freshwater fish constitute an important route of human exposure to banned pesticides. In 1994, the General Accounting Office expressed special concern about the ongoing presence of five different pesticides in fish, including DDT, chlordane, dieldrin, heptachlor, and Mirex.

It requires an ecological mind to understand why fish—as well as meat, eggs, and dairy—function as a major source of pesticides in our diet. The path between cause and effect can take us into the groundwater, down a river, along an air current, and through a complicated food chain involving organisms whose names we may have never heard. Ecologically speaking, a food chain consists of a series of organisms who pass chemical energy, unidirectionally, through each other. Each link of the chain is officially referred to as a trophic level. At the bottom are the producers—green plants that transform sunlight into food, thereby making chemical energy available to everyone else. The primary consumers feed on the producers directly, the secondary consumers feed on the primary consumers, and so forth. About 90 percent of the energy transfered from one trophic level to the next is dissipated as heat.

The phenomenon of biomagnification flows from this fact. Because so much food energy is lost, fewer organisms can be supported at each succeeding link. In order to survive, each individual must consume many individuals at lower trophic levels.

At this point, ecologists switch metaphors and refer to the pyramid of biomass: the total mass of all the secondary consumers is less than that of the primary consumers below them, which in turn is less than that of the producers. Thus, as organisms continue to feed on each other, any contaminant that accumulates in living tissues—such as an organochlorine pesticide—is funneled into the smaller and smaller mass of organisms at the top of the pyramid. There are fewer of them. Like a sauce simmering slowly, the poison concentrates.

With pyramids and chains firmly in mind, three mysterious facts become explainable. First, we see why infants are at special risk: residues of fat-soluble pesticides contained in the food eaten by nursing mothers are distilled even further in breast milk. In essence, breast-feeding infants occupy a higher rung on the food chain than the rest of us. In many cases, human milk contains pesticide and other residues in excess of limits established for commercially marketed food.

Second, we see why a diet rich in animal products exposes us to more pesticide residues than a plant-based diet, even though the plants are directly sprayed. For the most part, the flesh of the animals we eat contains more pesticides than the grasses and grains we feed them. Indeed, the largest contributors to total adult intake of chlorinated insecticides are dairy products, meat, fish, and poultry. In 1991, the National Research Council reported that more than half the cattle tested in a sample of Colorado ranches had detectable levels of pesticides in their blood serum. A commonly found contaminant was the banned pesticide heptachlor. Similarly, the FDA's Total Diet Study, which regularly monitors the concentration of contaminants in cooked, table-ready foods, continues to find traces of DDT in many food types, but particularly those of animal origin.

Third, we see why those of us born in the heyday of organochlorine pesticide use bear special risks. Passed along from one creature to the next, these persistent pesticides are especially subject to biomagnification. According to the long-running Total Diet Study, pesticide residues in table-ready American foods peaked in the 1960s and 1970s. Those of us whose formative years correspond to these decades received more pesticide residues in our childhood diets than any generation of American children before or since. Between 1965 and 1970, the average DDT intake was twenty-three times higher than it was in 1982. Levels of dieldrin residues in food were twenty times higher in the 1960s than in the 1980s. The recent decision to consider the vulnerabilities of infants and children when setting tolerance limits is not retroactive to the rest of us.

Rachel Carson once remarked how strange it was to live in an age where carcinogens were a basic element of our system of food production. This is still a strange notion. Our predicament, however, is hardly insoluble. In fact, partial answers and outright solutions exist all around us.

In Iowa, for example, a group of soybean farmers completely replaced a carcinogenic herbicide with tillage systems and planting techniques that use shade to control weeds. This group, calling themselves the Practical Farmers of Iowa, achieved yields higher than the state average and saved considerable money.

In Nebraska, the farmer and author Jim Bender slowly transformed his conventional corn and soybean acreage into a completely chemical-free operation. The secret to his success involves staggering planting times, diversifying and rotating crops, and reintegrating livestock. His book, *Future Harvest*, is a how-to manual for other Corn Belt farmers who want to step off the pesticide treadmill and still make a profit.

In 1989, the National Research Council investigated alternative agricultural operations such as these and reported that U.S. farming could be shifted to more natural forms without losses in yields or profits, without significantly higher food prices, and with significant gains in health and environmental protection. Subsequent studies have confirmed these conclusions. In the meantime, polls of farmers and consumers have shown that the majority of both are concerned about the health and environmental effects of pesticides: farmers say they would switch to less toxic methods if incentives existed to do so; one in three shoppers say they already seek out organically grown fruits and vegetables, and more indicate they would if the prices were lower. These intentions are reflected at the cash register. For each of the past two years, the sale of organic foods has jumped 20 percent.

Our progress in this direction can be hastened in a number of ways. One is to promote research into ecological methods of pest control. Another is to reward farmers for adopting these methods. A third is to encourage a collective recognition of the true expense of pesticides.

Based on direct crop returns, the use of conventional pesticides still appears profitable. But the costs of these poisons are far greater than the price the farmer pays the distributor. They include trucking in water to communities where the wells are contaminated, warning the public not to fish in certain lakes and rivers, coping with loss of honeybees and other pollinators, revising and enforcing thousands of tolerance limits for food sold in interstate commerce, cleaning up pesticide spills, monitoring pesticide residues in animal feed, and providing treatment for cancer patients. Most of these costs are ultimately borne by the public, and they amount to an indirect—and apparently unlimited—agricultural subsidy. When we look about us and assess the state of our food-production system, we need to take these costs into account—however difficult they are to quantify. The state of North Carolina, for instance, recently completed a five-year study of pesticides in its groundwater aquifers. The results were not reassuring: one of every five monitoring wells sampled in highly vulnerable areas was contaminated, and more than two dozen different pesticides were detected, including ten no longer legal for use. The cost of simply conducting this investigation exceeded $1 million. Pimentel places the total annual cost of pesticides at $8 billion, but as he himself asks, "What is an acceptable monetary value for a human life lost or for a cancer illness due to pesticides?"

Weaning American agriculture from chemical dependency will not be easy. In Illinois, for example, the get-big-or-get-out attitudes of the past have created farms so huge that simply substituting mechanical weed control for chemical herbicides would create serious erosion problems—leaving farmers the terrible choice of watching their soil blow away or spraying it with hazardous chemicals. And simply pulling a few hazardous herbicides off the market without offering alternatives may well accelerate herbicide resistance as farmers are left with fewer chemical weapons in their arsenal. Such problems are not intractable—as the success of many organic farmers attests—but solving them will require diligent effort on the part of many. Difficulty should not be used as an excuse to do nothing.

Out in the bean fields, John hollers over to me that Emily picked up a rock in her last pass on the combine. All its many working parts still seem to be working, however—the deafening decibel level of the engine certainly sounds the same. So after a brief inspection we climb into the cab together and swing the machine around. Usually, Emily runs the combine and her husband ferries the wagons back and forth to the storage bins, but they've switched places for a while.

The quiet of the cab gives us two cousins a chance to catch up with each other, gossip about the family, and laugh at our parents' various eccentricities. This kind of ride-along is a discouraged practice—guests in the cab are distracting and can compromise safety. John asks me to keep an eye out for rocks.

We're quiet for a while. The movement of the sickle bars is mesmerizing, like the churning paddlewheel of a large boat. If the weather holds off, Emily thinks they can get all their beans in by midnight. I look out at the many acres that remain—and the sea of corn beyond—and wonder how they will do it. A call comes in on the cellular phone from a friend near Bloomington who got rained out today. The price of beans has dropped a penny. The weather looks like it will pass to the north.

The combine's head lowers a bit as we descend into wetter ground. John points to the plants that rise above the beans—bright foxtail, glossy lambs-quarters, and the reedy flowering stalks of velvetleaf. He has a lot of respect for weeds. Since they're here to stay, he says, somebody ought to figure out a use for them. Besides, a few weeds means he's not using too many chemicals. John and Emily have five young kids. Most of their information on pest control comes from agrichemical distributors, but, John emphasizes, they want to become more environmental in their approach to farming.

An alarm sounds, indicating the hopper is full. John shakes his head. "Emily's going to ask me why I didn't dump this load on my last pass." We'll have to drive back the length of the field without harvesting any beans, which to Emily's thinking is not efficient farming.

All the crops they grow are for export. They sell to the river, as Emily puts it. Their entire corn harvest, for example, will be trucked to the Pekin docks. From Pekin it goes by barge to New Orleans; from there, who knows. Farmers have become as dissociated from those who eat their food as consumers have from those who grow it.

After John empties the hopper, Emily walks over to meet us. Instead of delivering the anticipated scolding, she smiles and hands me a plastic bag.

"These are tofu beans," she shouts over the machines. "We combined them yesterday. Right now, they're on their way to Japan."

I reach in and pull out a handful. They're a little bit bigger, rounder, and paler than the beans we just dumped in the wagon. John adds that he and Emily have never eaten tofu. Have I?

In fact, I have several vacuum-packed slabs in my Boston pantry. "What's it like?" John shouts. Standing in a rented field that's full of them, I try, at the top of my lungs, to describe the taste of soybeans.

EIGHT

air

Earth and the Sea feed Air; the Air those Fires Ethereal.
—JOHN MILTON, *PARADISE LOST*, BOOK V

The thing about air in Illinois is that there is so much of it. Air is a more conspicuous element here than in any other place I've lived. It seems deeper, wider, more present.

I first learned to see air when an art teacher began coming to our grade school. Of all the ideas she introduced, none made more sense to me than the concept of the vanishing point—an invisible place on the horizon where parallel lines mysteriously converge. Simply by choosing a point and drawing in perspective a house or a road, I could draw air. This was a great discovery. And the Illinois landscape was full of such points through which all objects seemed to be striving to disappear. Grain elevators along a railroad track. A plowed field. The silver towers and looping cables of high-power lines. Everything eventually vanished into air.

What I never figured out how to represent visually was air's transformative properties. In rural Illinois, how objects appear depends on how much air they are viewed through. A brown chip in the sky turns, a half mile closer, into a circling hawk. Fluttering black specks turn into black handkerchiefs and then into a flock of crows. The dead body on the side of the road eventually becomes a scrap of carpet.

In the fifth century B.C., the Greek philosopher and physiologist Empedocles declared that the atmosphere was not a void but a living substance. A thousand years later, his theory was animated by a Swiss alchemist, Paracelsus, who professed that air was inhabited by elemental beings he called sylphs. I am thinking of these two scholars while driving through central Illinois during a period of record-breaking cold. It is many degrees below zero—dangerously cold—equipment malfunctioning, cars seizing up, and hourly warnings against venturing outdoors issuing from the radio. I should not be out myself, but it seems important to see the landscape at its most unmoving, its constituent molecules vibrating at their slowest recorded speed, the earth with all its seeds frozen down to many feet, water only a memory.

Leaving the engine running, I step from the car onto the stone floor of a cornfield. Only the atmosphere seems alive. Every inhalation brings pain, every exhalation a puff of crystals. Whatever sylphs are, they instantly find the cracks between scarf and collar, glove and sleeve. In seconds, I am undressed, although still standing fully clothed. And yet never is this element more invisible. Even objects at the vanishing point seem distinct and permanent, untransformed by miles of air.

Far from Illinois, the White Mountains lie across upper New Hampshire like a crooked crown. Their westernmost peak, Mount Moosilauke, rises above a tract of trees familiar to every student of ecology: the Hubbard Brook Experimental Forest. Here, researchers have used large-scale, long-term field studies to trace the slow cycling of nutrients through an entire living community. Much of what we know about the ecological pathways of nitrogen, phosphorus, and calcium, for example, derives from work carried out in these woods. Some of the first investigations of acid rain were conducted here as well.

In 1993, a research team led by the biologist William H. Smith of Yale University discovered something else about the forest floor of the famed Hubbard Brook. Its newly fallen leaves and needles, as well as the composting earth beneath, contained detectable amounts of both DDT (0.8 pounds per acre) and PCBs (2.3 pounds per acre). Even more remarkably, neither of these long-banned chemicals had ever been used, distributed, or produced in the immediate area where the samples were collected.

Soil and leaf litter sampled from nearby Mount Moosilauke were found to be likewise contaminated—all the way up to its tundra-covered summit. Tellingly, the level of contamination rose with elevation and was greater on west-facing slopes. Such patterns are consistent with atmospheric deposition.

Smith and his colleagues thus postulated that molecules of DDT and PCBs were being carried into Hubbard Brook by prevailing winds. Their origin, however, remains obscure. Storm tracks that pass over major centers of agriculture and industry in the United States also pass frequently over New England. Quite possibly, regional air masses are spiriting these semivolatile, long-lived molecules from landfills, dump sites, and farm fields to this remote, pristine forest.

It is also possible that global air currents are the transporting medium. Like undocumented aliens, these chemicals may have been swept in from other countries, even other hemispheres. Studies conducted in rain-fed bogs across eastern North America support this possibility. These peculiar habitats receive all their input of pollutants from the atmosphere, and not from ground or surface water. Therefore, they function as a living map, revealing in detail the historical and geographical contours of atmospheric deposition. And because peat preserves organic compounds almost completely, they also function as living archives. Cores of peat extracted from these bogs show continuing accumulations of fresh, unmetabolized DDT. As DDT is no longer permitted for use within U.S. borders, researchers hypothesize that it is being carried in on air currents from Mexico, where DDT is still produced and used, and perhaps also from Central America.

A lake in northern England also has a interesting story to tell about the wind. Located far from industrial and residential areas, the lake bottom of Esthwaite Waters is nevertheless laced with DDT and PCBs. Sediment cores extracted from this basin and chemically dated provide a chronology of atmospheric desposition. For the most part, the changing levels of both contaminants mirror their historical rates of domestic production and use, with peak inputs occurring in the late 1950s and 1960s. However, samples dated between 1929 and 1954 also contain a broad range of PCBs. This is a curious finding because the United Kingdom had not begun manufacture of these chemicals during this time. The presence of PCBs in these old sediments indicates long-range atmospheric transport from either the United States or mainland Europe.

More evidence for the role of global circulation comes from the world's trees. A 1995 survey found traces of twenty-two different organochlorine pesticides—including DDT, chlordane, and endosulfan—in tree bark gathered from ninety different sites around the globe. Bark is a remarkably oily tissue and therefore readily absorbs oil-soluble pollutants from the air. That trees growing in the agricultural regions of the midwestern and eastern United States should bear residues of pesticides in common use there is thus hardly surprising. However, researchers also found pesticides that were thousands of miles away from where they had been sprayed. Trees in the Arctic, for example, were found to carry traces of insecticides used in tropical areas.

These results are explained by a form of chemical nomadism known as global distillation. When certain chlorinated pesticides and other persistent pollutants are released in warmer climates, they evaporate and are carried by winds to cooler areas, where they condense and descend back to earth. These trespassers overwinter in soil, snow, or water until the summer sun revaporizes them and air currents blow them further toward the poles. They then drift downward once again. During this process, the various chemical contaminants are spatially partitioned: those substances that evaporate at low temperatures—such as the more lightly chlorinated PCBs—are carried to farther-flung latitudes and higher altitudes, while substances that require higher temperatures to evaporate condense more quickly and are deposited closer to their source. The latter group includes the more highly chlorinated PCBs, as well as the carcinogen benzo[a]-pyrene.

The rising and falling movements of global distillation explain not only why chemicals used in rice paddies and cotton fields eventually end up in the skin of Arctic trees but also why the bodies of seals in Siberia's Lake Baikal—the world's oldest and deepest lake—contain the same two contaminants as the alpine soils of New Hampshire's Mount Moosilauke. Indeed, the levels of PCB and DDT in the blubber of these animals are high enough to render them illegal for consumption by U.S. standards.

Global distillation also explains why fish in the Yukon Territory's Lake Laberge became so full of carcinogenic toxaphene that the Canadian government was recently forced to ban angling there:

> Analyses of food chains and contaminant analyses of biota, water, and dated lake sediment show that the high concentrations of toxaphene in fishes from Laberge resulted entirely from the biomagnification of atmospheric inputs. A combination of low inputs of toxaphene from the atmosphere and transfer through an exceptionally long food chain has resulted in concentrations of toxaphene in fishes that are considered hazardous to human health.

❧

Transferred to the Yukon after the war, my father served with the army's Pipeline Distribution Unit in Whitehorse. In the summer of 1946, the story has it, he and his friends chartered a fishing boat on Lake Laberge and landed trout so large they could scarcely be subdued.

Eerie photographs of the lake—placid water, curtains of mist, black hills—form the final images in his wartime scrapbook. My father claims Lake Laberge as the most beautiful place he has ever seen and still lowers his voice when describing the Northern Lights reflected on its midnight surface.

Of all the component aspects of the environment, air is the one with which we, who inhale about a pint of atmosphere with every breath, are in most continuous contact. Even as the ongoing campaign against secondhand cigarette smoke has focused public attention on airborne carcinogens, air remains mysterious. Air is the element most diffuse, most shared, most invisible, least controllable, least understood.

The phenomenon of global distillation, along with more local forms of atmospheric deposition, shows that not all of the dangers from carcinogens in our air supply come from breathing. Some also come from eating. Poisons dumped and plowed into the earth are fed, molecule by molecule, into the air, where they redistribute themselves back to the earth and into our food supply.

This link is actually most profound in aquatic systems. Because its buoyancy helps organisms conserve energy, water supports longer food chains (taller pyramids of biomass, more trophic levels) than terrestrial ecosystems do; transfer of energy between levels is more efficient. Water thus provides more opportunity for pollutants that have sifted down from above to magnify in living tissues. Air deposition into the Great Lakes, for example, is now one of the major sources of chemical contamination there. In Lake Superior, somewhere between 76 to 89 percent of PCBs come from the air. Moreover, people who eat Great Lakes fish have higher body burdens of PCBs and other toxic substances than those who do not.

Air deposition into the Great Lakes is now officially recognized as posing cancer risks to the human inhabitants who live in their basin and consume the food they provide. According to the EPA's 1994 report on the subject, "Most of the chemicals of concern are probable human carcinogens, exposure to which is expected to increase the population incidence of cancer." A study of sport fishers in Wisconsin found consistent correlations between the number of "sport-caught fish meals" consumed each year and the levels of PCBs and DDT metabolites in their blood. In fact, serum levels were so impressively high that Wisconsin anglers, the authors concluded, would provide an excellent population for a study of PCB- and DDT-associated morbidity and mortality.

In short, because of air, we each consume suspected carcinogens released into the environment by people far removed from us in space and time. Some of the chemical contaminants we carry in our bodies are pesticides sprayed by farmers we have never met, whose language we may not speak, in countries whose agricultural practices may be completely unfamiliar to us. Some of the chemical contaminants we carry with us come from long-defunct products of industry—objects manufactured, used, and discarded by people of a previous generation. When we sit down to eat a meal of, for example, freshwater fish, we are linked to all these people through the medium of air.

Conversely, chemicals dumped and sprayed in our own neighborhoods, fields, and landfills have drifted to distant territories and found their way into the diets of the people who live there. I sometimes think of this multitude of connections while walking through Illinois corn and bean fields. In general, less than 0.1 percent of pesticides applied for pest control actually reach their target pests, leaving 99.9 percent to move into the general environment. Some runs into water, some binds to soil, and some rises into the air. While many of the pesticides used now are less environmentally persistent than their predecessors, I wonder where the chemicals sprayed in these fields when I was growing up here now reside. On what mountainside, in what forest or lake bottom, in whose bodies do they lodge now?

Air is by far the largest receptacle for industrial emissions. Of all of the toxic chemicals released by industry into the nation's environment each year, more than half is released into air. These emissions include about seventy different known or suspected human carcinogens. When vehicle exhaust and emissions from power plants are added to the mix, the number and amount of carcinogens in air rise further. According to the International Agency for Research on Cancer, ambient air in cities and industrial areas typically contains a hundred different chemicals known to cause cancer or genetic mutations in experimental animals. And while air pollution in the United States has markedly improved over the past quarter century, more than a hundred urban areas still fail to meet national air quality standards. In other words, nearly one hundred million Americans breathe air that is officially illegal.

These are facts not in dispute. How much airborne carcinogens actually contribute to human cancer, however, remains an elusive question. As two leading researchers have noted, airborne carcinogens create an "epidemiological dilemma": we know they exist, but we have no good method for linking them directly to disease.

Air can evade the rigors of scientific analysis through at least two means. First, its fluidity makes exposure very difficult to quantify. Wind speed and direction, as well as wind flow along river valleys, over hills, and around buildings, significantly alter the transporting path of airborne carcinogens. Residents of a single metropolitan area may all drink water from the same river and buy their food from the same supermarkets, but they may not all breathe the same air. Those who live downwind from the local industrial park may live in a very different atmosphere than those who live upwind. A centrally located air-monitoring system cannot account for differences in microclimate.

Secondly, air is a transmutational medium. As in an alchemist's flask, the atmosphere concocts new materials from the ingredients—droplets, fibers, vapors, and particles of various sizes and weights—placed into it. Recent evidence suggests that some of the major carcinogens in air are synthesized when organic chemicals released from various sources react with each other and are transformed into entirely new substances. Many may not even be identified yet. Thus, a simple laundry list of air emissions—such as the Toxics Release Inventory—cannot account for the presence of all the cancer-causing agents to which we are exposed.

Of the various pollutants that appear, literally, out of thin air, the most notorious is ozone. Recall from Chapter Three that this substance is created naturally up in the stratosphere from the interaction of ultraviolet radiation with oxygen. The resulting layer of ozone protects us from excessive UV exposure, and, with good reason, we concern ourselves with its gradual disappearance. At the earth's surface, however, ozone is a noxious, unnaturally occurring irritant to eyes and lungs. With equally good reason, we who live in cities during the summer concern ourselves with incremental rises in its daily parts-per-million concentration. (Molecules of ground-level ozone are too heavy to rise into the upper reaches of the atmosphere and take the place of their faltering compatriots, upon whose presence all life depends.)

Although it is a major ingredient of urban smog, ground-level ozone is emitted into the air by no known polluting source. Instead, it is created when sunlight catalyzes a reaction between two kinds of vapors: nitrogen oxides, which are emitted from tailpipes and smokestacks, and volatile organic compounds, which rise into the air when houses are painted, cars refueled, roads paved, and clothes dry-cleaned.

In the classic sense of the word, ozone is not a carcinogen. Nevertheless, in the complex unfolding of cellular events that typifies carcinogenesis, ozone seems to play the role of supporting actor. A powerful poison, ozone causes inflammation of the airways and thereby interferes with the body's ability to sweep foreign particles—some of which may be carcinogenic—out of the lungs. Ozone also hampers the activities of the lungs' macrophages. Provided by the immune system, these amoebalike scavengers offer a first line of defense against a variety of pathogens and foreign substances. In studies of laboratory animals, ozone appears to magnify the effect of other lung carcinogens and influence the carcinogenic process itself. Lung tumors in ozone-exposed mice have distinctly different genetic mutations than do those from mice that have breathed clean air.

In the attempt to understand how air may contribute to cancer, the issue of ozone raises some vexing questions. How do we assess exposure to airborne pollutants that emerge from chemical recombinations of other airborne pollutants? How do we quantify the cancer-causing potential of a substance that enhances the cancer-causing potential of other substances? How many cancer deaths do we chalk up to ozone? What is the body count?

With a five-year survival rate of only 13 percent, lung cancer is so swiftly fatal that we rarely hear the stories of its victims. While those diagnosed with breast cancer form support groups, write books, lobby Congress, and organize rallies, dress balls, and races-for-the-cure, those with lung cancer tend to vanish quietly from our midst. The small public presence afforded them is usually a posthumous one.

Guilt and blame also silence lung cancer patients, who are seen as having brought about their own misfortune.

Whether we hold individual consumers or corporate producers responsible, the primacy of tobacco in the epidemiological portrait of lung cancer is indisputable. Smoking is the dominant cause of lung cancer—all self-serving attempts by tobacco interests to cast doubt on the issue aside. Indeed, such attempts have inspired the coining of the phrase "cigarette science" to describe scientific study that ministers to the needs of industries beset by bad publicity.

Nevertheless, there is more to the story of lung cancer than cigarettes. If other factors seem minor in comparison, it is only because tobacco is such a major killer. But lung cancer among nonsmokers is responsible for more mortality in the United States than any other cancer except for colon, breast, and prostate. Not all of these deaths are cigarette independent. About 20 percent (three thousand deaths each year) are thought to be attributable to secondhand tobacco smoke. So shocking is this statistic that it has rightfully prompted substantive changes in laws governing smoking in workplaces, airplanes, restaurants, and other public domains. However, the majority of nonsmoking lung cancers remains unexplained. While air pollution is not the only possible cause, it is one that unavoidably affects us all, and it is one that may interact with and potentiate the effects of other factors. On these bases alone, this topic deserves thoughtful inquiry. Several lines of evidence suggest its role may be significant.

The first comes from doctors' offices. Oncologists who specialize in lung cancer are reporting increasing numbers of nonsmokers among their patients, as well as increasing cases of a specific kind of lung malignancy not strongly associated with tobacco. Called adenocarcinoma, it is distinguishable from oat cell and squamous cell carcinomas, both strongly linked to smoking. "Although better laboratory methods for classifying tumors have enabled doctors to identify more adenocarcinomas, scientists believe that environmental pollutants and carcinogens are also causing an increase in cases." So concludes a recent review article published by the Harvard Medical School.

Meanwhile, epidemiologists have been focusing on understanding the urban factor in lung cancer. Many ecologic studies, from countries around the world, have revealed urban lung cancer rates two to three times higher than those of surrounding countrysides.

However, city folk also tend to smoke more than their rural counterparts. When smoking habits are taken into account, the excess of lung cancer incidence in urban areas is much less—but is still higher than in rural areas. Areas with chemical plants, pulp and paper mills, and petroleum industries also show elevated rates of lung cancer. One recent cohort study of more than eight thousand U.S. adults showed that air pollution was positively associated with death from lung cancer after cigarette smoking, age, education, body mass, and occupational exposures were all accounted for. And a cohort study of five thousand chimney sweeps in Sweden found increased mortality from lung cancer and other tumors that was not explainable by smoking habits but that was related to exposure to carcinogenic soot.

Case-control studies are rarer. Because lung cancer patients expire so quickly, these studies tend to necessitate interviews with the subjects' next-of-kin. Some studies show no connection with air pollution; others find modest associations. In one such study conducted among Italian men, researchers painstakingly gathered data on smoking habits, occupation, social group, age, and place of residence. Even after all these factors had been taken into account, rates of lung cancer rose with rising levels of air pollution for all types of lung cancer, but especially for adenocarcinoma. "These results provide evidence," according to the authors, "that air pollution is a moderate risk factor for certain . . . types of lung cancer." In a second study, researchers corroborated these findings and reported significant relationships between lung cancer risk and proximity to specific air-polluting sources, such as an incinerator.

At first glance, these results would seem to be contradicted by case-control studies conducted in China, where lung cancer rates are highest among groups of rural women, few of whom smoke. Indeed, these women have some of the highest incidences of lung tumors in the world. However, these women all depend on burning coal for heating and cooking. As in Italy, researchers found a link between lung cancer and air pollution: women with the highest rates of cancer were those who burned smoky coal inside their homes.

As ports of entry, the long tunnels and spongy rooms of the lung are only the first place where airborne carcinogens meet human tissues. Those carcinogens absorbed across the lung's membranes are carried in the bloodstream and deposited throughout the body. Much less is known about the relationship of these contaminants to other forms of cancer, but the question is provoking increasing interest.

By-products from the burning of fossil fuels are under particular suspicion. Bladder cancer, for example, has been linked in several studies to exposure to diesel exhaust. Breast cancer, as we have seen in Chapter Four, was first linked to potential sources of air pollution in Long Island. In the laboratory, members of a chemical family of combustion by-products called aromatic hydrocarbons—of which benzo[a]pyrene is one—cause breast cancer in animals. According to researchers at Albert Einstein College in New York, aromatic hydrocarbons inhaled by the lungs can become stored, concentrated, and metabolized in the breast, where the ductal cells become targets for carcinogenesis.

Closing a lethal circle, air pollutants have also been implicated in promoting the spread of cancer from other organs *to* the lung. For example, melanoma-afflicted mice that breathed air polluted with nitrogen dioxide developed more tumors in their lungs than those that breathed clean air. They also died sooner. Not lung cancer per se, these tumors are secondary growths from cancer cells shed from the original tumor. They are carried by the blood to the lungs, where, like planted seeds, they take root. As cancer patients know, metastases to the lung are often the kiss of death. The first capillary bed encountered by blood leaving most other organs, the lung is the most common place for cancer metastases to take hold. And somehow—at least in mice—breathing nitrogen dioxide seems to facilitate this process.

The pathologist Arnis Richters, of the University of Southern California, believes that at least two sinister mechanisms are at work here. First, nitrogen dioxide impedes so-called killer T cells, whose function, among others, is to rid the body of wandering tumor cells. Second, nitrogen dioxide causes blisters to form deep in the lung's airy chambers, where such errant cells can then become trapped. "Since many cancer patients have circulating cancer cells," says Richters, "it is possible that noxious air pollutants may play a more important role in dissemination of cancer than is realized at the present time."

Thus, like its chemical offspring, ozone, nitrogen dioxide raises thorny questions about causality. Nitrogen dioxide is not, as far as we know, a carcinogen. And yet, for those of us who have had cancer, its presence in air may affect our chances of surviving our disease.

❧

I am driving through my favorite section of town, the old neighborhood north of the post office. It was Pekin's original town site, but I like it for a different reason. All the east–west streets have women's names, and not just ordinary ones: Cynthiana, Henrietta, Sabella, Caroline, Catherine, Matilda, Lucinda, Amanda, Charlotte, Susannah, Minerva, and the one I, as a child, most revered—Ann Eliza. All of these streets dead-end at the river.

I have my sister's two sons in tow, and we stop to pick up treats at Patsy's Bakery. It is a Sunday morning at the end of February, the first warm day in months. The church parking lots are all full. We head for the river, zigzagging as we go so we can hit all the girl streets. As we do, the smell that is almost always present here gets stronger. I'm half thinking about the research papers I've been studying.

> *These results are consistent with findings of previous analyses and provide further evidence that air pollution is a moderate risk factor for lung cancer.*

I have thought a lot about how to describe this smell, but I cannot. I can smell it more acutely now than I could when I lived here, although it is probably less potent than in earlier decades. It is still too familiar for words. A complicated smell, it seems to contain more than one odor. It is . . . pungent. After watching the barges and the fishing boats for a while, we swing onto Route 9 and head over the Pekin bridge.

> *Overall, the studies suggest that emissions from some types of industries may increase the lung cancer risk for the surrounding population.*

On the west bank of the Illinois, the floodplain spreads out like a dance floor in a pool hall. It's mostly corn and bean fields now—all the way up to the power plant and the coal piles and the access roads and the railroad yards and the bait shops. I love this river valley and the bluffs that rise above it, the backdrop of my childhood. I want the boys to love these landscapes, too—but with full knowledge rather than denial, in the terribly difficult way that one is asked to love alcoholic parents: not abandoning them to wretchedness, not enabling their self-destruction, not pretending there is no problem. I don't know how to explain this to my young nephews, but maybe I don't have to. Maybe we adults need only demonstrate an attitude of passionate attention about where we live.

"Let's check out the west bluff." I turn right onto Route 24 and then left onto a narrow side road, steep as any mountain grade, and downshift. Even in second gear the car stutters, and what's left of our doughnuts tumbles backward into our laps. Suddenly, we are in thick woods that draw a curtain on the river's floodplain behind us.

> *In conclusion, it is difficult to interpret the epidemiological evidence on ambient air pollution and lung cancer.*

At the top of the ramparts, we emerge back into the sunlight and find ourselves in Tuscarora, an unincorporated scattering of homes that follow the topographical contours of the bluff. These eventually give way to more corn and bean fields—a mirror image of the east bluff. Not so, claim my two companions, who believe the west bluff is hillier and therefore more fun to drive on. I'm not persuaded, but it's a good argument.

Back in the valley, I continue north on Route 24, the river now on our right. We pass the oil refinery, the steel and wire plant, the stockyards, another ethanol distillery, and a handful of other industries. I have spent the week reading medical literature on air pollution and picking through the TRI data on air emissions along this corridor. I am thinking about what all three of us grew up downwind from.

> *Many toxic chemicals are not routinely monitored in ambient air in Illinois, and little is known about ambient concentrations or the relative importance of various sources of these chemicals.*

In the nineteenth century, many good people—medical doctors and officers of the government among them—believed that infectious diseases were brought on by bad air. These were followers of the miasma theory of disease causation. Air was thought to be "corrupted" when it passed over decaying organic matter—swamps, sewage, dead bodies. Breathing the poisonous odors emanating from such places was, according to the miasma school of thought, injurious to the body and had the power to trigger terrible physical maladies.

The miasma theory was eventually supplanted by its rival, the germ theory. But before its drift into obscurity, miasma's disciples managed to usher in significant reforms in public health policy, namely, closed sewage systems, clean drinking water, and deep burial of the dead. These did much to deter epidemics even before the real, microscopic causes of disease were discovered. The miasma theory, although mistaken, saved many thousands of lives.

Old descriptions of the Illinois wilderness, including the valley I am now driving through, provide detailed observations of its air—how sweet or rank were its odors, how salubrious or potentially pestilent its breezes.

> *Whether the excess risk of lung cancer can be attributed to urban air pollution cannot be determined conclusively, but it is suggested that it at least contributes to the risk.*

I follow Route 24 east over the McClugage Bridge. Here the river is nearly three-quarters of a mile wide, and at the exact center we cross from Peoria back into Tazewell County. This border fascinated me as a child. A line in the middle of a river! How did they know exactly where it was? All stories about my childhood are fascinating to my current company, including my old fear of drawbridges and my enchantment with tugboats. And the fact of their fascination fascinates me. *This is where we come from.*

Still, I wonder if I should be bringing the boys down here at all. The younger one sitting next to me has asthma—as does his mother, who developed it as an adult. A secretary at the local grade school, Julie is astonished, she says, at the number of children with inhalers in their backpacks. In fact, the national incidence of asthma has jumped 40 percent in the past decade; it is now the number one cause of absenteeism for American schoolchildren. And asthma is also becoming more deadly: mortality is rising as swiftly as incidence, according to a recent report from the Centers for Disease Control. As with lung cancer, there is an urban factor at work. The majority of asthma sufferers in the United States live in places where the air does not meet federal standards.

On the days that she must send unusually high numbers of wheezing children home from school, my sister has begun to take note of what the weather is like, which way the wind is blowing, how the air smells, and how labored her own breathing is. Perhaps it is once again time, we both agree, to look at the environment to understand what ails us. Perhaps it is time to risk being right for the wrong reason—as did our predecessors who successfully prevented the spread of infectious disease by cleaning up pollutants in the absence of complete knowledge about the microbes they contained.

Increases in childhood asthma and the clustering of lung cancers around cities with dirty air are telling us something. Suppose we do nothing until the exact mechanisms are elucidated, until exposures are definitively ascertained, until the precise combination of air pollutants and their specific interactions with each other and with the tissues of our respiratory airways are exhaustively understood. Then are we not mimicking those who, at one time, could just as well have claimed that there was not sufficient reason—on the grounds that science had not yet identified any specific biological agent responsible for cholera—to keep human excrement out of the drinking water?

We begin the long climb up the bluff. At the top is the brokenhearted town of Creve Coeur, Pekin's smaller, meaner, drunker brother. On the other side of Creve Coeur is the road home. I celebrate by opening the sunroof.

"Aunt Sandy, when did you get this car?"

"Honey, I got it when a friend of mine in Boston was sick and needed to go to the doctor a lot. Do you remember when I told you about that?"

"She died, right?"

"Yes, she did."

"You had cancer, too, didn't you?"

In the absence of other data, it would be advisable to avoid excessive and prolonged exposure to such agents.

NINE

water

And the fish suspending themselves so curiously below
there and the beautiful curious liquid
And the water plants with their graceful flat heads, all
became part of him.

—WALT WHITMAN, "THERE WAS A CHILD WENT FORTH"

My mother grew up by the Vermilion River and my father alongside Lake Michigan. I have, therefore, no familial connection to the Illinois River, no handed-down tales to pass on. In getting to know this river I was raised beside, I've relied as much on library research as on my own observations. These sometimes tell two different stories.

In one archival photograph from the early 1900s, four men and two boys stand on the river's edge beside what looks to be an immense pile of stone butterflies. The solemn fellow in the foreground holds one of them, wings spread, in the palm of his upraised hand. The others in his outfit stand in the background, stiff and expressionless as fence posts. These men are, in fact, mussel gatherers showing off their catch. They will sell their heap of shells to one of fifteen button factories that line the shores of the lower Illinois River.

By 1948, the last one had closed. Pollution and overharvesting killed off the mussels, and plastic replaced mother-of-pearl in the production of shirt buttons. The species depicted in the photograph are as alien to me as the process of turning them into objects of human attire.

In 1948, diving ducks also began to disappear from the Illinois. Ring-necked duck, canvasback, ruddy duck, and lesser scaup: these are species I learned to identify from stuffed specimens and in distant field sites. I have been trained to recognize their patterns of coloration and differently pitched calls, but I do not recognize them as fellow inhabitants of the river system I grew up in—although it served for centuries as the flyway for their migrations.

My newlywed parents began building their house on the bluff in 1955. In this same year, the valley's population of scaups ("highly social . . . note purplish gloss on head . . . shows bold white stripe on secondaries . . . calls are short low croaks") plummeted to zero. Researchers attribute their disappearance to the synchronous demise of the river's fingernail clams. Likely poisoned by organochlorine contamination of the river's sediments, they had served as the ducks' major food source. The clams have never come back either.

Dabbling ducks, such as wigeons ("pale grey head and bluish bill") and gadwalls ("rarely congregates . . . call, very low and reedy"), feed on the seeds of aquatic plants. Their departure from the Illinois corresponds to the arrival of herbicides. As agriculture became increasingly mechanized and chemically dependent, the flow of silt and weed killers from surrounding fields created waters barren of all such vegetation. Wild celery, coontail, and sago: these species, according to old accounts, once flourished in the quiet, shallow waters of Peoria Lake. They vanished completely in the 1950s, along with the birds that ate them. I don't know how to recognize these plants.

The story of the fish begins fifty years earlier. At the turn of the twentieth century, over two thousand commercial fishers worked the Illinois and supplied their harvests to markets as far away as Boston. Special fishing trains also carried sport fishers to and from Havana, the river town just downstream of Pekin. As measured by pounds of fish caught per mile of stream, the Illinois was considered the most productive inland river in North America.

This remarkable fecundity was a gift of geology. Much of the Illinois flows through a floodplain left behind by the ancient Mississippi. This flat pan of ground allowed the river to spread out a far-flung web of interconnected backwaters—the perfect nursery, spawning grounds, and winter refuge for fish. In the summer, periodic droughts firmed up the bottom, improving conditions for vegetation. The plants, in turn, deterred the wind from stirring up sediment during periods of spring and autumn flooding, when the river poured itself into the twisting sloughs, swales, potholes, marshes, and subsidiary lakes that surrounded it.

Then came the Chicago Sanitary & Ship Canal. This part of the story is a kernel of central Illinois lore. The S&S Canal opened January 17, 1900, and effectively connected Lake Michigan to the Illinois River, creating a continuous navigational route down to New Orleans. This is the meaning of the second *S* ("ship"). The first *S* ("sanitary") refers to the flushing of Chicago's wastewater into this canal and, from there, through the Des Plaines River and into the Illinois. Consequently, the level of the Illinois rose considerably. Backwaters flooded and stayed flooded. Bottomland groves of pin oak and pecan trees died. A wave of industrial pollution moved slowly and inexorably south (reaching Pekin around 1915), and downstream residents protested vociferously. Finally in 1939, the U.S. Supreme Court was moved to reduce by one-half the diversion of water into the Illinois. In the meantime, locks and dams began shaping the river into a series of stepped navigational channels. By World War II, the river resembled its present configuration. Straightened, leveed, drained, and dammed, the Illinois River became a sewage canal for industry and a barge canal for shipping—S&S. A report published the year I turned seven features a photograph of Illinois River fish with open sores and fins eroded down to stumps.

The federal Clean Water Act of 1972 brought a modicum of improvement to the Illinois River. As annual amounts of industrial waste released into the river declined, water quality improved. The long-term ecological effects, however, are less clear. Like a cloth already frayed, the river shows signs of continued damage even at lower levels of stress. Recovery has been uneven, at best. Mussels have returned to some parts of the river, and fin erosion is a less common problem in fish. On the other hand, the level of pesticide contamination remains high, and aquatic plants have been unable to reestablish themselves.

In the Upper Illinois, fish advisories continue to caution sport anglers to severely limit—or eliminate—their consumption of fish known to contain high levels of cancer-causing chemicals. These warnings, most strict for children and women of reproductive age, are especially emphatic about the danger of eating large fish, in which the amplifying effects of biomagnification have had the longest time to operate. The bigger the fish, the more concentrated the poison.

This, then, is the river I know. Isolated from its floodplain by levees, factories, and farms, the Illinois River flows alone. Barge convoys, big as football fields, suck the river into their wakes and then send it crashing against the banks. The accompanying tugs churn the river like egg beaters, constantly resuspending toxic materials. These include vintage chemicals—PCBs, DDT, dieldrin, chlordane, heptachlor—as well as more contemporary pollutants contributed by industrial discharges, chemical spills, and farm runoff. The resulting waves slosh silt and poison into whatever fish-spawning backwaters still exist.

More than 350 different spills of hazardous substances into the waterway were reported between 1974 and 1989 alone. The continued absence of bottom-dwelling animals makes the aquatic biologist Doug Blodgett of the Illinois Natural History Survey in Havana suspect that such spills remain frequent. Killing as it goes, each spill creates a toxic pulse that moves through a given section of stream within hours. Once-a-month monitoring does little to detect most of these transient accidents. Such spills are, of course, in addition to routine industrial discharges.

The fastest way to get to the Illinois River from my parents' house is to follow Derby Street into Normandale. This was the route I used in high school—unbeknownst to my parents, who considered the riverfront dangerous.

Derby Street itself is an avenue of nostalgia and munitions. Storefronts with names like Karen's Kountry Kottage and Grandma's Feather Bed alternate down the block with various gun and ammo shops. One is notable both for the missiles on display in the parking lot and for the half of a Jeep (with GI mannequin positioned in the driver's seat) mounted trophy-style on the side of the building. Actually getting to the river from here is tricky. It requires a stroll through the subdivision, a climb over chain-link fencing, and a firm decision to ignore No Trespassing signs. Then, abruptly, there is the water—brown, familiar, blank.

Silence is comfortable here. The river embraces silence. The Illinois River seemed to me, as a teenager, not so much dangerous, or even endangered, as reassuring.

Standing here now, aware of all that is not here, I start to wonder whether I have become a natural historian of ghosts. Since 1908, twenty species of fish have disappeared from this river. One in every three native amphibian species have been completely, or almost completely, extirpated from the state. In their extinction, they join one in every five crayfish and more than half the species of mussels. The river reminds me of a poem by Robert Frost that asks, "What to make of a diminished thing?" and provides no answer.

❧

Water is regulated much the same way food is. Just as food has tolerances, drinking water has maximum contaminant levels. These represent the highest limits allowable by law of particular toxic substances in public water supplies.

In at least two respects, maximum contaminant levels for drinking water are a more stringent measure than food tolerances. Recall from Chapter Seven that only a very tiny slice of all the food shipped, sold, and consumed in the United States is actually tested for contaminants. In contrast, *all* public drinking water is monitored on a regular, ongoing basis. Furthermore, food tolerances govern only pesticides, whereas maximum contaminant levels in drinking water regulate pollutants from both industry and agriculture. For example, there is one maximum contaminant level for the herbicide atrazine (3 parts per billion) and another for the dry-cleaning fluid perchloroethylene (5 parts per billion). The maximum contaminant level for PCBs is 0.5 parts per billion, while those for the banned pesticide chlordane and the PVC feedstock vinyl chloride each stand at 2 parts per billion. The legal limit for the phthalate plasticizer DEHP is 6 parts per billion.

As with food tolerances, these numbers have been arrived at through a compromise between public safety and economics. Maximum contaminant levels are not a health-based standard. Instead, they take into consideration cost and the ability of available technology to reduce contaminants to particular levels. These then become the legal benchmark. For many chemicals, two numbers exist: the enforceable maximum and the health-based maximum-contaminant-level *goal,* officially defined as "a non-enforceable concentration of a drinking water contaminant that is protective of adverse human health effects and allows an adequate margin of safety." The enforceable values for the carcinogens benzene, vinyl chloride, and trichloroethylene, for instance, have been set at 5, 2, and 5 parts per billion, respectively. Their maximum-contaminant-level goals, however, are all zero.

Like an accountant who proficiently measures and records individual values but fails to sum the results, this system of regulating contaminants in water suffers from the same constricted one-chemical-at-a-time vision as the parallel system of regulating pesticide residues in food. It ignores exposures to combinations of chemicals that may act in concert. Radon gas and arsenic, for example, occur naturally in some aquifers tapped for public drinking water. Both are considered human carcinogens. Maximum contaminant levels have been established for each, and each is supposed to be regulated below those levels. However, if water containing these two elements is also laced with traces of herbicides, dry-cleaning fluids, and industrial solvents—even at concentrations well below their respective legal limits—the resulting mixture may well pose hazards not recognized by a laundry list of individual exposure limits. Exposure to one compound may decrease the body's ability to detoxify another, for example.

In other ways, maximum contaminant levels are a more lenient standard than tolerances. For one thing, there are far fewer of them. As of 1996, enforceable limits had been established for a mere eighty-four contaminants. Indeed, some pesticides strictly regulated in food are not regulated at all in drinking water. For example, no maximum contaminant level exists for the herbicide cyanazine, even though it has been registered since 1971 and even though concerns about its carcinogenic properties recently prompted a phaseout of its use. Cyanazine has been detected in wells in fourteen different states and in rivers and streams throughout the Corn Belt. In some Illinois drinking-water supplies, cyanazine detections continue to exceed health-based advisory limits. But because no enforceable standard exists, these detections do not constitute violations of the law. In 1991, the National Research Council expressed official concern about water contaminants without legal limits: "The absence of evidence of their risk is solely the result of the failure to conduct research; it should not be misconstrued that [unregulated pollutants] are without risk."

To the question, then, of whether drinking water is regulated on the basis of sound scientific knowledge, the answer is no. Perhaps most revealing of all is the fact that regulation for some contaminants is based on the annual average of four quarterly measurements. In other words, drinking-water standards are violated only when the yearly mean concentration of said contaminant exceeds its maximum contaminant limit. A one-time transgression does not automatically create a violation. This distinction is important in the Midwest, where herbicide concentrations in drinking water drawn from rivers and streams often reach hair-raising levels during the spring quarter, the months of planting and rain.

In 1995, in the first study of its kind, researchers sampled water from faucets in kitchens, offices, and bathrooms every three days from mid-May through the end of June in communities throughout the Corn Belt. Herbicides turned up in the tap water in all but one of twenty-nine towns and cities. Atrazine, the suspected breast carcinogen, exceeded its maximum contaminant level in five cities, including Danville, Illinois, where its concentration in water reached six times the legal limit. Danville is southeast of Pekin, near the Indiana border. My Uncle Jack grew up there.

Food shipments, of course, are seized on the basis of single violations. The issue of whether pesticide residues in milk or asparagus or animal crackers eventually settle into an acceptable average over time is not considered relevant. Nor should it be. Biologically speaking, we live only in the present. Our bodies do not respond to contaminants on the basis of averages; they must cope the best they can with the load of contaminants already received as well as with those streaming in at any given moment. If, during the period of April through June, a woman living in rural Illinois drinks enough weed killer to overwhelm her body's ability to detoxify it, and if, as some animal evidence suggests, these chemicals are capable of initiating and/or promoting genetic lesions in her breast tissue, then the damage has been done, regardless of what happens during the months of August, October, or January.

This issue is even more critical for infants and children. Many researchers believe that exposure to even minute amounts of carcinogens at certain points in early development can magnify later cancer risks greatly. What are the implications for the unborn child who happens to reach one of these key points at the same time a bevy of farm chemicals in the local water supply is reaching its peak? What are the implications for the adolescent girl whose breast buds start to form during this particular quarter of the calendar year?

However imperfect, the current system of monitoring and regulating drinking water does provide crucial information unavailable before this decade. A younger sibling of the Clean Water Act, the federal Safe Drinking Water Act became law in 1974 and brought all community water systems under federal and state regulation. It required the EPA to set legal limits for contaminants and placed the states in charge of enforcing these limits. Maximum contaminant levels for most organic chemicals were established only with the amendments of 1986, and maximum contaminant levels for many common insecticides and herbicides were promulgated as recently as 1991. To its credit, Illinois was the first state to comply with these new regulations and began routine monitoring of farm chemicals in drinking water in 1992. Illinoisans thus have a more complete chronicle of water contamination than do residents in many other states.

While it is shocking to contemplate how many decades have

passed between the widespread introduction of synthetic organic chemicals into the environment and the decision to quantify their presence in the water we drink, the data now available to us are valuable in the most intimate way: compliance monitoring data of finished drinking water describe the actual contaminants to which we are exposed whenever we turn on the faucet.

Happily, a recent right-to-know clause added to the Safe Drinking Water Act makes this information more accessible to the public. Under 1996 amendments, water utilities must tell customers, in their water bill and at least once a year, what pollutants have been detected in their drinking water and whether water quality standards have been violated. The law also mandates the creation of a national database of contaminants found in drinking water. Previously, national records did not tally contaminants unless they constituted actual violations.

Exposure to waterborne carcinogens is more commonplace than many people realize. In the same way that intake of airborne pollution involves the food we eat as well as the air we breathe, intake of contaminants carried by tap water involves breathing and skin absorption as well as drinking. These alternative routes are especially important for the class of synthetic contaminants called volatile organics—carbon-based compounds that vaporize more readily than water. The solvent tetrachloroethylene is a common one. Most are suspected carcinogens.

We have already seen in Chapter Eight how volatile organic compounds combine with nitrogen oxides to create poisonous ground-level ozone, a major air contaminant. As a contaminant of tap water, they present additional dangers. Volatile organics are easily absorbed across human skin and enter our breathing space when they evaporate. The higher the water temperature, the greater the rate of evaporation. Humidifiers, dishwashers, and washing machines all transform volatile waterborne contaminants into airborne ones, as does cooking. These sources of exposure are thought to be particularly worrisome for infants and women home all day engaged in housework.

The simple, relaxing act of taking a bath turns out to be a significant route of exposure to volatile organics. In a 1996 study, the exhaled breath of people who had recently showered contained elevated levels of volatile organic compounds. In fact, a ten-minute shower or a thirty-minute bath contributed a greater internal dose of these volatile compounds than drinking half a gallon of tap water. Showering in an enclosed stall appears to contribute the greatest dose, probably because of the inhalation of steam.

The particular route of exposure profoundly affects the biological course of the contaminant within the body. The water that we drink and use in cooking passes through the liver first and is metabolized before entering the bloodstream. A dose received from bathing is dispersed to many different organs before it reaches the liver. The relative hazards of each pathway depend on the biological activity of the contaminant and its metabolic breakdown product, as well as on the relative sensitivity of the various tissues exposed along the way.

The bathing studies raise additional questions about drinking-water standards. Once again, we see how narrow the purview of these regulations is. The environmental scientists Clifford Weisel and Wan-Kuen Jo, the authors of the 1996 study, pointedly explained:

> Traditional approaches for evaluating exposure to and adverse health effects from contaminants in tap water have assumed that ingestion is the major route of exposure. . . . Furthermore, the ingestion of two liters of water has been used to estimate the health risk associated with waterborne chemical contaminants and the establishment of drinking water standards without quantifying the doses received from other routes. This practice can lead to an underestimation of the potential health risk.

The work of Weisel, Jo, and others also helps explain an unusual finding in Rockford, Illinois. In 1984, an environmental investigation into the dumping practices of an electroplating company led to the incidental discovery that more than 150 private wells and one municipal well were polluted with volatile organochlorine solvents. Levels varied but in some cases exceeded five hundred parts per billion. Southeast Rockford was thus catapulted onto the Superfund National Priorities List.

In a study initiated five years later, researchers found elevated levels of these same chemicals in the air space of homes receiving water from the affected wells—and in the blood of their human occupants. Curiously, blood levels correlated more closely with household air levels than with actual water levels. Air levels, in turn, were roughly correlated with length of "shower run times." These results were based on a small study population and therefore have low statistical power. However, they support the notion that inhalation contributes more significantly to overall body burden of volatile organic compounds than drinking—even when water contamination is dramatic. Bottled water, by this accounting, is not the answer.

Not long ago, I came across a survey form used in 1918 by Illinois inspectors of wells and cisterns. Among its many questions, one required measuring the distance between the water source under inspection and all possible sources of pollution. Singled out for specific mention were feedlots, privies, stables, cesspools, and "dumping grounds for slops." The survey also inquired whether small animals could fall in at the top and, most important, whether any cases of typhoid fever had ever been attributed to use of this water.

The survey's approach to protecting drinking water seemed to me to show remarkable foresight. The thrust of its questions reflect an understanding about the relationship between the safety of drinking water and the kinds of activities that go on near the source of that water. "What care is taken in collecting and storing water?" "State general condition of health of those using water." "Is the drainage from all these places toward or away from the Well, Spring, or Cistern?" "If there is any other possible source of pollution, state it." Apparently, somewhere in the transition from the age of waterborne contagion to the age of chemical carcinogens, this type of consciousness was lost. Awareness was replaced by unthinkingness.

A lengthy report on groundwater quality in my hometown was released in 1993. It contains a detailed description of contamination in two of Pekin's seven drinking-water wells. Located near the river, both wellheads are close to various industrial sites, many underground storage tanks, and the local sewage treatment plant. The chemicals detected in the water—tetrachloroethylene and 1,1,1-trichloroethane—could have migrated in from any number of possible sources. Both are suspected carcinogens. The report's assessment team expresses specific concern about a site on Second Street once occupied by Valley Chemical and Solvents Corporation. Upon closing its doors in 1989, Valley Chemical left behind a shameful trail of soil and water contamination. Its property was, you might say, a dumping ground for slops.

The city of Pekin responded swiftly to this report. Committees were established and city ordinances proposed. A course on groundwater protection was even added to the public grade school curriculum. Before all this flurry of activity, however, the initial reaction was one of astonishment. "Nothing has been done through the years to protect that aquifer," the mayor of Pekin admitted in the newspaper. "Nobody really ever thought about it. We always had good water and nobody ever thought that would change."

"Have any cases of cancer ever been attributed to use of this water?" is a more difficult query than one asking about typhoid fever cases. No one is more knowledgeable about the nature of this difficulty than Kenneth Cantor, an environmental epidemiologist and senior scientist at the National Cancer Institute. Cantor has studied the relationship between water pollution and human cancer for much of his career.

In the introduction to a recent review, Cantor and his colleagues noted that discoveries of synthetic chemical contamination in drinking water are becoming increasingly common, while epidemiological investigations into their health effects remain few. Two reasons behind the collective reluctance to launch such studies should sound familiar: there is the limitation of focusing on small populations exposed to contaminant levels high enough to yield statistical significance, and there is the problem of ascertaining past exposures. Widespread chemical contamination of drinking water may be an unintentional, ongoing human experiment, but it is one that runs without the benefit of controls or experimental design.

Most of the studies that do exist are ecological in design, that is, they simply describe patterns of association between health problems and environmental problems. We have examined a few of these already. Recall from Chapter Four that bladder cancer mortality was elevated among men (but not women) living near Pennsylvania's infamous Drake Superfund site, an old chemical dumping ground full of known bladder carcinogens. On Cape Cod, high rates of bladder cancer and leukemia were associated with living in homes serviced by vinyl-lined water pipes that were leaching tetrachloroethylene. Recall also the nationwide study in which cancer mortality was found to be elevated in U.S. counties where drinking water was contaminated by leaking hazardous waste sites.

Similar studies have been conducted in other settings, both urban and rural. In New Jersey, researchers found associations between volatile organic compounds in municipal water and leukemia among women (but not men). In Iowa, lymphoma rates were elevated in counties where drinking water was drawn from dieldrin-contaminated rivers. In Massachusetts, childhood leukemias in the industrial town of Woburn were linked to a pair of water wells contaminated with chlorinated solvents. In North Carolina, a cancer cluster in the rural community of Bynum was linked to consumption of river water contaminated upstream with both agricultural and industrial chemicals. This study is particularly compelling because the sudden increase in cancer deaths that emerged in the 1980s corresponds closely with the time of peak exposure to known carcinogens in the river (1947 to 1976) once the normal latency period for cancer is factored in. Likewise in Woburn, the surge in childhood leukemias coincides with a period of known water contamination and abates a few years after the imputed wells closed down. (Public outcry about the plight of Woburn's children played a direct role in the creation of the Massachusetts Cancer Registry.)

Corroborating evidence also comes from abroad. In a study from China, liver cancer was strongly associated with drinking water from ditches containing agricultural chemicals. In Germany, excess cases of childhood leukemia in villages near uranium mines have been tentatively linked to radium-contaminated drinking water. And in Finland, high rates of non-Hodgkin's lymphoma were discovered in a rural community where water was contaminated by chlorophenols, probably from local sawmills. Used for treating lumber, chlorophenols are related chemically to the phenoxy herbicides, which are also linked to non-Hodgkin's lymphoma.

❧

In a practice that began early in the twentieth century, the city of Chicago began in 1908 to pour chlorine into wastewater before sending it downstream. In the same year, the waterworks of Boonton, New Jersey, became the first to add chlorine to water intended for drinking. Chlorination proved a cheap, effective means of halting waterborne epidemics during World War I. By 1940, about 30 percent of community drinking water in the United States was chlorinated, and at present, about seven of every ten Americans drink chlorinated water.

Over the past two decades, nearly two dozen studies have emerged that link chlorination of drinking water to bladder and rectal cancers and, in some cases, to cancers of the kidney, stomach, brain, and pancreas. These investigations include case-control and cohort studies in addition to ecologic studies. The collective evidence on water chlorination, affirms Kenneth Cantor, "supports concern over an elevated carcinogenic risk."

Upon hearing this news, many otherwise even-tempered individuals may feel tempted to throw up their arms in frustrated despair, as though they had just been asked to choose between death by cancer and death by cholera. Happily, this is not our predicament. Far less gloomy options are available. They will not be realized, however, unless we recognize the hazards created by the approach presently used to combat disease pathogens in our drinking water and, with this knowledge, insist on safer practices.

Chlorine gas is a noxious posion. However, the problem with chlorinated drinking water does not lie with chlorine itself. Rather, in a manner reminiscent of the way that air pollutants combine in the atmosphere to create new chemical species, the problem begins when elemental chlorine spontaneously reacts with organic contaminants already present in water. Their organochlorine offspring are known as disinfection by-products. Hundreds exist, and several are classified as probable human carcinogens. Trihalomethanes, a small subgroup of volatile disinfection by-products, are currently receiving the most scientific and regulatory attention. Chloroform is the most common one. As with any waterborne volatile compound, our route of exposure to trihalomethanes is threefold: ingestion, inhalation, and absorption. Indeed, trihalomethanes appear as one of the major chemical culprits in the bathing studies already discussed.

Volatile organic compounds in drinking water, then, can have variant life histories. Some may be escapees from landfills, waste dumps, farm fields, or industrial parks. These compounds arrive in our water supply ready-made, their chemical conformations intact. Others may be formed on-site at the waterworks. In this way, the chloroform present in finished tap water has at least two possible pedigrees: it could have leaked into the water supply as a contaminant, or it could have been created during the process of chlorination. All volatile organic compounds classified as trihalomethanes are regulated as a group, regardless of the precise genealogy or the individual components of the mixture. Their maximum-contaminant-level is 100 parts per billion. Their maximum-contaminant-level goal is zero. In the EPA's chart of drinking-water standards, along the row labeled "Total Trihalomethanes" and under the column titled "Potential Health Effects" is a single word: *cancer.*

Many studies all telescope into this one word. The early investigations were ecological in design and compared cancer rates in communities with and without chlorinated water. Conducted in Ohio, Louisiana, Wisconsin, Iowa, Norway, and Finland, these studies consistently found associations between water chlorination and cancers of the bladder and rectum. In a second wave of case-control and cohort studies, researchers then pursued the link between cancer and chlorination more intensely. These researchers interviewed individuals about the details of their tap-water habits, controlled for lifestyle confounders, used historical water records to estimate past exposures, and even gathered information about the sources of drinking water at previous residences. Carried out in Wisconsin, Illinois, Louisiana, Massachusetts, Maryland, North Carolina, Colorado, and Norway, these studies suggested an association between water chlorination and cancer, especially in regard to bladder and rectal cancers, and especially when drinking water is drawn from above-ground sources, such as rivers.

One of the most ambitious of these investigations was led by Kenneth Cantor himself. His research team personally interviewed nine thousand people living in ten different areas of the United States. Individual histories were then combined with water utility data to create a lifetime profile of drinking-water use for each respondent. In the final analysis,

> bladder cancer risk increased with the amount of tap water consumed, and this increase was strongly influenced by the duration of living at residences served by chlorinated surface water. . . . There was no increase of risk with tap water consumption among persons who had lived at places served by nonchlorinated ground water for most of their lives.

Giving people cancer in order to ensure them a water supply safe from disease-causing microbes is not necessary. Part of the solution lies in making wider use of alternative disinfection strategies. These include granular activated charcoal (which binds with contaminants and removes them) and ozonation (which bubbles ozone gas through raw water to kill microorganisms). Both techniques have been used successfully in many U.S. and European communities.

Part of the solution lies in directing a spirit of urgency, inventiveness, and ingenuity toward the development of other approaches. No doubt many technologies await discovery, requiring only the devotion of resources and a collaboration of creative minds to bring them into existence.

Finally, part of the answer lies in keeping carbon-based contaminants out of drinking water in the first place. This last dictum is doubly important. Less organic content means fewer trihalomethanes. Less organic content also knocks down the number of microorganisms, thereby reducing the amount of chlorine needed for disinfection. Tellingly, water from lakes, rivers, and reservoirs generates more trihalomethanes upon chlorination than does water drawn from aquifers. This is because, in general, surface water carries more organic matter than groundwater. Some of the progenitors of trihalomethanes are natural and unavoidable: decaying leaves, fallen feathers, and grains of pollen, for example. These all contribute to the total carbon load in a body of water. But many others are neither natural nor unavoidable: sewage, chemical spills, industrial discharge, soot and other fallout from air pollution, agricultural runoff, and motor oil, for instance. Drastically reducing these inputs would go a long way toward solving the problem of disinfection by-products—as well as other grosser forms of water contamination.

This part of the solution requires that water utilities and the water-consuming public become vigilant about the protection of watersheds and aquifers. Guarding water supplies means more than keeping swimmers out of reservoirs and erecting fences around wellheads. In some regions, this kind of protection will require new thinking about agriculture, which needs to substitute the techniques of organic farming for practices that pour soil and pesticides into river systems. In regions where cattle feedlots and hog farms periodically send lava flows of manure into watersheds, it will require new thinking about animal husbandry. In other areas, it will require new thinking about industry. Manufacturers must find safer alternatives to organic solvents and other synthetic carbon-based chemicals that are released into water directly, fall off barges in transit, waft into the air only to rain down elsewhere, or eventually worm their way into water via landfills and dump sites. Finally, in all regions, protection of water supplies will require new thinking on the part of individual citizens, who are asked to assume the frightening cancer risks that others have decided, on their behalf, are acceptable.

Back at the waterworks, additional improvements are possible. For example, making chlorination the last step of water treatment, rather than the initial one, lowers the amount of trihalomethanes generated, especially if the water is carefully filtered through granulated charcoal first. Artificial membranes can remove a slew of contaminants, including pesticides and solvents. Water can also be aerated to allow volatile organic compounds, including trihalomethanes, to vaporize. Because they transfer contaminants from drinking water to other environmental media, I consider these kinds of solutions less beneficial than a comprehensive strategy of primary prevention. Aeration sends waterborne organic compounds into the atmosphere where we can inhale them, and filters and membranes fill with toxic chemicals, which must go somewhere. Even while providing some immediate respite from exposure through tap water, these technological shell games keep carcinogens in circulation.

In 1910, a New Jersey court examiner declared that chlorination left "no deleterious substances in the water." He was wrong. Nevertheless, it is clear that the disinfection of drinking water with chlorine has prevented widespread contagion and death, even as it has also contributed to the burden of human cancers. I do not advocate a ban on the chlorination of drinking water. But neither do I believe we should blithely continue old disinfection practices as though our bodies and our water supplies still existed in the world of ninety years ago. I say this as an ecologist with a personal relationship to bladder cancer. In 1910, chloroform was not considered a deleterious substance. When its toxicity was later recognized, its use as a surgical anesthetic was phased out. We need not be forced to drink it now as the price for contagion-free water.

❧

At dead center, in the channel used by barges, the Illinois River is about as deep as the deep end of a swimming pool. If you dove to the bottom here, you would first pass through a flocculent layer of silt many feet deep. Underneath this fluffy mass is the clay trough of the riverbed. If you could somehow continue the descent, drilling down through this foundation, you would eventually find yourself once again in water—the water under the water—which is held between the glittering sand grains of the Sankoty Aquifer.

This underground basin not only lies beneath the river but stretches out for miles along its east flank and extends south toward Havana. It occupies what was once the valleys and snaky tributaries of the ancient Mississippi River—before they were bulldozed by glaciers. The Sankoty Aquifer is the source of Pekin's drinking water.

Technically speaking, an aquifer refers not to the groundwater it holds but to the collection of grit, gravel, clay, and rock the water flows through. The Sankoty ranges from 50 to 150 feet thick and consists mostly of quartz sand grains ranging in size from dust to marbles. They are said to be distinctly pink. Sankoty sand grains are also, I'm told, neatly sorted and stratified by size, indicating they were once carried along and then deposited by the flow of a melting glacier. The resulting outwash is referred to as a valley train. Aquifers can include any porous and permeable material, such as unsorted scrap gouged up and laid back down by the glacial ice itself—this would be called till—or layers of wind-blown silt, which geologists have named loess.

Regardless of materials, aquifers come in a few basic varieties. Bedrock aquifers are covered by a lid of impermeable materials. Artesian aquifers, often lying on a slant, are under hydrostatic pressure. Water-table aquifers are like uncovered pots with rain and melting snow periodically dribbling in through the overlying soil. Their surface, the water table proper, rises and falls with seasonal fluctuations in precipitation. The Sankoty is a very large water-table aquifer.

The chemical contamination of the Sankoty Aquifer is an ongoing story with no identifiable beginning, no defining catastrophic event, and nothing that could reasonably be called a resolution. Even before the 1993 assessment roused Pekin's residents into action, there were signs of trouble. As part of a 1989 survey, the Illinois Environmental Protection Agency discovered "a substantial level" of 1,1,1-trichloroethane in one of Pekin's drinking-water wells and low levels of benzene and tetrachloroethylene in another. A year later, the Illinois EPA issued two groundwater contamination advisories after accidents at loading docks sent into the public water wells of Creve Coeur a variety of gasoline additives and crude oil derivatives. Then, in the spring of 1991, high river levels contributed to the capture of chemical contaminants in a community drinking well in north Pekin.

This last discovery was particularly unsettling. The interchange between groundwater and surface water is normally a one-way affair with aquifers, recharged by rain, emptying themselves into the rivers and streams that lie across them. Barring a flood, flow of river water into groundwater is not supposed to happen. Abrupt increases in river volume or heavy pumping from wells can, however, alter the direction of these unseen currents and possibly divert surface water into the underground world of aquifers. Water levels inside certain Sankoty wells, for example, appear to fluctuate in tandem with lock-and-dam operations on the Illinois River, implying a more reciprocal communion between these two bodies of water than was once presumed.

In light of this and other ominous realizations, the results of the groundwater assessment actually appear quite mild. Industrial chemicals turned up in a few discrete locations, but the field team found no signs of aquiferwide contamination. Indeed, the authors of the study expressed surprise at not stumbling on an even bigger problem, especially after they reviewed the history of industrial practices within the local area: "Any contaminant that could have been produced, almost certainly was produced. And yet, we do not find widespread contamination of the ground-water environment."

It is an eerie paradox, and it is difficult to know how alarmed or reassured to feel. Certainly there is danger in breathing sighs of relief too soon. Just as the presence of a single cockroach in the kitchen sink speaks of the hundreds more behind the wall, periodic detections of contaminants in groundwater aquifers are often harbingers of widespread contamination yet to come. Groundwater flows both leisurely—sometimes only inches per year—and smoothly as it moves along the pores and cracks of the underground landscape. Without speed or turbulence, dispersion of chemical contaminants is also slow. Over time, an intermittent detection in one well can eventually become a constant detection in several. Moreover, even the merest trace of contamination can portend a serious problem if what is being detected is the bottom edge of a falling curtain of chemicals slowly moving its way down the aquifer's overlying substrate.

As a general rule, contamination in lowland areas of discharge—where aquifers give up their water to rivers and streams—is considered a lesser problem than contamination in the upland areas of recharge where aquifers receive rain and snow from the atmosphere. Areas of recharge are the headwaters of aquifers, and contamination here can fan out and fill the whole. In either case, once groundwater becomes contaminated, little can be done to remedy the problem. In contrast to the alfresco run of surface water, groundwater has no oxygen to hasten the breakdown of chemical contaminants nor open air to facilitate the evaporation of solvents and other volatile organics. Contamination lingers in the still, watery vaults of aquifers.

With great pomp and flair, the city of Pekin passed its proposed groundwater protection ordinance in 1995. It has since been hailed as a model for the state. Essentially, the statute regulates land use in three recharge areas, each a narrow stretch of ground a mile or so long where the groundwater underneath flows into the city's wells. Additionally, the ordinance draws a two-thousand-foot ring of protection around each of the wellheads. Inside these seven circles, the city restricts, and in some cases prohibits, the siting of new businesses that handle large quantities of hazardous materials. Existing businesses are largely unaffected, although some have promised to make improvements.

Ever since the ordinance was drafted, personal knowledge about recharge, discharge, depth to water table, glacial deposits, and other details of hydrogeology has become a matter of civic pride in Pekin. Groundwater maps overlaid with a grid of the city's streets have appeared on the front page of the newspaper so that residents can locate themselves vis-à-vis the aquifer. The west end of Derby Street lies within a recharge area, for example, as do sections of Sabella, Charlotte, and Henrietta Streets. The east bluff does not. The owner of a gas station on Fourth Street, upon discovering himself inside one of the protected zones, pledged to install double-walled tanks to prevent leaks. He even praised the ordinance after attending a public hearing. "It's great, it should have been done years ago, and nobody paid attention to it."

Pekin's ordinance is a candle in the dark. It has sparked open discussion about the relationship between health and the environment, and it has lit in people's hearts new respect for the body of water they walk over and drink from. Its ability to safeguard the Sankoty Aquifer against the toxic activities that continue to go on ninety feet above it, however, is not at all clear. Toxic runoff from storm sewers empties into several creeks and at least one lake that overlay the recharge zone. And as the district superintendent of the water company points out, the rain itself contains pollutants. No local zoning laws can legislate against pesticide-laced raindrops or solvent-contaminated snowflakes falling in a recharge zone. Solutions to these problems need to be hammered out in chambers larger than small-town city councils.

In the meantime, industrial chemicals and pesticides persist in making cameo appearances in Pekin's drinking-water wells—sometimes briefly exceeding their maximum contaminant levels, sometimes remaining well below the legal limit. They include benzene, perchloroethylene, 1,1,1-trichloroethane, the phthalate plasticizer DEHP, and a couple of lawn chemicals. This brings us to the present moment. We know the story of Sankoty Aquifer begins with a glacier. We know that someone, sooner or later, ends up drinking whatever poisons are spread on the earth above it. What happens next is the part of the story that is still unwritten.

About one-third of Americans draw their water from aquifers. The rest drink from rivers, lakes, and streams. Of course, ecologically speaking, everyone drinks from aquifers: all running surface water was at one time groundwater, aquifers being the mothers of rivers. As Rachel Carson pointed out, contamination of groundwater is, therefore, contamination of water everywhere.

Groundwater provides no archival photographs to consult. It offers no shores to walk along, no reflective surfaces to peer into, no fish, bivalves, grasses, or game birds to inquire about. Our relationship to aquifers is deeply biological, but it is not visual.

I once descended to the bottom of a well in order to look at groundwater in its natural habitat. This happened in Hawaii, where drinking water is drawn from a flattened lens of rain that is trapped under the island between the volcanic rock it trickled through and the Pacific Ocean beneath. (Freshwater floats on salt.) At the Halawa pumping station, I rode a cable car down a three-hundred-foot shaft to arrive in a blasted-out cavern filled with water. It was very dark and very quiet.

Illinois, I believe, would provide a more exotic underground landscape. In the descent, one could examine the mashed remains of preglacial forests, bluffs, dunes, islands, and cliffs. The bedrock floor would be inlaid with the trenches of ancestral riverbeds. The water table's undulating ceiling would offer a subdued reflection of the overlying topography.

Cultivating an ability to imagine these vast basins beneath us is an imperative need. What is required is a kind of mental divining rod that would connect this subterranean world to the images we see every day: a kettle boiling on the stove, a sprinkler bowing over the garden, a bathtub filling up. Our drinking water should not contain the fear of cancer. The presence of carcinogens in groundwater, no matter how faint, means we have paid too high a price for accepting the unimaginative way things are.

TEN

fire

this world in peace
this laced temple of darkening colors
it could not have been made for shambles
—JOHN KNOEPFLE, "CONFLUENCE"

"It's different country down here," my mother says, and I agree.

We're following the river valley south into Mason County on a winding unmarked road everyone around here calls the Manito Blacktop. To catch it, you take Route 29 past Normandale, continue on by the distillery, and turn right at the federal prison. From here the blacktop appears to be an access road for the power plant, but once you drive through the silver forest of transformers and towers, with the Illinois River on your right, you know you're on a real road heading out of town.

The houses along the blacktop have a scattered, haphazard feel to them—as though the river gave permission to strew one's possessions about. Propane tanks, extra cars, and satellite dishes are parked in sloping yards alongside signs advertising the sale of garden produce. The soil here is very porous, a fact made evident by the irrigating center pivots that lie over the fields like the skeletons of enormous bats. These draw water up from the Sankoty Aquifer and spray it in all directions. They grow a lot of specialty crops here—green beans, peas, sweet corn, cucumbers, pumpkins, and melons—as well as standard-issue field corn and soybeans. The little village of Manito announces itself as the Imperial Valley of the Midwest, and then the road banks and curves away to the right. None of this resembles life on the right-angled Illinois prairie east of here where farmhouses are spaced across the landscape like battleships assigned to their own black-dirt square of sea and farmers depend only on rain to get them through.

I'm assuming, anyway, these differences are the ones my mother is referring to. We continue south and west toward Havana. It is September 1994. A convergence of various forces—political, historical, and personal—form the purpose for this trip. They all have to do with garbage incinerators.

The political reasons start with an obscure law that was passed at the end of the 1980s. Called the Illinois Retail Rate Law, this legislation requires electric power companies to purchase—at the retail rate charged to customers—any and all electricity generated by trash-burning incinerators. The utilities then recoup their loss of profit through tax credits. Overnight, garbage incinerators became the most heavily subsidized development project in Illinois history.

In their aim to attract incinerator builders and investors to the state, the lawmakers had succeeded splendidly. Previously, Illinois was home to one operating garbage incinerator (in Chicago), but as of 1994, twelve others were in various stages of study, siting, or construction. Six were under consideration in central Illinois, one of them in Havana, where siting had been approved by the city council the previous December. Specifically, the plan was to build a fifteen-acre incinerator—handling eighteen hundred tons of garbage per day—in a popcorn field adjacent to the railroad's coal docks. Trash would be hauled in by boxcar from Chicago. In order to turn the steam-driven, electricity-generating turbines, three million gallons of water a day would be drawn up from the aquifer, requiring a doubling of Havana's pumping capacity.

The historical circumstances surrounding the summer and fall of 1994 ran counter to the immediate political ones. On a national level, successful recycling efforts had taken the pressure off of landfills, and recyclers were now competing with incinerators for trash. Incinerators themselves were proving costly. Columbus, Ohio, was preparing to shut down its energy-generating incinerator in the wake of intractable economic and environmental problems. Albany, New York, had become plenty vexed at its own garbage burner the previous January. Already facing expensive upgrades in order to meet air pollution standards, the incinerator had covered a fresh snowfall with a black layer of soot after a series of pollution control mishaps. It was quickly mothballed.

Meanwhile, environmental research indicated that trash incinerators routinely release troubling amounts of toxic and carcinogenic pollutants, including the most potent of all the organochlorines: dioxin. In addition, several new studies had demonstrated that dioxin is harmful at far lower exposures than anyone ever suspected. Even at a few parts parts per *trillion*, dioxin is capable, it seems, of profoundly altering biological processes.

Also in the fall of 1994, the EPA released a three-thousand-page draft reassessment of dioxin and was now soliciting public commentary and reaction. Three years in the making, the study reaffirmed dioxin's classification as a probable human carcinogen. The draft report also announced three other findings. First, dioxin's effects on the immune system, reproduction, and infant development are much more significant than previously thought. Second, there is no safe dose below which dioxin causes no biological effect. Third, quantities of dioxin and dioxinlike chemicals present in most people's bodies are already at or near levels shown to cause problems in animals. Finally, the report identified incineration—of both medical waste and common household garbage—as the leading source of dioxin emissions in the United States and food (meat, dairy, and fish) as the immediate source of 95 percent of the dioxin found in the bodies of the general population.

The release of this draft had been rumored for months. I had not yet received my copy of the six-volume set, but Dorothy Anderson, a pediatrician and the president of the Mason County Board of Public Health, had. It is to her house my mother and I are headed.

Our personal interest in the incinerator issue is simple. First, Pekin is directly downwind of Havana. Second, a plan for a trash incinerator nearly identical to this one—introduced by the same developer and backed by the same set of corporate investors—is under consideration by the village board of Forrest, eighty miles northeast of Pekin, in Livingston County. This facility is to be built not in the village itself but out in Pleasant Ridge township, three miles north of town. My mother knows precisely the section of cornfield they have in mind. It's exactly one mile south and three-quarters of a mile east of her brother's farm. She also knows that Roy has thrown his hat in with a group of farmers organizing to oppose the siting.

❧

No matter how you look at it, scooping garbage into an oven and setting it afire is an equally primitive alternative to digging a hole in the ground and burying it. The former contaminates air; the latter, groundwater.

The relative popularity of these two options has waxed and waned over the decades. In 1960, about one-third of the nation's trash was burned in incinerators. Because of serious air pollution, these were later phased out in favor of landfills. In the 1980s, incinerators, now sporting high-tech pollution-control devices and designed to generate electricity, staged a comeback, their promoters referring to them as "waste-to-energy" or "resource recovery" plants.

No matter how improved or what they are called, incinerators present two problems that landfills do not. First, incinerators only transform garbage; they don't provide a final resting place for it. There remains the question of where to put the ashes. Second, these cavernous furnaces create, out of the ordinary garbage they are stoked with, new species of toxic chemicals. In addition to producing electricity, they generate hazardous waste.

The first problem flows from a primary law of physics. Most of us at one point or another in our education probably had it memorized: matter can neither be created nor destroyed. Every single atom fed into an incinerator survives. If 1,800 tons of garbage per day go in, 1,800 tons per day also come out, albeit in a chemically altered form. Some of this matter rises as gas or tiny particles and is released into the air as stack emissions. (Much of the gas is carbon dioxide.) The rest of it is captured as ash, which requires disposal.

In 1993, John Kirby, the developer responsible for the downstate Illinois incinerator proposals, showed newspaper reporters a jar of ashes weighing 3.7 pounds. This is all that remains, he boasted, of an average person's weekly 40 pounds of garbage after it has been run through a waste-to-energy facility. Landfilling 3.7 pounds, Kirby pointed out, creates less of a volume problem than burying all 40. In this he is certainly correct, but by extension, the containment of 3.7 pounds means 36.3 pounds of garbage are sent up into the sky. Incinerator advocates cannot have it both ways: fewer ashes means less required landfill space but more air emissions; more ashes means less air pollution but creates a bigger disposal problem. The indestructability of matter reigns supreme.

Moreover, the process of burning concentrates into the ash whatever hazardous materials are present in the original refuse. Heavy metals, such as mercury, lead, and cadmium, for example, are not destroyed by fire. Occurring as ingredients in household batteries, lightbulbs, paints, dyes, and thermometers, they are absolutely persistent. Air pollution control depends on the ability of an incinerator's cooling chambers to condense these metals onto fine particles, which are then caught in special filters.

Once again, the irony of trade-offs becomes readily apparent: the less air pollution, the more toxic the ash. An incinerator burning eighteen boxcars of trash per day, for example, produces about ten truckloads of ashes per day. The trucks must then rumble out onto the highways, hauling their poisonous cargo through all kinds of weather. Once ensconced in special burying grounds, incinerator ash, of course, presents a hazard to groundwater.

The second problem is more an issue of chemistry than of physics. Somewhere between the furnace and the top of the stack, on the papery surfaces of fly ash particles, in the crucible of heating and cooling, carbon and chlorine atoms rearrange themselves to create molecules of dioxins and their closely related organochlorine allies, the furans.

There are many dozens of dioxins and furans, but, as with snowflakes, their individual chemical configurations are all variations on a theme. Recall that benzene consists of a hexagonal ring of carbon atoms. This ring can then be studded with chlorine atoms. Two chlorinated benzene rings bonded directly together form a polychlorinated biphenyl, a PCB. By contrast, two chlorinated benzene rings held together by a single atom of oxygen and a double carbon bond are called a furan. A pair of chlorinated benzene rings linked by two oxygen atoms form a dioxin. There are 135 furans and 75 dioxins, each with a different number and arrangement of attached chlorines.

Dioxins and furans behave similarly in the human body, and they all to some degree elicit the set of biological effects described earlier. The most poisonous by far, however, is the dioxin known as TCDD. This particular molecule bears four chlorine atoms, each bonded to an outer corner. Because these points of attachment are located on the carbon atoms numbered 2, 3, 7, and 8, its full name is a mouthful: 2,3,7,8-**t**etra**c**hloro**d**ibenzo-*p*-**d**ioxin. Imagine looking down from an airplane window at a pair of skydivers in free fall, both hands joined together. Their geometry provides a reasonable impersonation of a TCDD molecule: the divers' linked arms represent the double oxygen bridge, their bodies the benzene rings, and their splayed, outstretched legs the four chlorine atoms.

TCDD is scary because it is so stable. The symmetrical arrangement of its chlorine legs prevents enzymes—ours or any other living creature's—from breaking TCDD apart. In human tissues, TCDD has a half-life of at least seven years. As we shall see, this particular geometry also allows TCDD admission into a cell's nucleus and access to its DNA.

Incineration is not the only source of dioxins and furans. They can also form spontaneously during the manufacture of certain pesticides—especially phenoxy herbicides and chlorophenols—and during the bleaching of paper products, for example. What all three of these processes have in common is chlorine. Dioxin is synthesized when certain types of organic matter are placed together with chlorine in a reactive environment. Such conditions are created by combinations as banal as newspapers plus plastic wrap plus fire.

In the inferno of an incinerator, many common synthetic products may serve as chlorine donors for the spontaneous generation of dioxins and furans: paint thinners, pesticides, household cleaners. A major source of chlorine is PVC (polyvinyl chloride), which can take the form of discarded toys, appliances, shoes, or construction debris.

However unselective, the conditions required for the formation of dioxins and furans are largely limited to those created by contemporary human activities. Compared to incineration of synthetic materials, forest fires produce trivial amounts—and these traces may represent the release of dioxin molecules from soil and vegetation contaminated by previous aerial deposits rather than de novo synthesis. Sediment cores show no extensive contamination with dioxins until the 1920s and 1930s, corresponding to the advent of organochlorine production. People living in industrialized nations have higher dioxin body burdens than those living in unindustrialized areas. We also carry far greater levels of dioxin in our tissues than do 2,800-year-old human mummies or 400-year-old frozen Eskimos, which scarcely have any.

Dioxins and furans are not the natural-born children of fire. They are the unplanned, unwanted offspring of modern chlorine chemistry.

Dorothy feeds us slabs of freshly baked bread, garden tomatoes, and great hunks of cheese. She and my mother have already figured out they belong to the same church, a discovery that creates a spirit of common purpose between them. I've already plowed through most of the reassessment's executive summary (*Estimating Exposure to Dioxin-like Compounds*) and have taken a few notes from the first volume

(*Health Assessment Document for 2,3,7,8-Tetrachlorodibenzo-p-dioxin [TCDD] and Related Compounds*).

Now the three of us are sitting around her kitchen table discussing the ambitions of the man named John Kirby—the incinerator developer, state lobbyist, and entrepreneur—whose plans affect us all so deeply. The bright haze of an Indian summer afternoon fills the windows. Trees buzz with cicadas. Despite the issue that has brought us together, I feel completely at peace.

Dorothy's perspective is straightforward. She is a practicing Methodist, a practicing physician, and the head of the board of public health. Opposing the Havana incinerator is both an affirmation of her spiritual beliefs and an act of preventive medicine.

"If nothing else, I have a charge to keep."

Places like Havana, she asserts, are especially vulnerable to the designs of incinerator developers, who entice rural communities with promises of jobs and lucrative "host fees." These often exceed a small town's entire annual budget.

Dorothy ticks off the statistics: Mason County has 15 percent unemployment, lots of teenage pregnancy, and high rates of infant mortality. It is one of the state's poorest counties. One in four children live in poverty.

"So Kirby offered Havana a million dollars," she shrugs, as if to say, end of story.

We are quiet for a while.

Dorothy brings out apples and paring knives. My mother can core and quarter an apple faster and more perfectly than anyone else I know. Doctor Dorothy, blonde mother of four, is pretty swift at apple dissection herself, I notice. Somehow I am botching mine. I silently convince myself that I suffer from such problems only when in the presence of my mother. A chunk with a stem attached shoots across the table. Mom takes what's left in my hand and makes quick work of it. Then she begins to speak.

Forrest, she says, is in similar straits. Included in the package of temptations presented to residents out there is the promise of a new school library. A school referendum to do the same had just failed. As in Havana, Kirby's proposal had torn the tight-knit community asunder. Teachers found themselves on the opposite side of the issue from parents, farmers from grain elevator operators, ministers from their parishioners, village board members from village board members. The farmer who had given the incinerator developers the option on his land was now sunk in remorse. His brother had joined the opposition. I hear in her voice her worry for Roy, who is a tax assessor for the township and a farmer.

"Neighbors aren't talking to neighbors. Even some family members aren't on speaking terms anymore. It's the money talking."

Dorothy's theory is that a small, but possibly fatal, slipup by Kirby in Havana may explain his activities in Forrest. In December 1993, after a rancorous siting hearing, the city council of Havana voted five to two in favor of building the incinerator. The group opposing the project, of which Dorothy is a part, then appealed the council's decision to the Illinois Pollution Control Board.

Obvious improprieties abounded, their lawyer argued. From the start, Kirby and the mayor had been working hand in hand on this project. Furthermore, one of Kirby's several companies had flown city council members to Boston, where they toured an incinerator located near Cape Cod that was to serve as a model for the one in Havana. Incinerator opponents were not invited. Hence, they were denied crucial knowledge obtained by trip participants.

The newspaper had represented the tenor of the Pollution Control Board hearing with this piece of testimony on the details of the infamous Boston trip:

> **LAWYER:** Was it a big plane, little plane?
>
> **COUNCIL MEMBER:** Oh, any plane would have been big to me. I never been on one.
>
> **LAWYER:** And did they take you to supper, Mr. Thomas?
>
> **COUNCIL MEMBER:** Oh, yes, they had everything you ever wanted to eat there. I had crab something one night and lobster the next. It was real, real, real nice. You know, I've never had lobster before.
>
> **LAWYER:** So you were treated first class everywhere you went?
>
> **COUNCIL MEMBER:** Well, it wasn't just me. All of us were treated that way.

Dorothy places this clipping on the glass surface of the copy machine parked in her dining room and pushes the button. She also makes copies for me of articles relaying more recent news: in June, the Pollution Control Board ruled in favor of incinerator opponents, overturning the city council's decision—at least temporarily. Kirby was in the process of appealing. In the meantime, he may be pursuing Forrest as a way of hedging his bets, Dorothy suggests. We know that he recently ferried a delegation from Forrest out to Massachusetts, but this time he wisely extended his invitation to include incinerator opponents.

I close my eyes. The board's decision brings to mind an image of an unarmed man standing before a line of tanks. How long can the ruling hold? The Havana incinerator proposal had been officially declared dead twice before—once in 1991 when the Mason County board voted it down and again in 1993 when Kirby's corporate backers pulled out—and both times it had come roaring back. He shifted his proposal from county board to city council. He found new investors.

Still, Kirby doesn't always prevail. In 1990, he brought an incinerator proposal, with all its glittering promises of money and jobs, to the city council of another struggling town in central Illinois. Opponents promptly submitted a petition with three thousand attached signatures and staged a candlelight vigil the night of the vote. At the last possible moment, Kirby withdrew his application.

This happened in Pekin. The citizens who organized to oppose the siting stated publicly their firm belief that incineration is not the answer to the landfill dilemma. And they won.

❧

The image of a giant incinerator sitting out in the silence of Livingston County cornfields, fleets of ash trucks and refuse-filled railcars forever coming and going, was disorienting, virtually impossible to accept. Understanding the effects of incinerator emissions on agriculture—the soil, the crops, the food chain—is equally baffling. However, we must examine the probable effects, because so much of our exposure to dioxins and furans comes from eating food.

Even the newest, fanciest incinerators send traces of dioxins and furans into the air. These molecules cling to bits of dust and sediment. As they move downwind, they sink back to earth or are washed out with rain. Here they coat soil and vegetation—grass, clover, corn, beans, hay, watermelons, whatever. These chemical contaminants are then consumed by us directly or are first concentrated in the flesh, milk, and eggs of farm animals. A number of European studies have documented elevated levels of dioxin in the milk of cows grazing in pastures near municipal incinerators, for example.

Thankfully, dioxin lacks a few of the nastier traits of other organochlorines. It tends not to migrate to groundwater and is not very volatile. Water and air are not, therefore, major routes of exposure for us. Dioxin does accumulate in river sediments and in the bodies of fish, and it collects in soil. It is not, however, easily absorbed by the roots of most crops. The main problem for us comes when dioxin-contaminated particles are deposited onto the leaves, stems, and flowers of crop and pasture plants, thus initiating the ballooning process of biomagnification. Foraging farm animals can also accumulate dioxin from ingesting soil directly.

By contrast, incinerator-distributed cadmium contaminates crops both from aerial deposition and from root uptake. Unlike dioxin, cadmium can be concentrated in plant tissues at levels higher than those in soil. Cadmium intake by humans is mostly from direct consumption of contaminated plants, especially fruits and vegetables.

Cadmium is classified as a probable human carcinogen. In animals, it is associated with sarcoma, lung cancer, and prostate cancer. High rates of lung, prostate, and testicular cancers have also been reported in workers who inhale cadmium on the job—but the question of incineration, cadmium ingestion, and cancer risk remains unexplored. Somewhere between 50 to 75 percent of the cadmium in the waste stream—about thirteen hundred tons—comes from discarded batteries. "Dead batteries, if incinerated, do pose a cancer risk," reasoned one epidemiologist after analyzing the numbers.

Fruits, vegetables, chickens, hogs, grain destined for animal feed: from Mason to Livingston counties, we had them all. Every day, someone somewhere sits down to a pork chop, a pile of peas, or a bowl of popcorn that originated from the central Illinois countryside. The question of whether or not to construct an incinerator in the midst of all this agriculture seemed like a national issue—one the whole country should vote on. Instead, the decision was being made by a handful of small-town city councils desperate to shore up their communities' economies.

Who could blame them? Who could be against jobs or school libraries? From what I could see, most of those promoting Kirby's plan were well-intentioned souls who saw themselves upholding, rather than betraying, the public trust. Still, the politics were being played out within the smallest of jurisdictions while the potential biological impact loomed beyond all city, county, and state boundaries.

Ascertaining dioxin's contribution to human cancers is one of the more frustrating challenges for public health researchers. Because dioxin is so potent at such vanishingly small levels, exposure is expensive to measure. Because it is so widely distributed, there remain no populations to serve as unexposed controls. Because dioxin so often rides the coattails of other carcinogens, confounding factors abound. U.S. military personnel exposed to Agent Orange in Vietnam, for example, were simultaneously exposed to 2,4-D and dioxin-contaminated 2,4,5-T, as noted in Chapter Three.

Animal studies provide a complex set of clues. In the laboratory, dioxin is an unequivocal carcinogen. As the dioxin researcher James Huff once noted, "In every species so far exposed to TCDD . . . and by every route of exposure, clear carcinogenic responses have been found." These include cancers of the lung, mouth, nose, thyroid gland, adrenal gland, lymphatic system, and skin. Dioxin also causes liver cancer in rats and mice, but it does so more often in females. Female rats whose ovaries have been removed, however, tend *not* to develop liver cancer when exposed to dioxin. On the other hand, they are far more likely to succumb to lung cancer. Clearly, an organism's own internal hormones modulate dioxin's carcinogenic powers, but through some unclear means.

Sensitivity to dioxin also varies among species. A thousand times more dioxin is required to kill a hamster than a guinea pig. Such differences raise the question of where humans fall along the dioxin/cancer sensitivity spectrum. Most researchers place us in the middle.

Epidemiologists studying dioxin have focused on human populations exposed in the workplace or through chemical accidents. Several report an association between dioxin exposure and overall cancer incidence, but, with the exception of soft-tissue sarcomas (tumors that arise in muscles, fat, blood vessels, or fibrous tissues), no one particular cancer stands out. A 1991 study of five thousand TCDD-exposed workers employed at twelve U.S. plants, for example, showed significant elevations in overall cancer mortality.

Several intriguing studies have come from Germany. In a 1990 study, researchers found excess cancer mortality among workers known to be heavily exposed to TCDD during a 1953 explosion at a German chemical plant. In another cohort study, researchers found elevated cancer deaths among workers at a dioxin-contaminated chemical plant in Hamburg. When compared to other workers, chemical plant employees with twenty or more years of employment suffered twice the cancer mortality, and women workers showed elevated breast cancer mortality. Likewise, a 1996 cohort study revealed significant increases in cancer mortality among more than twenty-four hundred German workers involved in manufacturing herbicides known to be contaminated with TCDD. In both studies, cancer risk rose with level of exposure.

One of the largest studies to date is still in progress. In July 1976, an explosion at a pesticide-manufacturing plant in Seveso, Italy, released a dioxin-suffused cloud of chemicals into the air. Within a few days, leaves fell from trees, birds and other animals died, and children developed skin lesions. Since then, the epidemiologist Pier Alberto Bertazzi and his colleagues have been monitoring the health of some two thousand families in and around Seveso.

As of 1993, Bertazzi had found excesses of certain cancers among inhabitants in Zone B, the second-most dioxin-contaminated area. Compared to the general population, Zone B residents had three times the rate of liver cancer. Rates of leukemia, multiple myeloma, and certain soft-tissue sarcomas were also elevated. In contrast to the findings in Germany, breast cancer incidence among the women of Zone B was actually *lower* than normal. So were their rates of uterine cancer. Statistics in Zone A, the most contaminated area, were problematic, since many of its residents fled the area immediately.

❧

I spent a lot of time in the fall of 1994 driving around the back roads of Illinois to various incinerator meetings. Some of these took place in school gymnasiums, and others in farmhouse kitchens. The assorted communities considering Kirby's proposal—from Forrest to Beardstown—were in various stages of deliberation. From what I could see, the same conversations were happening everywhere. It was as though someone had handed each of these towns identical scripts. Some were just beginning Act I (where civic leaders circle cautiously around the proposal, and a few members of the constituency issue admonishments along the lines of "beware of Greeks bearing gifts"). Some were already at the end of Act II (where the town becomes a house divided, predictions of ruin fly from both camps, and lifelong friendships dissolve into enmity). The whole drama had a foreordained feel to it, but I couldn't begin to guess how it would end.

Havana had been at it the longest, and the plot there had thickened more than once. The feasibility study, for example, was supposed to clarify once and for all what the incinerator's health risks would be. Instead, it further deepened the lines of fear, distrust, and contempt. Conducted by an independent team of scientists but paid for by Kirby's company—the city of Havana couldn't afford it—the study was released in the spring of 1992. The team concluded that dioxin would be emitted at acceptable levels, predicted no significant effect on human health or wildlife, and recommended project approval.

Two rebuttals swiftly followed—one commissioned by private citizens, and the other by the county farm bureau. The first, also authored by an independent, university scientist, criticized the feasibility study for downplaying food consumption as a route of exposure, for miscalculating wind direction, and for ignoring the fact that people already have background levels of dioxin in their tissues. The second one claimed the feasibility study exaggerated the incinerator's economic benefits.

The county farm bureau and the Central Illinois Irrigated

Growers Association officially declared themselves opposed. Their position was given new credence when a national popcorn buyer announced that it would reconsider its long-standing commitment to Mason County if the incinerator threatened adversity to the popcorn crop.

By midsummer of 1992, Ban the Burn signs had sprouted up in lawns, and a convoy of tractors pulling anti-incinerator floats joined the annual Fourth of July parade. The sign proclaiming God Recycles and the Devil Burns caught the most attention and disapproval—both because of its religiosity and because the chamber of commerce was presumed proincinerator and they were the sponsors of the parade.

The specific arguments and counterarguments were peculiar to each community, but even these shared common elements. There was, for example, the Hypocrisy Argument. Incinerator opponents, according to incinerator advocates, had no right to object to the possible environmental risks when in fact a) their own unrecycled garbage was being carted off to some other community for disposal and b) they were heavy users of pesticides. From a letter to the editor in Havana:

> I want to write and say please give it a chance. You say garbage from other places will be brought in there to burn. May I ask you where is our garbage being hauled? . . . I challenge any of you that is against the incinerator to prove to me that all the toxics you say will go in the air and pollute our air is any more toxic than all the sprays, plow-downs, herbicides and all the hundreds of chemicals farmers are using now days.

Incinerator opponents also cried hypocrite. Incinerator advocates, they countered, had no right to claim they were only looking out for the well-being of the community when a) they were actually seeking to line their own pockets and b) would allow ruination to befall a neighbor's farm in the process. From a letter to the editor in Forrest: "What has become of our Christian virtues? What has happened to 'Love thy neighbor as thyself'? Why has the wonderful, Christian community of Forrest sold out to the god of money?"

There was also the Trust Argument. Proincinerator folks emphasized that village board and city council members had been freely elected for the purpose of pursuing opportunities for community growth. The community, therefore, needed to trust them to do their job. If the naysayers were unhappy, let them run their own slate of candidates next time. Opponents located trust elsewhere. They pointed out that future generations were dependent on the current generation to safeguard the environment, with failure to do so being a betrayal of trust. Moreover, public servants, even freely elected ones, were susceptible to the seductive influences of a developer who stood to make a personal fortune if the project went through. The goings-on of the whole crew required rigorous surveillance. One busy citizen even wrote in to report that Kirby could hardly be counted a trustworthy steward of air quality, as he, along with the mayor, *was a smoker.*

Finally, there was the Question of Risk. Incinerator proponents tended to couch risk as opportunity and risk-taking as bravery. Without risk, the community would die: "Will Forrest reject a move forward for the more complacent choice of slow death? Nothing new is without problems. The dead have no worries."

Those opposed saw risk as recklessness. One pollution-control mishap, one Seveso-style explosion, one overturned ash truck and a brisk wind—and the community would regret its decision forever. Even if accidents never happened, they argued, the emissions that were now said to pose little or no risk might someday be revealed as hazardous. Nobody really knew what was in garbage anyway. How could developers be so sure what would come out the stack?

Dioxin is an agent provocateur. It works its evil in part by inciting cells to certain actions that increase their susceptibility to damage by other carcinogens. One of dioxin's known tricks is to induce cells to step up production of a group of enzymes called cytochromes P450, which serve the important function of metabolizing toxic substances. Sometimes, however, the first step of this conversion transforms a harmless chemical intruder into something truly dangerous. As we have seen, it is very often the metabolic breakdown product, rather than the parent material, that goes on to wreak carcinogenic havoc.

By virtue of its shape, dioxin is protected from deconstruction by these same enzymes. Hence, it remains powerful at faint concentrations.

Its effect on cytochromes P450 may also explain why dioxin is linked with so many different kinds of cancer. If, as it now seems, dioxin aids and abets a whole assortment of carcinogens—some associated with one set of cancers and some with another—then different people should develop different ailments, depending on their specific prior exposures, hormonal status, and stage of life. Dioxin may bring on liver cancer in some dioxin-exposed individuals, for example, and hasten the progression of lymphoma in others.

We also know with some confidence how dioxin stimulates P450 production in the first place. Once a dioxin molecule leaves the bloodstream and slips into the interior of a cell, it binds to a naturally occurring protein called the Ah (aryl hydrocarbon) receptor. This complex is subsequently shuttled into the nucleus, or walled-off chamber within every cell that contains the DNA. Once here, the trio attaches to and turns on a particular set of genes. These activated genes then send out instructions for the manufacture of particular enzymes, namely, cytochromes P450.

Genes coding for P450 enzymes are not dioxin's only target, but less is known about the others. They seem to include genes that regulate growth, as well as genes responsible for the regulation of and sensitivity to certain hormones. Each is relevant to the expression of cancer. Interestingly, dioxin sometimes acts as an antiestrogen. More to the point, it makes cells less responsive to the body's own estrogen hormones. This observation may explain the odd fact that dioxin exposure is sometimes associated with lower rates of breast cancer.

Seven of the 75 dioxins, 10 of the 135 furans, and 11 of the 209 PCBs have the ability to bind with the efficient little Ah receptor. (TCDD does so most tightly.) Apart from allowing industrial chemicals access to our genes, what is its function? Why does it exist? With what naturally occurring agent is an Ah receptor supposed to make contact? No one actually knows, but we do have some inklings from a 1995 study.

In this study, researchers developed a strain of mouse completely lacking Ah receptors. They did so by knocking out the gene that codes for this protein. The result? Overcome by infections, defective in immunity, and debilitated by liver problems, these mice either died shortly after birth or soon became too sick for further experimentation. Obviously, the receptors play a key role in the development of immunity and liver functioning. Probably they are also part of some little-understood detoxification system.

This experiment sheds a ray of light on dioxin's other shadowy habits. Dioxin is known to depress immunity, an outcome that may promote a variety of cancers. Sometimes it is associated with cancers that originate in the immune system, such as lymphoma. Dioxin also influences thyroid functioning, blood glucose levels, sexual development, and testosterone production. In rats, it interacts synergistically with PCBs to alter certain liver functions, and in monkeys it has been linked to the painful uterine disorder endometriosis. Whether all these effects are mediated through the Ah receptor or involve some other unidentified pathway remains to be seen.

In all the driving around I did that fall, I never ran into Mr. Kirby. He would be in one town while I was in another. We probably passed each other on the highway. Everyone I spoke with said he was a nice guy. The pleasantness of Kirby's personality was probably the only issue on which there was complete agreement. Published photographs showed a sizable man with snow-white hair, a farmer's leathery complexion, and a tendency toward colorful dress.

He was in fact a farmer—or at least he had grown up on a farm and had once owned one. His poultry operation was so successful that the Springfield paper once featured it in a full-page profile. Someone else actually ran the farm, but it was Kirby's children who had sorted and packed the eggs. These kinds of details impressed people. Kirby was a Korean War veteran and had been handpicked by his father to go to college. At one time he had been a school principal.

He also was what my mother called a wheeler-dealer. He entered politics by becoming assistant to the state school superintendent. He ran (unsuccessfully) for state auditor and then (briefly and unsuccessfully) in the Republican primary for U.S. senator. Kirby became influential with the governor and counted Ronald Reagan and Everett Dirkson among his friends. In the 1970s, Kirby was a special consultant to the developers of a racetrack. The newpapers called him brilliant. The racetrack was never built.

Apparently, Kirby got into the trash business the old-fashioned way—by buying trucks, landfills, and transfer stations and then selling them for a profit. Later, he focused on putting together agreements for other trash haulers and then incinerator operators. Here is where I lost the thread of exactly what Kirby did for a living. It involved some combination of lobbying government, negotiating with insurance brokers, acquiring permits, attracting venture capital, and otherwise dealing with revenue bonds and investment companies. I wasn't the only one who felt mystified by his profession. The *Pekin Daily Times* asked him who he thought John Kirby was, and John Kirby replied, "I think I'm a guy that pretty well knows what he wants to do."

He also said, "There's no question there'll be, by the turn of the century, five or six 1,800-ton-a-day incinerators" in Illinois.

❧

Hanging above the stairway in my parents' house is an aerial photograph of the Maurer family farm, circa 1950. You must walk by it without even a sidelong glance of interest unless you truly want to hear my mother's accompanying disquisition. This involves explaining—in the reverent, authoritative tone she reserves for talking about the farm—what each and every outbuilding was used for, how all three generations who lived here had six children apiece, and why Roy and Pop slept out in the milkhouse during the scarlet fever outbreak. (The house was quarantined; they needed to sell the milk.)

But of course you express curiosity about the photograph—which appears to depict an entire village rather than a single farm—and so my mother is duty-bound to point out the following. First, the house. Built in 1908, it features two stairways and eight bedrooms. All the kids were born in the downstairs one. To the north, out to the road, is the orchard.

To the south is the garden, the chicken house, the toolshed, the smokehouse, the cob house, and the threshing-machine shed. The garden was always organic and still is today. The toolshed housed the discs, harrows, plows, manure spreaders, and planters. The cob house held corncobs and coal for the stove. The threshing-machine shed is the one with the tall doors to accommodate the massive contraption that separated oats from straw. The chicken house had a special room for setting hens.

The rest of the animals were quartered on the east side, where lie the barnyard, the barn, the milk house, the shop, the pig house, and the corncrib. To the west are another orchard, a special pasture for calves, and a windbreak of catalpa trees. Beyond all that are the 160 acres of the Maurer spread in Pleasant Ridge township—four miles north of Forrest, just west of Route 47, along County Road 1200 N, in southeast Livingston County.

Even I have a hard time perceiving the ridge in Pleasant Ridge township, which is about as flat as they come. The open earth lies like a black sail from horizon to horizon.

It is October 1994. In two days I will fly back to Boston, but right now I am driving home to Pekin from an evening meeting with a group of Pleasant Ridge farmers, and I am full of pie. One does not leave a convocation of Illinois farmers without eating pie.

Out here I am known as "Kath-urn's girl" (my mother's name being Kathryn). This fact makes even more peculiar the realization that I just spent the last two hours in spirited conversation about furans, Ah receptors, the flow pattern of the Vermilion River—as well as the particulars of a certain Massachusetts incinerator—with people my mother knew in high school. Tomorrow this group, organized as Citizens for a Healthy Community, is hosting a critical incinerator teach-in. This, they hope, will balance the rosy presentation that Kirby's local collaborator, the Forrest Development Corporation, sponsored the month before. All this education anticipates the referendum in November.

The opposition group in Havana succeeded in getting a similar referendum on the ballot in a previous election. The majority of Havanans voted against the incinerator, but the city council had gone ahead and approved it anyway. I am not at all sure that this one will end more happily, but as someone fighting the same proposal up in Summit said, "If this incinerator comes in, and I never lifted a hand to stop it, I just couldn't live with myself." That statement seemed to represent the collective sentiment around the table tonight.

It's late and I'm just at the crossroads, the junction of Routes 47 and 24, when the radio begins playing the opening strains of a symphony. It's modern and orchestral, but the feeling is that of an enormous choir singing. What is it? I realize immediately where I need to be to listen to this music and so execute a U-turn and head back north.

Three miles later, I pull the car over, turn on the flashers, turn up the volume, unroll the windows, and walk straight out onto the section of land containing the eighty-acre rectangle—optioned, annexed, disputed, despised—where the incinerator is to be sited. The music follows me.

Earlier tonight, I spoke with the farmer who works this field. The drainage is troublesome, he said, and he mentioned another problem, too—the specifics of which I have already forgotten. Now, he rues the day he ever complained about it. "I'd give anything just to keep plowing that field forever."

The music plays on. It is sad but somehow glorifying. (Months later I will learn that it is Vaughan Williams's "Fantasia on a Theme by Thomas Tallis.") I walk as far as I can without straying out of earshot and then lie down, mindful of broken cornstalks. I realize for the first time how much the events of the last month—all the research, the strategizing, the attempts to predict the future—have exhausted me.

The music's loveliness makes me realize other things, too. Whatever the facts about the incinerator were, the truth was that it was obscene. And so was arguing about exactly how many picograms of dioxin could acceptably contaminate these fields, the bodies of the people who plow them, and the flesh of their hogs, turkeys, and garden vegetables. So was discussing exactly how many thousands of gallons of water a day could be pumped out of this land in order to burn garbage. So was manufacturing substances that are poisonous when incinerated and undegradable when buried. And so was the rapacity of subsidizing incineration over recycling in the first place.

Between 170 and 190 incinerators operate at any given time in the United States. They handle about 17 percent of the nation's trash. Any respectable recycling program would easily put them all out of business. I think again about what the folks in Pekin said when they successfully ran the incinerator train out of town: incineration is not the answer to the landfill dilemma. Knowing what is *not* the solution is sometimes the most important first step in confronting a problem.

Perhaps incinerator emissions are no worse than the injection of agricultural chemicals into these same fields—incinerator proponents certainly had a point there—but two obscenities do not cancel each other out. This land is bordered on one side by the north fork of the Vermilion River, and on the other by the south fork. The incinerator will pollute the watershed no matter which way the wind blows. The fish of the Vermilion are already contaminated with chlordane, DDT, dieldrin, heptachlor, and aldrin. Still, one of its tributaries manages to support a small population of river redhorse, an endangered fish species.

The music lifts and falls like a human voice.

I try to imagine it out there somewhere, the river redhorse—whatever it looks like—resting quietly in a current of water, even as I am resting here on the earth. A scud of clouds covers over the stars and then blows by. A melody carried from one instrument to another is transformed and then submerged.

Some of the men I ate pie with tonight are no doubt responsible for the contamination of Vermilion River fish, but in a larger sense, so are we all. Now we had the chance to work together to prevent a problem rather than debate the damage later. It was a start.

I send my thoughts out to our farm, which lies a mile to the north. It is a greatly abridged version of its earlier self. Except for the odd sheep or two, the animals are all gone, and so is the metropolis of barns, shops, sheds, and corncribs. But the fields are still there, and the house is still there, flanked by a few silver grain bins and my Aunt Ann's organic garden. When the music ends, I'll be able to see its lights as I walk out of this field.

On November 9, 1994, the results of the incinerator referendum in Forrest showed 466 against and 406 for. Some members of the Forrest Development Corporation vowed to proceed anyway, but Kirby demurred. "We're apprehensive about committing to the project if that support for it is soft. We don't want to have to fight a battle every time we want a sewer extension."

The following September, an appellate court in Springfield, Illinois, upheld unanimously the decision of the Illinois Pollution Control Board regarding the unfair siting approval of the incinerator in Havana. The judges cited both a Massachusetts trip paid for by Kirby's corporation and the improper influence of that corporation on the hearing officer.

On January 11, 1996, the Illinois General Assembly repealed the Retail Rate Law. According to the governor, "Most communities do not want the incinerators. And it is time we stopped asking our taxpayers to subsidize them."

On January 25, 1996, John Kirby died of malignant mesothelioma—a form of lung cancer—in a Springfield, Illinois, hospice.

ELEVEN

our bodies, inscribed

Among forest trees, size and age can be remarkably dissociated. Seedlings germinating in deep shade are often swiftly overtaken by those sprouting up in light-filled spaces nearby. Saplings browsed by a passing deer lose vertical height relative to neighbors less palatable. By these and other means, senior members of a forest community sometimes grow old beneath a canopy of younger trees.

Field ecologists, therefore, rely on tree-ring analysis to reconstruct the history of forests. I once spent a summer in Minnesota engaged in this kind of work, which begins with pressing the bit end of a hand borer against the bark of a tree at chest height, leaning against it with all one's weight, and slowly turning the handle until the steel threads have chewed into the flesh beneath and have wound themselves straight into the tree's exact center. A slender wand of cool, damp wood is then extracted with the narrowest of spatulas, sealed in an envelope, and, along with an assortment of other tree cores, taken back to the laboratory to be read.

These cores are banded with colored rings, each representing a season of growth. An experienced dendrochronologist (which I am not) can identify in the subtler patterns of these circles not only age but also periods of changing light levels, insect plagues, drought, flood, or fire. An individual tree carries within its own body an ecological chronicle of the entire community.

In this, people are not so different. Our bodies, too, are living scrolls of sorts. What is written there—inside the fibers of our cells and chromosomes—is a record of our exposure to environmental contaminants. Like the rings of trees, our tissues are historical documents that can be read by those who know how to decipher the code.

❧

Body burden refers to the sum total of these exposures and encompasses all routes of entry (inhalation, ingestion, and skin absorption) and all sources (food, air, water, workplace, home, and so forth). In the case of fat-soluble, persistent chemicals, body burdens provide a measure of cumulative exposure. For example, 177 different organochlorine residues can be detected in the body of an average middle-aged American man. Some of these exposures occurred in infancy, others in adolescence, and still others in adulthood. In the case of chemicals quickly metabolized and excreted, the body burden is an index more akin to a press release than a biography. It reports on the status of immediate and ongoing exposures to particular contaminants at single points in time.

The problem with body burdens is that they require sampling each and every fluid and compartment of tissue. This task can be accomplished during an autopsy, but for living people, total exposures are more often derived from measurements taken from a specific source. Blood, urine, breast milk, exhaled air, fat, semen, hair, tears, sweat, and fingernails have all been used for this purpose.

Different tissues work more or less well for different contaminants. The blood inside umbilical cords, for example, may identify compounds that pass through the placenta and enter the bodies of developing fetuses. Their presence provides clues to the causes of childhood cancers. So far, these include PCBs and an array of pesticides. Urine, on the other hand, is a good medium for looking at water-soluble contaminants, such as organophosphate and carbamate pesticides. Sampling urine, researchers have estimated that the bodies of most members of the U.S. population contain detectable levels of the insecticide chlorpyrifos, a common ingredient in pet flea collars, lawn and garden pest control products, indoor foggers, and roach, ant, and wasp poisons.

PCB levels in blood have been demonstrated to correlate roughly with their overall body burden, once differences in fat content have been accounted for. Hence, a simple blood draw can provide an estimate of lifetime PCB exposure. (Blood contains a certain fraction of fat.) Nevertheless, complications arise even here. Different organs seem to sequester differing proportions of each of the 209 chemical varieties. If PCB molecules were all created equal, this partitioning process would matter less. However, members of the PCB family differ in their persistence, potency, and carcinogenic potential. Furthermore, PCBs are broken down into different metabolic products that distribute differently though human tissues. If greater amounts of the more toxic varieties differentially settle into the lung, kidney, and uterus rather than the liver, breast, and adrenal glands, for instance, then a simple measure of total PCB concentration in blood plasma may not tell the whole story.

A sponge for oil-soluble chemicals, body fat is considered an especially sensitive indicator of exposure to persistent environmental contaminants. In Japan, researchers examined a variety of industrial contaminants in preserved fat collected from men who had died between 1928 and 1985. The highest concentrations of DDT, PCBs, and chlordane were found in samples collected during their respective periods of maximum production, import, and use. In a 1996 study conducted in Mexico, researchers found that levels of DDT in living human tissues varied predictably across geographic space: residue levels in both abdominal fat and breast fat were highest in areas of intense agriculture and in tropical regions where DDT was used for malaria control.

Breast milk has a lexicon all its own. About 3 percent fat, it contains high concentrations of fat-soluble contaminants. These pollutants are carried by the blood into the breast from fat reserves scattered throughout the body and probably including the breast fat itself. Since 1951, surveys of human milk in the United States have consistently shown contamination by an array of persistent, chlorinated chemicals. The issue of insecticides in breast milk received close attention from Rachel Carson in 1962. A dozen years later, 99 percent of breast milk sampled in the United States was also shown to contain PCBs. About one of every four of these samples contained PCB concentrations exceeding the legal limit (2.5 parts per million), above which level commercial formula is pulled from the shelves. Or, to express this another way: by 1976, roughly 25 percent of all U.S. breast milk was too contaminated to be bottled and sold as a food commodity.

The cancer risks assumed by these mothers and their nursing infants—now adults, some with children of their own—remain to be seen. The possible relationship between carcinogens in breast milk and breast cancer (or cancer in offspring) has not been systematically investigated.

A study of more than eight hundred nursing mothers in North Carolina has uncovered three patterns that make this question an urgent one. Researchers found that the concentration of organochlorine chemicals in breast milk increased with the age of the mother, increased with the amount of sport fish consumed, and decreased dramatically over the course of lactation and with the number of children nursed. The first trend indicates that our bodies are still amassing fat-soluble contaminants faster than we can eliminate them. The second attests to the ongoing contamination of our rivers, streams, and lakes.

The third fact is the most ominous one. Organochlorine contaminants are not easily expunged from our tissues. Their sharp decline in concentration over the course of breast-feeding, therefore, represents the movement of accumulated toxins from mother to child. It signifies that during the intimate act of nursing, a burden of public poisons—insect killers, electrical insulating fluids, industrial solvents, and incinerator residues—is shifted from one generation into the tiny bodies of the next.

Happily, concentrations of a few of the most pernicious contaminants of breast milk are stabilizing or even beginning to drop. Long-term monitoring of human milk in Germany, for example, showed slight declines during the early 1990s in levels of dioxins, furans, organochlorine pesticides, and PCBs. Similarly, pooled samples of human milk archived in the Mothers' Milk Centre in Stockholm, Sweden, show declines in many PCB and DDT metabolites from 1972 to 1992. These trends indicate that efforts to shut down known sources of these chemicals are finally beginning to have an effect on their respective body burdens.

❧

The human body is an endless construction site where demolition and renovation occur simultaneously and continuously. Different tissues carry on this work at different rates; the lining of the stomach is entirely overhauled every few days, while a complete restoration of the bones' internal scaffolding requires years. All tissues replace themselves through the orderly process of cell division—mitosis—in which one cell splits in half and becomes two. Damaged and aged cells slated for removal undergo a programmed form of death known as apoptosis. All this activity is coordinated through an elaborate system of communication that cell biologists are just beginning to understand.

A certain amount of supervision is provided by a cell's own DNA, which sends out from the nucleus periodic messages instructing the cell to begin (or cease) dividing. We know also that chemical signals from neighboring cells can alter the pace of this process. And we know that marching orders sometimes arrive from distant headquarters. These often take the form of hormones, as when estrogen from a woman's ovaries causes the cells in her breasts to begin dividing.

However scant our knowledge about its regulation, the actual feat of mitosis, its procession of precise, elegant steps, is becoming increasingly clear. Mitosis begins inside a circle within a circle: the nucleus of the cell where the DNA is quartered.

The first step is the doubling of each of the strands of DNA, the chromosomes. Their duplication will enable both daughter cells to receive a complete set. For this task, a crew of enzymes creates an exact replica of each original chromosome (which is split in half lengthwise and used as a template for its own duplication). Lying side by side, the two identical strands are then cuffed together and come to resemble a gangly letter *H* or sometimes a stout *V*.

Humans possess forty-six individual chromosomes, each consisting of a curly DNA ladder and each bearing many thousand genes. Once all forty-six gene-studded chromosomes have been so copied, a dance begins. The nuclear membrane disintegrates. The chromosomal couples move to the center of the cell and form a vertical line. Fine threads called spindle fibers extend horizontally from opposing ends of the cell and attach to each member of a pair. The fibers contract. Simultaneously, the twinned chromosomes pull apart, their midpoint connections giving way as the left and right halves of the *H*s and the *V*s are towed through the watery protoplasm to opposite poles. Just as the cell begins to pinch in half, a membranous curtain closes around each new grouping of single-stranded chromosomes, and they are once again cloistered within a nucleus. They will remain there, directing the synthesis of proteins, until the mitotic cycle begins anew and once again releases them.

Cancer is mitosis run amuck. Instead of reproducing in careful, methodical fashion, cancer cells carry on replication and division despite a myriad of directives designed to restrain such activity. Cancer cells are dancers deaf to the choreographer. They are builders in flagrant disregard of zoning ordinances and architectural blueprints. They are defiant, disobedient, and in the view of many cancer biologists, almost purposeful in the ways they disrupt cellular biochemistry.

Besides a propensity for unrelenting growth, a cancer cell is known for two other traits: invasiveness and primitivism. The ability to invade other tissues distinguishes cancer from other freakish outgrowths, such as warts. This facility operates at both a local level—cancer also ignores property lines—and a distant one, as when cancer cells are shed from the primary tumor and seeded throughout the body as metastases. Destroying healthy tissue and clogging vital passageways, both habits make cancer life-threatening.

By *primitive*, biologists mean that the tissues created by cancer appear to have reverted back to some earlier, cruder, unformed stage of development. They no longer bear much resemblance to the differentiated structures of which they were originally a part. Typically, the hard lump in the breast that turns out to be a malignancy is a direct descendent of one of the smooth, flat cells that wallpaper the interior surfaces of the slender mammary ducts. But, microscopically, the tumor's mass of cells no longer looks anything like the benevolent sheets of breast epithelial tissue it came from. In general, the less a tissue resembles its previous, respectable, specialized self, the more virulent the cancer. Along with runaway growth and the propensity to spread, this tendency to devolve into an immature, unrecognizable state is the result of a long accumulation of genetic injuries.

A cancer cell, then, is made, not born. Cancer arises through a series of incremental changes to chromosomal DNA. Some of these DNA alterations can be inherited, but the vast majority are acquired during the lifetime of an individual when genes perfectly healthy at the time of conception become damaged. This process can happen through numerous pathways. Routine errors made during DNA replication are one. Sabotage by carcinogens is another. About 100,000 different genes are strung along our chromosomes. To contribute to cancer, at least some of these encounters between carcinogens and genes must involve the handful that help govern cell division.

These growth-regulating genes come in two basic varieties. The first group are called oncogenes. In their normal state, these bits of DNA convey messages that encourage cell division. When mutated, however, oncogenes become hyperactive and ratchet up the rate of growth. Working on exactly the opposite principle are the tumor suppressor genes. Normally, they dampen the rate of cell division. In some circumstances—as when signs of DNA damage are about—they actually halt mitosis altogether and thereby nip in the bud the possible genesis of cancerous growth. Loss or inactivation of tumor suppressor genes may contribute to the birth of a tumor. If a mutant oncogene is a stuck accelerator pedal, then damaged tumor suppressor genes are faulty brakes. Either problem can result in runaway cell growth.

Different kinds of cancers are associated with different kinds of mutations. The cells of most colon tumors, for example, turn out to contain both hyperactive oncogenes and nonfunctional tumor suppressor genes. One specific tumor suppressor gene located on chromosome 17 has been fingered in several big-ticket malignancies, including cancers of the lung, breast, colon, esophagus, bladder, brain, and bone. Indeed, alterations of this gene, named p53, may be involved in as many as half of all human cancers. Much as a gunshot wound indicates what kind of firearm was used in an assault, the particular nature of the p53 mutation often suggests the type of carcinogen responsible for the damage. Cigarette smoke leaves one kind of lesion, ultraviolet radiation another, and exposure to vinyl chloride a third. The mutational spectrum of this gene is so broad that the lung tumors from uranium miners can sometimes be distinguished from the lung tumors of smokers simply by looking at the specific location of the mutation. Breast tumors frequently display p53 mutations in a spectrum resembling that seen in lung tumors and varying across geographic regions.

Harm can befall growth regulator genes through a whole variety of pathways. Benzo[a]pyrene can adhere to a section of chromosome and, in so doing, create a DNA adduct. Like bits of chewing gum stuck to a strand of hair, adducts can cause mistakes to be made during the next cycle of DNA replication. Other carcinogens disable the spindle fiber apparatus, causing chromosomes to pull apart improperly. By these and other means, daughter cells can end up receiving mutated oncogenes and/or missing or impaired tumor suppressor genes. Alterations in other kinds of genes can abet the process. For example, DNA repair genes normally function to fix chromosomes vandalized by mutating agents or damaged accidentally during the normal course of mitosis. An injury to a repair gene is, therefore, a treacherous event, as it can lead to the accumulation of genetic lesions of all kinds. Fortunately, the carcinogenic process is lengthy and complicated, often requiring decades to unfold. It is also capable of being arrested at many points along the way.

In the language of cancer biology, the making of a cancer cell involves three overlapping stages: initiation, promotion, and progression. To become a full-blown malignancy, a cancer cell must pass through them all.

The first rite of passage, initiation, is characterized by small structural alterations to the cell's DNA strands. Arising spontaneously or resulting from an encounter with a carcinogen, these modifications—like tiny tattoos—are swift, permanent, and subtle. A small hole here. An inconspicuous inversion there. Cells so affected remain, to the human eye, indistinguishable in shape and appearance from their undamaged counterparts. Nevertheless, many initiated cells meet an early demise through the winnowing action of apoptosis. Any agent, then, that interferes with cell death can contribute to cancer by permitting damaged cells to continue along the pathway to tumor formation.

The immune system also plays a role in the selective destruction of incipient cancer cells, which presumably reveal their hand by exhibiting biochemical traits recognizable as abnormal. At what specific stage immune cells begin to mount a reaction is not entirely clear. It is known that certain environmental contaminants, including dioxin, suppress human immunity and that immune suppression is associated with several kinds of cancers, most notably leukemias and lymphomas. Recent studies from the former Soviet Union have shown clear relationships between exposure to certain pesticides and depression of the immune system's T cells.

Initiated cancer cells that escape detection advance to the next stage, promotion, which requires additional exposures to cancer-stimulating substances. Unlike initiation, promotion unfolds over a long period and may involve no actual mutations. In general, cancer promoters encourage cells to divide not by altering the physical structure of genes but by altering the expression of their chemical messages. Genes that are normally quiescent, for example, may become activated. Estrogen, in some cases, acts as a cancer promoter. As demonstrated in lab animals, so do many organochlorine compounds. The good news is that these effects wane when such agents are removed from the body.

Quite often, cancer promoters perturb an intricate communications pathway known as signal transduction. This system consists of a team of proteins relaying messages back and forth between the perimeter of the cell and the heartwood of the nucleus. By mechanisms barely elucidated, signal transduction proteins play a key role in the timing and coordination of cell division. Promoting agents can affect the production and behavior of these courier molecules without permanently damaging the genes that code for their manufacture. The result is an expanded cluster of abnormal cells.

Like initiation but unlike promotion, the progression stage involves exposures that inflict physical injury to the DNA molecule. Mutations pile up. Chromosomes fall into disrepair and become increasingly unstable. Ironically, substances that act at this stage bestow on the cells they cripple some of cancer's most fearsome abilities: the capacity to spread and invade, enhanced sensitivity to hormones, and a knack for attracting blood vessels to the growing mass of tumor cells. Some researchers believe that arsenic, asbestos, and benzene can each function as cancer progressors, under certain conditions.

Agents that contribute to cancer do not all fall neatly into the categories of initiator, promoter, and progressor. Some, like radiation, are complete carcinogens that can play all three roles by themselves. Others, such as dioxin, appear to behave as promoters at low doses and complete carcinogens at higher levels, and they may also interfere with apoptosis. Still others initiate at low doses and promote and progress when their concentration in the body rises.

These shifting biological possibilities bring with them many social implications. First, they explain why no safe dose of a carcinogen exists. They also explain why similar exposures can pose very different degrees of danger to different people. The trace presence of a cancer-promoting pesticide in drinking water, for example, may represent absolute hazard to those whose breast, prostate, colon, or bladder tissue has already been initiated by some prior event (perhaps during childhood or because of occupation) or to those rare few born with a mutated gene that predisposes them to cancer. Individuals whose genetic material has suffered less previous damage may more successfully ward off the effects of promoting agents—as would those lucky persons who happen to possess a set of metabolism genes that allows for especially efficient detoxification and excretion of promoting substances.

The implications become even broader when we consider the dozens of known and suspected carcinogens to which we are routinely exposed and which may work alone, in concert, or cumulatively anywhere along the cancer continuum. In rats, for example, DDT acts to accelerate tumors induced by an agent called 2-acetamidophenanthrene, even though neither one alone is capable of causing tumors to progress to a detectable level.

In the words of the veteran cancer biologist Ross Hume Hall, "Too often cancer research has focused on finding the last straw. It's time we looked at all the straws."

❧

They have been compared to footprints, fingerprints, graffiti, and stigmata. They have also been hailed as the jewel in the crown of molecular epidemiology and described as decoding tools by which to read the body. They are biological markers, and, defined most plainly, they are indicators of physical damage caused by the interplay between human genes and environmental carcinogens. As such, biological markers serve as both signals of past exposure and predictors of future cancers.

Adducts, formed by mutation-inducing chemicals that adhere to DNA, are one type of marker. As discussed in Chapter Six, the tissues of beluga whales living in contaminated stretches of the St. Lawrence River display high concentrations of benzo[a]pyrene adducts. Similarly, in laboratory animals, researchers consistently find tight correlations between exposure to chemicals known to cause cancer and the concentration of adducts in the DNA of certain tissues. In humans, the relationship between adduct levels and cancer risk has not been worked out as definitively. However, some compelling evidence is now emerging from one of the most polluted regions on earth: Silesia, Poland.

Hard up against Poland's southern border, Silesia is blanketed with chemical plants, foundries, smelters, steel mills, coal mines, and cokeries (the great ovens that distill coal into coke for steelmaking). The cancer death rate is also impressively high here, persuading the molecular epidemiologist Frederica Perera of Columbia University to examine Silesian DNA closely. Her pioneering work has uncovered consistent associations between toxic exposures and adduct formation, on the one hand, and adduct formation and cancer risk, on the other.

Perera and her coworkers focused on polycyclic aromatic hydrocarbons, such as benzo[a]pyrene, which are released into Silesia's air in great abundance, mostly as by-products of coal and coke burning. Simply measuring their airborne concentration turns out not to be a reliable indicator of individual human exposure because polycyclic aromatic hydrocarbons are not only available for inhalation but also stick to skin (and are absorbed) and insinuate themselves into food (and are ingested). Moreover, these carcinogenic contaminants are handled differently by different people, depending on genetic and other factors that affect metabolism and detoxification.

The proof is in the cells' pudding. Perera found that the DNA of Silesian coke workers and Silesian city dwellers bore similar loads of polycyclic aromatic hydrocarbon adducts. These levels were two to three times higher than among rural folk. Perera also discovered a pronounced seasonal effect: the number of adducts rose during the winter months, when coal burned for domestic heating adds to the burden of aromatic hydrocarbons contributed by industry. Moreover, the level of adducts was correlated with the presence of chromosomal mutations thought to be affiliated with lung cancer. Together with studies showing that people with lung cancer carry higher burdens of polycyclic aromatic hydrocarbon adducts on their DNA than people without the disease, Perera's findings "strongly suggested that severe air pollution could indeed help induce lung cancer."

As Perera observed, DNA adducts provide us with a molecular link between environmental exposure and genetic injuries relevant to cancer. But they are not the only biological marker to do so. Alterations in certain proteins can also signal that villainy is afoot. For example, as a result of rearranging the genetic code, the carcinogen vinyl chloride triggers the production of a defective signal transduction protein. The presence of this protein in blood serum is therefore an unmistakable marker of vinyl chloride exposure. Alterations in DNA repair enzymes indicate other kinds of foul play, as do elevated levels of enzymes used for metabolizing foreign substances. The premier example here is cytochrome P450 enzymes, levels of which, as we have seen in Chapter Ten, rise rapidly in response to the presence of dioxinlike molecules.

Mutations themselves have a story to tell. For example, abnormally high levels of chromosomal breakages and genetic rearrangements have been identified in Minnesota fumigant and pesticide applicators. Some of these alterations consistently affect certain areas of chromosomes 14 and 18, and these mutations are of particular interest to researchers because they are the ones most commonly observed in non-Hodgkins lymphoma patients.

Certain mutational patterns are indicators of free radical exposure. A free radical, not part of any one classifiable chemical group, is any atom or molecule with just one electron in its outermost orbital. Electrons prefer to circle in pairs. When one is missing, the particle to which they belong becomes reactive—quick to surrender or absorb an electron from nearby molecules. If these molecules are chromosomes, mutations may result.

As part of the normal process of breaking apart food and hormone molecules, the cells of our bodies are constantly generating free radicals (and these undoubtedly contribute to our load of acquired DNA mutations). Fortunately, we possess several means of protecting our chromosomes from the resulting electron scramble—including the use of dietary vitamins to soak up free radicals as they are produced. Research by the molecular epidemiologist Donald Malins indicates that certain environmental contaminants generate free radicals when the body attempts to detoxify and metabolize them. Malins and his colleagues are currently attempting to determine whether specific patterns of free radical damage in the DNA of the human breast could provide a means of predicting breast cancer risk. Breasts may be particularly susceptible to free radical damage, even in the absence of toxic exposures. The process of metabolizing estrogen is itself a free radical–generating operation. Foreign chemicals that add to this burden—or that compromise DNA repair systems designed to counteract the ravages of routine free radical damage—may amplify the risk of breast cancer. In other words, while free radical generation is a normal but unfortunate consequence of fueling ourselves with chemical energy, preliminary evidence—from both animal and human tissue studies—suggests that chronic exposure to certain toxic substances can, in some circumstances, overwhelm the body's multilayered defense system against free radical stress and thereby accelerate the rate at which we accumulate genetic injury. More research along this line of inquiry is essential.

The first clue that estrogen might play a role in breast cancer came in 1896 when a British surgeon reported that removal of the ovaries sometimes caused breast tumors to shrink. Many exhaustive studies conducted over the years since then have clearly indicated that a woman's chances of developing breast cancer are related in some way to her lifetime exposure to estrogen. Early first menstruation, late menopause, and late or no childbirths all raise a woman's lifetime exposure to estrogen and all are considered established risk factors for breast cancer—as is having a mother or a sister with the disease. Even so, taken together, such factors still account for only a minority of breast cancer cases.

Because the origin of most breast cancers remains unexplained and because there exists an apparent connection between breast cancer and naturally occurring estrogen, scientific attention has begun to turn to the possible role of xenoestrogens—chemicals foreign to the human body that, directly or indirectly, act like estrogens. We have already examined the evidence on xenoestrogens from epidemiological studies, animal data, and human cell cultures in Chapters Five and Six. I focus here on the specific pathways by which these hormone mimics leave their signatures within the cell.

But first, a bit of background on estrogen itself. Manufactured from cholesterol by a woman's ovaries each month, estrogen circulates in the blood, passes freely in and out of all organs and tissues, is eventually metabolized by specific enzymes, and, with the help of the liver, is eliminated from the body through the gut. Most cells are completely unaffected by all this activity. The cells of certain tissues, however, contain receptors that latch on to estrogen molecules as they float through. The estrogen-receptor complex then goes to work inside the nucleus. Some genes are activated, while others are switched off. Different messages are sent out from the nucleus and, hence, different proteins manufactured. For tissues possessing estrogen receptors, the net effect of these various alterations is an increase in cell proliferation. The cells of the vagina, the uterus, and the breast all contain large numbers of estrogen receptors. In the presence of estrogen, they divide. Ovulation, breast development, menstruation, and pregnancy are all made possible by estrogen's actions.

Estrogen comes in several chemical configurations, each with its own name. By far, the most potent one is estradiol. Its particular structure allows it easy passage from blood into surrounding cells. To regulate this movement, estradiol is not permitted to travel about unescorted. Instead, most estradiol molecules are attached to serum proteins that slow down their entry into target tissues and thereby blunt their dramatic effects.

Like estradiol, xenoestrogens slip from blood serum into the interior of cells, attach themselves to estrogen receptors, and, by tinkering with particular genes, elicit growth-promoting changes within target tissues. The ability of certain synthetic chemicals to mimic estrogen in these regards has been known for some time, but until recently, many researchers had assumed that any breast cancer risk created by this sort of mischief paled in comparison to the sovereign power of a woman's own hormones. This assumption was based on several observations. First, few synthetic chemicals closely resemble the ornately designed estrogen molecule, and estrogen is the key that must fit into the receptor's lock in order to ignite the whole process. Second, assays show that foreign estrogens are much less potent than naturally occurring estradiol. Indeed, most are thousands, even millions, of times weaker. Third, xenoestrogens exist in much lower concentrations in the body than naturally occurring estrogens, which surge to impressive levels during the first half of a woman's menstrual cycle. Also, many of the plants we eat, such as soy, contain naturally occurring plant estrogens, which are far more commonly encountered by our cells than their synthetic counterparts, such as pesticide residues. In short, xenoestrogens have been presumed rare, ineffective, and dilute.

Several recent findings have cast doubt on such reassuring suppositions. It turns out, for example, that close physical resemblance is not required for successful estrogen impersonation. As a lock, the estrogen receptor accepts many keys, some widely divergent in shape and size. Organic compounds that look nothing like estradiol—from pesticides to plastics to surfactants—can possess estrogenic properties. Xenoestrogens are far more common than anyone had imagined.

Additionally, xenoestrogens appear able to compensate for their lack of individual potency through remarkable synergistic interactions: together they can exert estrogenic effects hundreds and even thousands of times higher than any one working alone—at least under certain laboratory conditions. For example, the pesticides dieldrin, endosulfan, toxaphene, and chlordane, when combined, were shown to exert estrogenic effects in cultured yeast cells that were 160 to 1,600 times higher than any one acting alone.

Futhermore, many artificial estrogens compensate for their low numbers through longevity and enhanced availability. As we have seen, synthetic xenoestrogens are not easily metabolized and excreted. They linger, sometimes for decades. Recent studies have also shown that some xenoestrogens, including DDT, are not as tightly bound to blood proteins as estradiol. They can, therefore, enter target cells more quickly and at lower concentrations; they are more available.

Xenoestrogens not only mimic natural estrogens directly but also can indirectly enhance their effects. Some, for example, appear to stimulate the manufacture of more estrogen receptors. More receptors means an amplified response to estradiol. Still others influence how estradiol is metabolized and eliminated from the body. This second effect has been the subject of several recent studies led by the biochemical endocrinologist Leon Bradlow at the Strang Cornell Cancer Research Center in New York and his collaborator Devra Davis.

As explained by Bradlow, estradiol molecules can be broken apart by metabolic enzymes in one of two ways. The first one alters carbon atom number 2. The second alters carbon atom number 16. Which of these two pathways estradiol takes turns out to be critical. The 16-metabolite is still estrogenic; it is easily reabsorbed across the gut and is capable of binding to estrogen receptors just like its parent, estradiol. More menacingly, the 16-metabolite can directly damage DNA. It is believed capable of both initiating and promoting breast cancer. Indeed, many researchers consider the level of this metabolite a potential marker for breast cancer risk. In contrast, the 2-metabolite is minimally estrogenic and nontoxic to DNA, and it may even protect the breast against cancerous changes. According to Bradlow and his colleagues, a low 16-to-2 ratio is desirable.

Unfortunately, many contaminants push the ratio in the other direction. In cultured cells, the pesticides DDT, atrazine, and endosulfan—as well as benzene and certain PCBs—all skew the balance away from 2 and toward the 16 pathway. In essence, these environmental contaminants turn the natural hormone estrogen into a weapon that is aimed at the breasts it caused to grow in the first place.

❧

I had bladder cancer as a young adult. If I tell people this fact, they usually shake their heads. If I go on to mention that cancer runs in my family, they usually start to nod. *She is from one of those cancer families*, I can almost hear them thinking. Sometimes, I just leave it at that. But, if I am up for blank stares, I add that I am adopted and go on to describe a study of cancer among adoptees that found correlations within their adoptive families but not within their biological ones. ("Deaths of adoptive parents from cancer before the age of 50 increased the rate of mortality from cancer fivefold among the adoptees. . . . Deaths of biological parents from cancer had no detectable effect on the rate of mortality from cancer among the adoptees.") At this point, most people become very quiet.

These silences remind me how unfamiliar many of us are with the notion that families share environments as well as chromosomes or with the concept that our genes work in communion with substances streaming in from the larger, ecological world. What runs in families does not necessarily run in blood. And our genes are less an inherited set of teacups enclosed in a cellular china cabinet than they are plates used in a busy diner. Cracks, chips, and scrapes accumulate. Accidents happen.

My Aunt Jean died of bladder cancer. Raymond and Violet both died of colon cancer. LeRoy is currently under treatment. These are my father's relatives. About Uncle Ray I remember very little, except that he, along with my dad, was one of the less loud of the concrete-pouring, brick-laying Steingraber brothers. Aunt Jean laughed a lot and once asked me to draw a pig so she could tape it to her refrigerator door. Red-haired Aunt Vi cooked magnificent dinners, was partial to wearing pink, and was married to a man truly untempted by silence. Together, she once remarked, the two of them sure knew how to enjoy themselves. Her widowed husband, my Uncle Ed, is now being treated aggressively for prostate cancer. Nonetheless, at last report, he was busy building a shrine to his wife out in the backyard. When it comes to expressions of grief, my father's side of the family tends toward large-scale construction projects.

The man who was to be my brother-in-law was stricken with intestinal cancer at the age of twenty-one. He cleaned out chemical drums for a living. Three years before Jeff's diagnosis, I was diagnosed with bladder cancer, and three years before my diagnosis, my mother learned she had metastatic breast cancer. That she is still alive today is a topic of considerable wonder among her doctors. Mom is matter-of-fact about this, although she will, if prompted, shyly point out that she has outlived her oncologist and three of her other doctors, two of whom died of cancer.

My mother was first diagnosed in 1974, a year that is considered an anomaly in the annals of breast cancer. Graphs displaying U.S. breast cancer incidence rates across the decades show a gently rising line that suddenly zooms skyward, falls back, then continues its slow ascent. The story behind the blip of '74 has been deemed a textbook lesson in statistical artifacts.

In this year, First Lady Betty Ford and Second Lady Happy Rockefeller both underwent mastectomies. The words *breast cancer* entered public conversation. Women who might otherwise have delayed routine checkups or who were hesitant to seek medical opinion about a lump were propelled into doctors' offices. The result was that a lot of women were diagnosed with breast cancer within a short period of time, my mother among them.

When I, at age fifteen, inquired why my mother was in the hospital, the answer was "Because she has what Mrs. Ford has." When my mother, at age forty-four, questioned whether a radical mastectomy was necessary, she was told, "If it's good enough for Happy, it's good enough for you."

Back at home, a new fixture appeared on the dresser in my parents' bedroom: a bald Styrofoam head. It had come with the wig—which it dutifully wore when my mother wasn't—and it remains in my mind as the most vivid image of her illness. Its features were peculiar. It lacked ears. Its closed eyes and too-small nose were half formed, as though worn smooth by water. It wore the serene, expressionless face of someone drowned or unborn.

Not that the rest of us were any more demonstrative. My father vanished into his workshop. I became the heroine of homework and long walks. My twelve-year-old sister wrote protracted, angry manifestos—and then tore them up into small fragments. These were secretly reassembled and read by our mother, who steadfastly believed that an atmosphere of normalcy was health promoting.

Some twenty years later, Mom and I sit out on my Boston balcony, drinking iced tea. I describe some medical decisions that I am facing. She provides calm, thoughtful advice—as I knew she would. Finally, I ask her about all those years of chemotherapy, surgeries, and bad news. Did she feel supported during that time?

She looks away. "Too much sympathy would have weakened me." It isn't exactly an answer to my question, and I want to ask what she means. But I don't.

My sister and I sit out in her backyard, drinking beer and watching her boys chase fireflies. I realize—as though for the first time—that she had seen her mother, sister, and fiancé all in treatment for cancer by the time she was old enough for college. I ask her about this.

"It just kept happening." Julie says, ticking off the chronology of diagnoses we both have memorized. "You and I quit talking for a while. Dad and Mom quit talking. We all got very quiet."

"That's how I remember it, too. Everybody lost their vocabulary." I want to ask her about Jeff's death and about the Styrofoam head. But I don't.

TWELVE

ecological roots

In 1983, I took the train home to Illinois for the holidays—and an appointment at the hospital.

The scheduling of cancer checkups is always an elaborate decision. The calendar date must sound auspicious. Monday or Tuesday appointments are best; otherwise, one risks waiting through the weekend for the results of a laggard lab test or delayed radiology report. It's also best if these appointments fall within a hectic, deadline-filled month so that frenetic activity can preclude fretfulness. During the years I was a graduate student, this meant the ends of semesters, which explains why some half dozen Christmas carols now remind me of outpatient waiting rooms. This particular appointment was destined to turn out fine. What I remember most clearly is my journey there by train.

Something about the landscape changes abruptly between

northern and central Illinois. I am not sure what it is exactly, but it happens right around the little towns of Wilmington and Dwight. The horizon recedes, and the sky becomes larger. Distances increase, as though all objects are moving slowly away from each other. Lines become more sharply drawn. These changes always make me restless and, when driving, drive faster. But since I am in a train, I close the book I am reading and begin impatiently straightening the pages of a newspaper strewn over the adjacent seat.

That is when my eye catches the headline of a back-page article: SCIENTISTS IDENTIFY GENE RESPONSIBLE FOR HUMAN BLADDER CANCER. Pulling the newspaper onto my lap, I stare out the window and become very still. It is only early evening, but the fields are already dark, a patchwork of lights quilted over and across them. They have always soothed me. I look for signs of snow. There are none. Finally, I read the article.

Researchers at the Massachusetts Institute of Technology, it seems, had extracted DNA from the cells of a human bladder tumor and used it to transform normal mouse cells into cancerous ones. Through this process, they located the segment of DNA responsible for the transformation. And by comparing this segment to its unmutated form in noncancerous human cells, they were able to pinpoint the exact alteration that had caused a respectable gene to go bad.

In this case, the mutation turned out to be a substitution of one unit of genetic material for another in a single rung of the DNA ladder. Namely, at some point during DNA replication, a double-ringed base called guanine was swapped for the single-ringed thymine. Like a typographical error in which one letter replaces another—*snow* instead of *show*, *block* instead of *black*—the message sent out by this gene was utterly changed. Instead of instructing the cell to manufacture the amino acid glycine, the altered gene now specified for valine. (Nine years later, other researchers would determine that this substitution alters the structure of proteins involved in signal transduction—the crucial line of communication between the cell membrane and the nucleus that helps coordinate cell division.)

Guanine instead of thymine. Valine instead of glycine. I look away again—this time at my face superimposed over the landscape by the window's mirror. If, in fact, this mutation was involved in my cancer, when did it happen? Where was I? Why had it escaped repair? I had been betrayed. But by what?

Thirteen years later, I possess a bulging file of scientific articles documenting an array of genetic changes involved in bladder cancer. Besides the oncogene just described, two tumor supressor genes, p15 and p16, have also been discovered to play a role. Their deletion is a common event in transitional cell carcinoma, the kind of cancer I had. Mutations of the famous p53 tumor suppressor gene, with guest-star appearances in so many different cancers, have been detected in more than half of invasive bladder tumors. Also associated with transitional cell carcinomas are surplus numbers of growth factor receptors. Their overexpression has been linked to the kinds of gross genetic injuries that appear near the end of the malignant process.

The nature of the transaction between these various genes and certain bladder carcinogens has likewise been worked out in the years since a newspaper article introduced me to the then new concept of oncogenes. Consider, for example, that redoubtable class of bladder carcinogens called aromatic amines—present as contaminants in cigarette smoke; added to rubber during vulcanization; formulated as dyes for cloth, leather, and paper; used in printing and color photography; and featured in the manufacture of certain pharmaceuticals and pesticides. Aniline, benzidine, naphthylamine, and *o*-toluidine are all members of this group. The first reports of excessive bladder cancers among workers in the aniline dye industry were published in 1895. (Recall also Wilhelm Hueper's dogs, described in Chapter Six.) More than a century later, we now know that anilines and other aromatic amines ply their wickedness by forming DNA adducts in the cells of the tissues lining the bladder, where they arrive as contaminants of urine.

We also now know that aromatic amines are gradually detoxified by the body through a process called acetylation. Like all such processes, it is carried out by a special group of detoxifying enzymes whose actions are controlled and modified by a number of genes. People who are slow acetylators have low levels of these enzymes and are at greater risk of bladder cancer from exposure to aromatic amines. Members of this population can be readily identified because they bear significantly higher burdens of adducts than fast acetylators at the same exposure levels. These genetically suspectible individuals hardly constitute a tiny minority: more than half of Americans and Europeans are estimated to be slow acetylators.

Very likely, I am one. You may be one, too.

We know a lot about bladder cancer. Bladder carcinogens were among the earliest human carcinogens ever identified, and one of the first human oncogenes ever decoded was isolated from some unlucky fellow's bladder tumor. More than most malignancies, bladder cancer has provided researchers with a picture of the sequential genetic changes that unfold from initiation through promotion to progression, from precursor lesions to increasingly more aggressive tumors.

Sadly, all this knowledge about genetic mutations, inherited risk factors, and enzymatic mechanisms has not translated into an effective campaign to prevent the disease. The fact remains that the overall incidence rate of bladder cancer increased 10 percent between 1973 and 1991. Increases are especially dramatic among African Americans: among black men, bladder cancer incidence has risen 28 percent since 1973, and among black women, 34 percent.

Somewhat less than half of all bladder cancers among men and one-third of all cases among women are thought to be attributable to cigarette smoking, which is the single largest known risk factor for this disease. As we saw in Chapter Three, the lung cancer rate among white men in the United States is now falling, reflecting—at long last—the significant decline in smoking among members of this demographic group. If a parallel decline in bladder cancer incidence among white men should follow, we would have reason to finger tobacco as one possible explanation for the 1973–1991 increase. So far, it has not, but perhaps bladder cancer simply has a longer lag time than lung cancer. In the meantime, the question still remains: What is causing bladder cancer in the rest of us, the majority of bladder cancer patients for whom tobacco is not a factor?

I also possess another bulging file of scientific articles. These concern the ongoing presence of known and suspected bladder carcinogens in rivers, groundwater, dump sites, and indoor air. For example, industries reporting to the Toxics Release Inventory disclosed environmental releases of the aromatic amine *o*-toluidine that totaled 14,625 pounds in 1992 alone. Detected also in effluent from refineries and other manufacturing plants, *o*-toluidine exists as residues in the dyes of commercial textiles, which may, according to the *Seventh Annual Report on Carcinogens,* expose members of the general public who are consumers of these goods: "The presence of *o*-toluidine, even as a trace contaminant, would be a cause for concern." A 1996 study investigated a sixfold excess of bladder cancer among workers exposed years before to *o*-toluidine and aniline in the rubber chemicals department of a manufacturing plant in upstate New York. Levels of these contaminants are now well within their legal workplace limits, and yet blood and urine collected from current employees were found to contain substantial numbers of DNA adducts and detectable levels of *o*-toluidine and aniline. Another recent investigation revealed an eightfold excess of bladder cancer among workers employed in a Connecticut pharmaceuticals plant that manufactured a variety of aromatic amines. This study was reported as having national implications because the main suspect, dichlorobenzidine, has been widely used throughout the United States.

What my various file folders do *not* contain is a considered evaluation of all known and suspected bladder carcinogens—their sources, their possible interactions with each other, and our various routes of exposure to them. As we have seen, trihalomethanes—those unwanted by-products of water chlorination—have been linked to bladder cancer, as has the dry-cleaning solvent and sometime-contaminant of drinking-water pipes, tetrachloroethylene. I possess individual reports on each of these topics. What I do not have is a comprehensive description of how all these substances behave in combination. What are the risks of multiple trace exposures? What happens when we drink trihalomethanes, absorb aromatic amines, and inhale tetrachloroethylene? Furthermore, what is the ecological fate of these substances once they are released into the environment? What happens when dyed cloth, colored paper, and leather goods are laundered, landfilled, or incinerated? And why—almost a century after some of them were so identified—do powerful bladder carcinogens such as amine dyes continue to be manufactured, imported, used, and released into the environment in the first place? However improved the record of effort to regulate them, why have safer substitutes not replaced them all? These questions remain, to my knowledge, largely unaddressed by the cancer research community.

Several obstacles, I believe, prevent us from addressing cancer's environmental roots. An obsession with genes and heredity is one.

Cancer research currently directs considerable attention to the study of inherited cancers. Most immediately, this approach facilitates the development of genetic testing, which attempts to predict an individual's risk of succumbing to cancer, based on the presence or absence of certain genetic alterations. These efforts may also reveal which genes are common targets of acquired mutation in the general population. (Hereditary mutations are present at the time of conception, and they are carried in the DNA of all body cells; acquired mutations, which accumulate over an individual's lifetime, are passed only to the direct descendents of the cells in which they arise.)

Hereditary cancers, however, are the rare exception. Collectively, fewer than 10 percent of all malignancies are thought to involve inherited mutations. Between 1 and 5 percent of colon cancers, for example, are of the hereditary variety, and only about 15 percent exhibit any sort of familial component. The remaining 85 percent of colon cancers are officially classified as "sporadic," which, confesses one prominent researcher, "is a fancy medical term for 'we don't know what the hell causes it.'" Breast cancer also shows little connection to heredity (probably between 5 and 10 percent). Finding "cancer genes" is not going to prevent the vast majority of cancers that develop.

Moreover, even when rare, inherited mutations play a role in the development of a particular cancer, environmental influences are inescapably involved as well. Genetic risks are not exclusive of environmental risks. Indeed, the direct consequence of some of these damaging mutations is that people become even more sensitive to environmental carcinogens. In the case of hereditary colon cancer, for example, what is passed down the generations is a faulty DNA repair gene. Its human heirs are thereby rendered less capable of coping with environmental assaults on their genes or repairing the spontaneous mistakes that occur during normal cell division. These individuals thus become more likely to accumulate the series of *acquired* mutations needed for the formation of a colon tumor.

Cancer incidence rates are not rising because we are suddenly sprouting new cancer genes. Rare, heritable genes that predispose their hosts to cancer by creating special susceptibilities to the effects of carcinogens have undoubtedly been with us for a long time. The ill effects of some of these genes might well be diminished by lowering the burden of environmental carcinogens to which we are all exposed. In a world free of aromatic amines, for example, being born a slow acetylator would be a trivial issue, not a matter of grave consequence. The inheritance of a defective carcinogen-detoxifying gene would matter less in a culture that did not tolerate carcinogens in air, food, and water. By contrast, we cannot change our ancestors. Shining the spotlight on inheritance focuses us on the one piece of the puzzle we can do absolutely nothing about.

Risks of lifestyle are also not independent of environmental risks. And yet public education campaigns about cancer consistently accent the former and ignore the latter. I collect the colorful pamphlets on cancer that are made available in hospitals, clinics, and waiting rooms. When I was teaching introductory biology and also spending many hours in doctors' offices, I began to compare the descriptions of cancer in the tracts displayed in the skinny, silver racks above the magazines with the chapter on cancer provided in my students' textbook. Here are some of my findings.

On the topic of how many people get cancer, a pink and blue brochure published by the U.S. Department of Health and Human Services offers the following:

> Good News: Everyone does not get cancer. 2 out of 3 Americans never will get it.

Whereas, according to *Human Genetics: A Modern Synthesis:*

> One of three Americans will develop some form of cancer in his or her lifetime, and one in five will die from it.

(Since these materials were published, the proportion of Americans contracting cancer has risen from 30 to 40 percent.)

On the topic of what causes cancer, the brochure states:

> In the past few years, scientists have identified many causes of cancer. Today it is known that about 80% of cancer cases are tied to the way people live their lives.

Whereas the textbook contends:

> As much as 90 percent of all forms of cancer is attributable to specific environmental factors.

In regard to prevention, the brochure emphasizes individual choice and responsibility:

> You can control many of the factors that cause cancer. This means you can help protect yourself from the possibility of getting cancer. You can decide how you're going to live your life—which habits you will keep and which ones you will change.

The genetics book presents a somewhat different vision:

> Because exposure to these environmental factors can, in principle, be controlled, most cancers could be prevented. . . . Reducing or eliminating exposures to environmental carcinogens would dramatically reduce the prevalence of cancer in the United States.

The textbook goes on to identify some of these carcinogens, the routes of exposure, and the types of cancer that result. In contrast, the brochure emphasizes the importance of personal habits, such as sunbathing, that raise one's risk of contracting cancer. Thus, in my students' textbook, vinyl chloride is identifed as a carcinogen to which PVC manufacturers are exposed, whereas in the brochure, occupations that involve working with certain chemicals are called a risk factor. The textbook declares that "radiation is a carcinogen." The brochure advises us to "avoid unnecessary X-rays." Both emphasize the role of diet and tobacco.

In its ardent focus on lifestyle, the Good News brochure is typical of the educational pamphlets in my collection. By emphasizing personal habits rather than carcinogens, they frame the cause of the disease as a problem of *behavior* rather than as a problem of *exposure* to disease-causing agents. At its best, this perspective can offer us practical guidance and the reassurance that there are actions we as individuals can take to protect ourselves. (Not smoking, rightfully so, tops this list.) At its worst, the lifestyle approach to cancer is dismissive of hazards that lie beyond personal choice. A narrow focus on lifestyle—like a narrow focus on genetic mechanisms—obscures cancer's environmental roots. It presumes that the ongoing contamination of our air, food, and water is an immutable fact of the human condition to which we must accommodate ourselves. When we are urged to "avoid carcinogens in the environment and workplace," this advice begs the question. Why must there be known carcinogens in our environment and at our job sites?

The experience of the anthropologist Martha Balshem is revealing here. In the late 1980s, Balshem served as a health educator in an industrial, working-class community near Philadelphia where cancer rates were discovered to be unusually high. In response, the cancer control program of which she was part launched a public outreach campaign urging residents to adopt healthier lifestyles. The residents themselves suspected environmental causes and reported to the educational team that many neighborhood dogs were also afflicted with cancer: Did their pets have faulty personal habits as well? In her book *Cancer in the Community* Balshem recalls:

> As representatives of the cancer center, we sought to deflect this concern and stressed lifestyle changes to reduce cancer risk. Privately, we acknowledged our own feelings or suspicions that the profound pollution we observed in the community was somehow linked to the high cancer rates. We said to each other that this did not present us with a moral dilemma, because in any case, people were well advised to quit smoking, improve their diets, and get regular cancer tests.

In the end, Balshem came to believe the lesson she was transmitting—"accept authority and accept blame"—was the wrong one.

Cancer is certainly not the first disease to inspire this kind of message. In 1832, at the height of an epidemic, the New York City medical council announced that cholera's usual victims were those who were imprudent, intemperate, or prone to injury by the consumption of improper medicines. Lists of cholera prevention tips were posted publicly. Their advice ranged from avoiding drafts and crude vegetables to abstaining from alcohol. Maintaining "regular habits" was also said to be protective. Decades later, improvements in public sanitation (as mentioned in Chapter Eight) would bring cholera under control, and the pathogen responsible for the disease would finally be isolated by the bacteriologist Robert Koch in 1883. Of course, the behavioral changes urged by the 1832 handbills were not all without merit: uncooked produce, as it turned out, was an important route of exposure, but it was a fecal-borne bacteria—and not a salad-eating lifestyle—that was the cause.

The orthodoxy of lifestyle today finds its full expression in the public educational literature on breast cancer. In scores of cheerful pamphlets, women are exhorted to exercise, lower the fat in their diets, perform breast self-examinations, ponder their family history, and receive regular mammograms. "Delayed childbirth" (after age twenty) is frequently mentioned as a risk factor. (I have never seen "prompt childbirth" in the accompanying list of cancer prevention tips—undoubtedly because such advice would be tantamount to advocating teenage pregnancy.)

All by itself, a lifestyle approach to preventing breast cancer is inadequate. First, the majority of breast cancers cannot be explained by lifestyle factors, including reproductive history. We need to look elsewhere for the causes of these cancers. Second, mammography and breast self-examinations are tools of cancer detection, not acts of prevention. The popular refrain "Early detection is your best prevention!" is a non sequitur: Detecting cancer, no matter how early, negates the possibility of preventing cancer. At best, early detection may make cancer less fatal, allowing us, as the epidemiologist Robert Millikan puts it, "to live in a toxic soup without breasts or prostates, et cetera."

Finally, the adage that high-fat Western diets are the cause of breast cancer has not yet been supported by data. Dietary fat has long been a centerpiece of study in the investigation of breast cancer risk.

And yet, several long-term, heavily funded studies have indicated that dietary fat is unlikely to play a major role by itself. Rather than continuing to focus singlemindedly on the absolute quantity of fat consumed, several researchers have called for a more refined, ecological approach to diet. Two obvious starting points would be to assess the link between breast cancer and diets high in animal fat and to launch a definitive investigation into the extent to which various kinds of fats are contaminated by carcinogens. We already know with certainty that animal-based foods are our main route of exposure to organochlorine pesticides and dioxins. It's time to look at the whole picture.

Even reproductive choices have environmental implications. Breasts, for example, do not complete their development until the last months of a woman's first full-term pregnancy. During this time, the latticework of mammary ducts and lobules differentiate into fully functioning secretory cells. This process of specialization permanently slows the rate of mitosis, dampens the response to growth-promoting estrogens, and renders DNA less vulnerable to damage. According to the leading hypothesis, a full-term pregnancy early in life protects against breast cancer precisely because it reduces a woman's vulnerability to carcinogens and other cancer promoters, such as estrogens.

One of the principle proponents of this hypothesis, the Harvard epidemiologist Nancy Krieger, has urged its further testing. She has also urged a redirection of breast cancer research toward environmental questions. Investigators have repeatedly confirmed that reproductive history contributes to breast cancer risk. We need to know now, Krieger argues, whether women with similiar reproductive histories but divergent exposure to carcinogens have marked differences in breast cancer incidence. This need is made urgent by the results of animal studies showing that exposure to certain organochlorines hastens the onset of puberty. As we have noted, early first menstruation—along with late parenthood—is considered a risk factor for breast cancer in women.

Within the scientific community, grand arguments have ensued from the attempt to classify and quantify cancer deaths due to specific causes. Traditionally, the final result of this task takes the visual form of a great cancer pie sliced to depict the relative importance of different risk factors. "Smoking" is always a big wedge, monopolizing about 30 percent of the circle. "Diet" is also a sizable helping. Depending on who's doing the apportioning, an array of other lifestyle factors—"alcohol," "reproductive and sexual behavior," and "sedentary way of life"—divvy up the remainder, along with "occupation" and "pollution."

The quarreling begins immediately. How do we account for malignancies, such as certain liver cancers, to which both drinking and job hazards contribute? Or lung and bladder cancers where both job hazards and smoking conspire? Should the effects of pesticides be tallied under "pollution" or under "diet"? What about pollution's indirect effects—such as hormonal disruption, inhibition of apoptosis (cell death), and immune system suppression—that act to augment the dangers of risk factors across the board? What about formaldehyde, which seems to bind with DNA in such a way that it prevents repair of damage induced by ionizing radiation, possibly raising the cancer risk from medical X-rays?

Interactions between risk factors aside, how can the environment's death toll be calculated at all when the vast majority of industrial chemicals in commerce have never been tested for their ability to cause cancer?

The futility of what the cancer historian Robert Proctor calls "the percentages game" has not deterred public health agencies from using this kind of simplistic accounting to formulate cancer control policies and educational programs. Lifestyle is the bull's-eye of cancer prevention efforts, while targeting of environmental factors, perceived as a small contribution to the cancer problem, is seen as inefficient. Moreover—the rationale continues—not enough is known about environmental risks to make specific recommendations. (Incomplete and inconsistent evidence about the role of dietary fat in contributing to breast cancer is, on the other hand, not an obstacle to advising women to change their diets.)

In my own home state, a recent county-by-county cancer report reproduced an old cancer pie chart, published originally in 1981, that relegated environmental factors to a single, tiny slice and depicted tobacco and diet as major risk factors. The report concluded, "Many persons could reduce their chances of developing or dying from cancer by adopting healthier lifestyles and by visiting their physicians regularly for cancer-related checkups." The fact that Illinois is a leading producer of hazardous waste, a heavy user of pesticides, and home to an above-average number of Superfund sites is neither mentioned nor considered. No attempt is made in this report to correlate cancer statistics with Toxics Release Inventory data. No attempt is made in this report to determine whether cancer might follow industrial river valleys, rise in areas of high pesticide use, or cluster around contaminated wells.

Lifestyle and the environment are *not* independent categories that can be untwisted from each other: to talk about one is to talk about the other. A discussion about dietary habits is necessarily also a discussion about the food chain. To converse about childbirth and breast cancer is also to converse about changing susceptibility to carcinogens in the breast. And to advise those of us at risk for bladder cancer to "void frequently" is to acknowledge the presence of carcinogens in the fluids passing through our bodies.

During the last year of her life, Rachel Carson discussed before a U.S. Senate subcommittee her emerging ideas about the relationship between environmental contamination and human rights. The problems addressed in *Silent Spring*, she asserted, were merely one piece of a larger story—namely, the threat to human health created by reckless pollution of the living world. Abetting this hidden menace was a failure to inform common citizens about the senseless and frightening dangers they were being asked, without their consent, to endure. In *Silent Spring*, Carson had predicted that full knowledge of this situation would lead us to reject the counsel of those who claim there is simply no choice but to go on filling the world with poisons. Now she urged recognition of an individual's right to know about poisons introduced into one's environment by others and the right to protection against them. These ideas are Carson's final legacy.

The process of exploration that results from asserting our right to know about carcinogens in our environment is a different journey for every person who undertakes it. For all of us, however, I believe it necessarily entails a three-part inquiry. Like the Dickens character Ebenezer Scrooge, we must first look back into our past, then reassess our present situation, and finally summon the courage to imagine an alternative future.

We begin retrospectively for two reasons. First, we carry in our bodies many carcinogens that are no longer produced and used domesticially but which linger in the environment and in human tissue. Appreciating how, even today, we remain in contact with banned chemicals such as PCBs and DDT requires a historical understanding. Second, because cancer is a multicausal disease that unfolds over a period of decades, exposures during young adulthood, adolescence, childhood—and even prior to birth—are relevant to our present cancer risks. We need to find out what pesticides were sprayed in our neighborhoods and what sorts of household chemicals were stored under our parents' kitchen sink. Reminiscing with neighbors, family members, and elders in the community where one grew up can be an eye-opening first step.

This part of the journey is, in essence, a search for our ecological roots. Just as awareness of our genealogical roots offers us a sense of heritage and cultural identity, our ecological roots provide a particular appreciation of who we are biologically. It means asking questions about the physical environment we have grown up within and whose molecules are woven together with the strands of DNA inherited from our genetic ancestors. After all, except for the original blueprint of our chromosomes, all the material that is us—from bone to blood to breast tissue—has come to us from the environment.

Going in search of our ecological roots has both intimate and far-flung dimensions. It means learning about the sources of our drinking water (past and present), about the prevailing winds that blow through our communities, and about the agricultural system that provides us food. It involves visiting grainfields, as well as cattle lots, orchards, pastures, and dairy farms. It demands curiosity about how our apartment buildings are exterminated, clothing cleaned, and golf courses maintained. It means asserting our right to know about any and all toxic ingredients in products such as household cleaners, paints, and cosmetics. It requires a determination to find out where the underground storage tanks are located, how the land was used before the subdivision was built over it, what is being sprayed along the roadsides and rights-of-way, and what exactly goes on behind that barbed-wire fence at the end of the street.

Acquiring a copy of the Toxics Release Inventory for one's home county, as well as a list of local hazardous waste sites, is a simple place to begin (see the Afterword that follows). Such information is not available for the years prior to 1987 and so tells us less about our formative years than it does about the present decade. Nevertheless, these documents often contain clues to the past as well: the toxic chemicals loitering around an abandoned Superfund site, for example, can reveal what kinds of activities occurred there decades earlier.

In full possession of our ecological roots, we can begin to survey our present situation. This requires a human rights approach. Such an approach recognizes that the current system of regulating the use, release, and disposal of known and suspected carcinogens—rather than preventing their generation in the first place—is intolerable. So is the decision to allow untested chemicals free access to our bodies, until which time they are finally assessed for carcinogenic properties. Both practices show reckless disregard for human life.

A human rights approach would also recognize that we do not all bear equal risks when carcinogens are allowed to circulate within our environment. Workers who manufacture carcinogens are exposed to higher levels, as are those who live near the chemical graveyards that serve as their final resting place. Moreover, people are not uniformly vulnerable to effects of environmental carcinogens. Individuals with genetic predispositions, infants whose detoxifying mechanisms are not yet fully developed, and those with significant prior exposures may all be affected more profoundly. Cancer may be a lottery, but we do not each of us hold equal chances of "winning." When carcinogens are deliberately or accidentally introduced into the environment, some number of vulnerable persons are consigned to death. The impossibility of tabulating an exact body count does not alter this fact. A human rights approach to cancer strives, nonetheless, to make these deaths visible.

Suppose we assume for a moment that the most conservative estimate concerning the proportion of cancer deaths due to environmental causes is absolutely accurate. This estimate, put forth by those who dismiss environmental carcinogens as negligible, is 2 percent. Though others have placed this number far higher, let's assume for the sake of argument that this lowest value is absolutely correct. Two percent means that 10,940 people in the United States die each year from environmentally caused cancers. This is more than the number of women who die each year from hereditary breast cancer—an issue that has launched multi-million-dollar research initiatives. This is more than the number of children and teenagers killed each year by firearms—an issue that is considered a matter of national shame. It is more than three times the number of nonsmokers estimated to die each year of lung cancer caused by exposure to secondhand smoke—a problem so serious it warranted sweeping changes in laws governing air quality in public spaces. It is the annual equivalent of wiping out a small city. It is thirty funerals every day.

None of these 10,940 Americans will die quick, painless deaths. They will be amputated, irradiated, and dosed with chemotherapy. They will expire privately in hospitals and hospices and be buried quietly. Photographs of their bodies will not appear in newspapers. We will not know who most of them are. Their anonymity, however, does not moderate this violence. These deaths are a form of homicide.

A human rights approach to cancer would also speak out against other deprivations besides gross loss of life. The dispossession of Chattanooga Creek is one example. In 1993, the U.S. Agency for Toxic Substances and Disease Registry dispatched a group of representatives to Chattanooga, Tennessee, expressly to teach schoolchildren to stay away from the local creek, which happens to be surrounded by no less than forty-two hazardous waste sites. In the agency's words: "Training workshops highlighted the dangers of fishing, swimming, and playing in the creek and of eating fish from the creek. . . . Children were encouraged to take the information home and share it with their parents."

No one can quantify what the loss of a creek means to a child in Tennessee or measure the grief of parents who must forbid their son or daughter from exploring along its banks. But I think we can say with assurance that the transformation of a popular swimming hole into a cancer hazard and child's play into a cancer risk factor is a terrible diminishment of our humanity. And we can say that the agency's gesture of educational responsibility is indicative of a vast national *ir-*responsibility.

According to the most recent tally, forty possible carcinogens appear in drinking water, sixty are released by industry into ambient air, and sixty-six are routinely sprayed on food crops as pesticides. Whatever our past exposures, this is our current situation.

After having carefully appraised the risks and losses that we have endured by tolerating it, we can begin to imagine a future in which our right to an environment free of such substances is respected. It is unlikely that we will ever rid our environment of all chemical carcinogens. However, as Rachel Carson herself observed, the elimination of a great number of them would reduce the carcinogenic burden we all bear and thus would prevent considerable suffering and loss of human life. Three key principles can assist us in this effort.

One is the idea that public and private interests should act to prevent harm before it occurs. This is known as the *precautionary principle*, and it dictates that indication of harm, rather than proof of harm, should be the trigger for action—especially if delay may cause irreparable damage. Central to the precautionary principle is the recognition that we have an obligation to protect human life. Our current methods of regulation, by contrast, appear governed by what some frustrated policymakers have called the dead body approach: wait until damage is proven before action is taken. It is a system tantamount to running an uncontrolled experiment using human subjects.

Closely related to the precautionary principle is the *principle of reverse onus.* According to this edict, it is safety, rather than harm, that should necessitate demonstration. This reversal essentially shifts the burden of proof off the shoulders of the public and onto those who produce, import, or use the substance in question. The principle of reverse onus requires that those who seek to introduce chemicals into our environment first show that what they propose to do is almost certainly *not* going to hurt anyone. This is already the standard we uphold for pharmaceuticals, and yet for most industrial chemicals, no firm requirement for advance demonstration of safety exists. But chemicals are not citizens. They should not be presumed innocent unless proven guilty, especially when a verdict of guilt requires some of us to sicken and die in order to demonstrate the necessary evidence.

Finally, all activities with potential public health consequences should be guided by the *principle of the least toxic alternative*, which presumes that toxic substances will not be used as long as there is another way of accomplishing the task. This means choosing the least harmful way of solving problems—whether it be ridding fields of weeds, school cafeterias of cockroaches, dogs of fleas, woolens of stains, or drinking water of pathogens. Biologist Mary O'Brien advocates a system of alternatives assessment in which facilities regularly evaluate the availability of alternatives to the use and release of toxic chemicals. Any departure from zero should be preceded by a finding of necessity. These efforts, in turn, should be coordinated with active attempts to develop and make available affordable, nontoxic alternatives for currently toxic processes and with systems of support for those making the transition—whether farmer, corner dry-cleaner, hospital, or machine shop. Receiving the highest priority for transformation should be all processes that generate dioxin or require the use or release of any known human carcinogen such as benzene and vinyl chloride.

The principle of the least toxic alternative would move us away from protracted, unwinnable debates over how to quantify the cancer risks from each individual carcinogen released into the environment and where to set legal maximum limits for their presence in air, food, water, workplace, and consumer goods. As O'Brien observed, "Our society proceeds on the assumption that toxic substances *will* be used and the only question is how much. Under the current system, toxic chemicals are used, discharged, incinerated, and buried without ever requiring a finding that these activities are necessary." The principle of the least toxic alternative looks toward the day when the availability of safer choices makes the deliberate and routine release of chemical carcinogens into the environment as unthinkable as the practice of slavery.

Sitting at my desk in my Boston apartment, I am skimming through a journal article about hormone disruption in young female rats. The study is unusual because the animals were exposed not to a single chemical but to a real-life, low-level mixture of substances derived from the dust, soil, and air from a dioxin-contaminated landfill site. After only two days, the test animals exhibited abnormal changes in their livers, reproductive organs, and thyroid glands. Even rats exposed only to air from the landfill experienced significant changes in their development. These results indicate, the authors concluded, that current methods used for calculating health risks from chemical mixtures may "underestimate certain biological effects."

Flipping back to the beginning of the report, my eye catches on a familiar word: *Illinois.* The contaminated dust, soil, and air mixtures used in this study were collected from an old, inoperative landfill in Illinois.

Dust. Soil. Air. The year after my cancer diagnosis, I signed up for a field ecology class and learned to identify plant species in the rarest of rare Illinois habitats: the black soil prairie. Its remnants are almost completely confined to a few old pioneer gravesites. Hunkered down between headstones, I cupped the unfamiliar plants in my hands and tried to will into existence thousands of acres of these grasses and herbs, the sound of animals running, wildfires, birdsong.

As I became ever more enchanted with the Illinois prairie, I found that I was, nevertheless, unable to banish from my heart its remaining enemies—the nonnative invading species. Queen Anne's lace, ox-eye daisy, chicory, foxtail, goat's beard, teasel: all European immigrants, these are the familiar weeds of roadsides and fallow fields. My mother taught me the names of most of them. I am especially fond of teasel. It represents a special threat to prairie plants because mourners brought bouquets of it into the old prairie cemeteries, where it set seed and spread. In the winter, its stiff wands stand in the snow like pinecones on the ends of antennas. I keep a few stalks near my desk to remind me of home. I keep a scientific monograph of prairie plants on the shelf for the same reason.

After finishing the article on the health hazards of trace chemical mixtures, I look at the brown, spiny flowers and then out the window at the city I live in. Dust. Soil. Air. What I see are the contours of home.

AFTERWORD

exercising your right to know

There are essentially three ways of acquiring environmental data available to the public under the Emergency Planning and Community Right-to-Know Act (EPCRA). One is to request them from the state or federal government. Another is to search for them electronically using a computer. The third is to contact one of a number of public interest groups with expertise in providing such services.

If you are unaccustomed to dealing with environmental data—or unsure about exactly what information you want—the last option is probably the least frustrating. Organizations I have found very helpful in this regard are the Working Group on Community Right-to-Know (218 D Street SE, Washington, DC 20003; 202-544-9586); the Right-to-Know Network at the Unison Institute (1731 Connecticut Avenue NW, Washington, DC 20009; 202-234-8494); the John Snow Institute's Center for Environmental Health Studies (44 Farnsworth Street, Boston, MA 02210; 617-482-9485); and CCHW: Center for Health, Environment and Justice (P.O. Box 6806, Falls Church, VA 22040; 703-237-2249). Additionally, the Environmental Research Foundation (P.O. Box 5036, Annapolis, MD 21403; 410-263-1584) provides technical assistance to those who want to research specific chemicals or industries. The Mountain Association for Community and Economic Development in Berea, Kentucky, has assembled an excellent step-by-step handbook that describes how to become an environmental detective in your home community: D. F. Harker and E. U. Natter, *Where We Live: A Citizen's Guide to Conducting a Community Environmental Inventory* (Washington, D.C.: Island Press, 1995). Citizen groups who wish to conduct their own health survey can find the necessary tools in a book edited by two toxicologists: M. S. Legator and S. F. Strawn, *Chemical Alert! A Community Action Handbook* (Austin: University of Texas Press, 1993).

The U.S. Environmental Protection Agency is the branch of the federal government charged with administering EPCRA and providing public outreach. The EPA can be contacted in Washington (401 M Street SW, Washington, DC 20460) or at any one of its regional headquarters. Paper copies of the Toxics Release Inventory and other EPCRA documents can be obtained free of charge by calling EPA's EPCRA hotline (800-535-0202) or the TRI User Support (TRI-US) Service (202-260-1531). Upon request, TRI-US will also conduct short computer searches specific to one's home county and provide referrals to state agencies and local libraries where further information may be available. Hazardous waste is regulated separately under the Resource Conservation and Recovery Act (RCRA). To identify hazardous waste facilities that may be located in your community, call EPA's RCRA hotline (800-424-9346). For information about drinking water contamination, call EPA's Safe Drinking Water Hotline (800-426-4791). For information about the health effects of environmental contaminants, call EPA's Pollution Prevention Clearinghouse (202-260-1023).

A number of other federal agencies are important sources of information. Detailed toxicological profiles for hazardous substances commonly found at hazardous waste sites and thought to pose significant threats to public health are available from the U.S. Agency for Toxic Substances and Disease Registry (1600 Clifton Road NE, Atlanta, GA 30333; 404-639-0501). The Clearinghouse on Environmental Health Effects (800-643-4794) at the National Institute of Environmental Health Sciences answers questions about illnesses in which the environment may play a role, conducts on-line computer searches, and researches specific inquiries. Questions about workplace carcinogens can also be directed to the National Institute of Occupational Safety and Health (800-35-NIOSH) or to the Occupational Safety and Health Administration (800-321-OSHA).

For those so inclined, right-to-know information is available through a growing variety of electronic media and on-line services. The U.S. Government Printing Office makes the Toxics Release Inventory, as well as selected information about the health effects of TRI chemicals, available on CD-ROM and computer diskette. Contact the U.S. Government Printing Office Superintendent of Documents, P.O. Box 371954, Pittsburgh, PA 15250; 202-512-1800. The National Library of Medicine makes the Toxics Release Inventory—as well as information about chemical carcinogens—accessible through the on-line service TOXNET. Many local libraries also have access to this database. Contact the TRI Representative, National Library of Medicine, Specialized Information Services, 8600 Rockville Pike, Bethesda, MD 20894; 301-496-6531. RTK NET, made available for no charge through the Right-to-Know Network (see page 273), is an on-line telecommunications link that puts TRI data together with other environmental data and provides a mapping program. Increasingly, TRI data and other right-to-know documents are available through the Internet. The EPA manages a full-service Internet site, offering cost-free access via the World Wide Web and other servers.

As of this writing, pesticide sprayings are not subject to public disclosure, except in the states of California and New York, both of which have passed pesticide registry laws. In New York, for example, the law tracks the sale and use of pesticides, down to ZIP-code level, in a right-to-know database. The first statewide report is due for release in July 1998. In the absence of a federal registry, the National Center for Food and Agricultural Policy (1616 P Street NW, Washington, DC 20036; 202-328-5048) has compiled a national pesticide-use database. Specific to agriculture and drawing information from federal and state surveys, this document provides estimates on the total number of acres treated in each state and on the total pounds of active ingredient used by each state and each crop. The EPA, in cooperation with Oregon State University, provides a hotline to field questions on the ecological and health effects of pesticides: the National Pesticide Telecommunications Network (800-858-7378).

Several nonprofit groups provide extensive pesticide information services to the public on topics ranging from the export of banned pesticides to nontoxic treatments for head lice. Those I have found helpful include the National Coalition Against the Misuse of Pesticides (701 E Street SE, Suite 200, Washington, DC 20003; 202-543-5450); the Northwest Coalition for Alternatives to Pesticides (P.O. Box 1393, Eugene, OR 97440; 541-344-5044); the Pesticide Action Network North American Regional Center (116 New Montgomery Street, Suite 810, San Francisco, CA 94105; 415-541-9140); the Pesticide Education Center (P.O. Box 420870, San Francisco, CA 94142; 415-391-8511); and the Rachel Carson Council (8940 Jones Mill Road, Chevy Chase, MD 20815; 301-652-1877).

In Canada, right-to-know legislation varies by province and territory. In most jurisdictions, a comissioner makes information collected under these laws publicly available. Access is also provided by many Canadian libraries. For a guide to access-to-information laws in Canada, as well as to the specifics of the National Pollution Release Inventory, individuals may contact the Information Commissioner of Canada (112 Kent Street, Ottawa, Ontario K1A 1H3; 800-267-0441; 613-995-2410). Additional assistance and advocacy is provided by Democracy Watch (P.O. Box 821, Station B, Ottawa, Ontario K1P 5P9; 613-241-5179) and the Canadian Environmental Law Association (517 College Street, Suite 401, Toronto, Ontario M6G 4A2; 416-960-2284). For information about the Canadian Cancer Registry, contact Health Canada (Tunney's Pasture, Ottawa, Ontario K1A 0L2; 613-957-0327) or the Statistics Canada Reference Centre nearest you.

SOURCE NOTES

ABBREVIATIONS USED IN NOTES

ACS	American Cancer Society
AEH	*Archives of Environmental Health*
AJE	*American Journal of Epidemiology*
AJPH	*American Journal of Public Health*
ATSDR	Agency for Toxic Substances and Disease Registry
CDC	Centers for Disease Control
EDF	Environmental Defense Fund
EHP	*Environmental Health Perspectives*
EPA	U.S. Environmental Protection Agency
FDA	Food and Drug Administration
GAO	General Accounting Office
GPO	U.S. Government Printing Office
IARC	International Agency for Research on Cancer
IASS	Illinois Agricultural Statistics Service
IDA	Illinois Department of Agriculture
IDC	Illinois Department of Conservation
IDENR	Illinois Department of Energy and Natural Resources
IDPH	Illinois Department of Public Health
IEPA	Illinois Environmental Protection Agency
IFB	Illinois Farm Bureau
INHS	Illinois Natural History Survey
ISGS	Illinois State Geological Survey
ISGWS	Illinois State Geological and Water Surveys
ISWS	Illinois State Water Survey
JAMA	*Journal of the American Medical Association*
JNCI	*Journal of the National Cancer Institute*

JTEH	*Journal of Toxicology and Environmental Health*
MDPH	Massachusetts Department of Public Health
NCI	National Cancer Institute
NEJM	*New England Journal of Medicine*
NIH	National Institutes of Health
NIOSH	National Institute for Occupational Safety and Health
NRC	National Research Council
NRDC	National Resources Defense Council
OSHA	Occupational Safety and Health Administration
PDT	*Pekin Daily Times*
PJS	*Peoria Journal Star*
SSJR	*Springfield State Journal Register*
USDA	U.S. Department of Agriculture
USDHHS	U.S. Department of Health and Human Services
WHO	World Health Organization

Note: Organized by page number, the citations provided below represent the primary sources I consulted and are not intended to serve as a comprehensive review of the scientific literature. Some of the articles, monographs, and texts cited here are difficult to obtain, and some are highly technical in nature. Whenever I was aware of them, I also provided references to articles appearing in popular publications (*Science News* and the *New York Times*, for example) that can be found in most public libraries and that, I hope, may be more accessible to lay readers.

PROLOGUE

xiii Illinois prairies: S. L. Post, "Surveying the Illinois Prairie," *The Nature of Illinois* (winter 1993): 1–8.

xv–xvi direct link between smoking and lung cancer: M. F. Denissenko, "Preferential Formation of Benzo[a]pyrene Adducts at Lung Cancer Mutational Hotspots in p53," *Science* 274 (1996): 430–32; D. Stout, "Direct Link Found between Smoking and Lung Cancer," *New York Times*, 18 Oct. 1996, pp. A-1, A-11; "Smoking Leaves Fingerprints on DNA," *Science News* 150 (1996): 284.

xvi the duty to inquire: See, for example, J. D. Sherman, *Chemical Exposure and Disease: Scientific and Investigative Techniques* (Princeton, N.J.: Princeton Scientific Publishing, 1994), 4–13.

xvi "Population Health Looking Upstream" (editorial), *Lancet* 343 (1994): 429–30.

ONE *trace amounts*

2 Mahomet River: J. P. Kempton and A. P. Visocky, *Regional Groundwater Resources in Western McLean and Eastern Tazewell Counties with an Emphasis on the Mahomet Bedrock Valley*, Cooperative Groundwater Report 13 (Champaign: ISGWS, 1992); J. P. Kempton et al., "Mahomet Bedrock Valley in East-Central Illinois: Topography, Glacial Drift Stratigraphy, and Hydrogeology," in N. Melhorn and J. P. Kempton (eds.), *Geology and Hydrology of*

the Teays-Mahomet Bedrock Valley System, Special Report 258 (Boulder, Colo.: Geological Society of America, 1991); J. P. Gibb et al., *Groundwater Conditions and River-Aquifer Relationships along the Illinois Waterway* (Champaign: ISWS, 1979); M. M. Killey, "Do You Live above an Underground River?" Geogram 6 (Urbana: ISGS, 1975).

2–3 the ancestral Mississippi River valley: M. A. Marino and R. J. Schicht, *Groundwater Levels and Pumpage in the Peoria-Pekin Area, Illinois, 1890–1966* (Champaign: ISWS, 1969), 3; S. L. Burch and D. J. Kelly, *Peoria-Pekin Regional Groundwater Quality Assessment*, Research Report 124 (Champaign: ISWS, 1993), 6.

3 Illinois farm statistics: IFB, *Farm and Food Facts* (Bloomington, Ill.: IFB, 1994).

3 the disappearance of the Illinois prairie: IDENR, *The Changing Illinois Environment: Critical Trends*, summary report and vol. 3, ILENR/RE-EA-94/05 (Springfield, Ill.: IDENR, 1994); S. L. Post, "Surveying the Illinois Prairie," *The Nature of Illinois* (winter 1993): 1–8; R. C. Anderson, "Illinois Prairies: A Historical Perspective," in L. M Page and M. R. Jeffords (eds.), *Our Living Heritage: The Biological Resources of Illinois* (Champaign: INHS, 1991).

5 current pesticide application in Illinois: L. P. Gianessi and J. E. Anderson, *Pesticide Use in Illinois Crop Production* (Washington, D.C.: National Center for Food and Agricultural Policy, 1995), table B-2. Fifty-four million represents pounds of active ingredient. This figure is an extrapolation derived from small-scale surveys. Other than California and New York, both of which maintain state pesticide registries, no state or federal agency keeps track of pesticide use (unless the pesticide is classified as restricted). In Illinois, records are kept only on the number of acres sprayed, not amounts sprayed per acre. Moreover, many newer pesticides are more potent at low dosages. Decreases in pounds or ounces sprayed per acre do not necessarily indicate decreases in pesticide reliance or toxicity. See IDENR, *Changing Illinois*, summary report, 81.

5 percentage of corn treated with pesticides in 1950: IDENR, *Changing Illinois*, vol. 3., 78.

5 percentage of corn treated in 1993: IASS, *Agricultural Fertilizer and Chemical Use: Corn—1993* (Springfield, Ill.: IDA, 1994).

5 pesticide drift: C. M. Benbrook et al., *Pest Management at the Crossroads* (Yonkers, N.Y.: Consumers Union, 1996); C. A. Edwards, "The Impact of Pesticides on the Environment," in D. Pimentel et al. (eds.), *The Pesticide Question: Environment, Economics, and Ethics* (New York: Routledge, 1993), 13–46; D. E. Glotfelty et al., "Pesticides in Fog," *Nature* 325 (1987): 602–5.

5 pesticides in Illinois surface streams: A. G. Taylor and S. Cook, "Water Quality Update: The Results of Pesticide Monitoring in Illinois' Streams and Public Water Supplies" (paper presented at the 1995 Illinois Agricultural Pesticides Conference, Univ. of Illinois, Urbana, 4–5 Jan. 1995).

5 pesticides in Illinois groundwater: A. G. Taylor, "The Effects of Agricultural Use on Water Quality in Illinois" (paper presented at the 1993 American Chemical Society Agrochemicals Division Symposium, "Pesticide Management for the Protection of Ground and Surface Water Resources," Chicago, Ill., 25–26 Aug. 1993); S. C. Schock et al., *Pilot Study: Agricul-*

tural Chemicals in Rural, Private Wells in Illinois, Cooperative Groundwater Report 14 (Champaign: ISGWS, 1992).

5 atrazine's link to cancer: EPA, *The Triazine Herbicides: Atrazine, Simazine, and Cyanazine*, Position Document 1, Initiation of Special Review, OPP-30000-60-4919-5 (Washington, D.C.: Office of Pesticide Programs, 1994); A. Pinter et al., "Long-term Carcinogenicity Bioassay of the Herbicide Atrazine in F344 Rats," *Neoplasma* 37 (1990): 533–44; A. Donna et al., "Triazine Herbicides and Ovarian Epithelial Neoplasms," *Scandinavian Journal of Work Environment and Health* 15 (1989): 47–53; A. Donna et al., "Ovarian Mesothelial Tumors and Herbicides: A Case-Control Study," *Carcinogenesis* 5 (1984): 941–42.

5–6 hazardous waste in Illinois: C. W. Forrest and R. Olshansky, *Groundwater Protection by Local Government* (Champaign: IDENR and IEPA, 1993); W. H. Allen, "Hazardous Waste: Past, Present, Future," *The Nature of Illinois* (winter 1992): 13–16. Updated estimates were obtained from the IEPA's Office of Chemical Safety in Jan. 1997.

5–6 number of waste sites in Illinois: IDENR, *Changing Illinois*, summary report, 29, 68; R. D. Brower and A. P. Visocky, *Evaluation of Underground Injection of Industrial Waste in Illinois*, Joint Report 2 (Champaign: Illinois Scientific Surveys, 1989). Updated estimates were obtained from the IEPA's Office of Chemical Safety in Jan. 1997.

5–6 import and export of hazardous waste: IEPA, *Summary of Annual Reports on Hazardous Waste in Illinois, 1991 and 1992: Generation, Treatment, Storage, Disposal, and Recovery* (Springfield, Ill.: IEPA, 1994), v; IEPA, *Illinois Nonhazardous Special Waste Annual Report for 1991* (Springfield, Ill.: IEPA, 1993). Updated estimates were obtained from the IEPA's Office of Chemical Safety in Jan. 1997.

6 legal releases of toxic chemicals: IEPA, *Sixth Annual Toxic Chemical Report*, IEPA/ENV/94-151 (Springfield, Ill.: IEPA, 1994), v.

6 metal degreasers and dry-cleaning fluids: IDPH, *Chlorinated Solvents in Drinking Water* (Springfield, Ill.: IDPH, Division of Environmental Health, n.d.).

6 quote from a recent state assessment: IDENR, *Changing Illinois*, summary report, 6.

6 universal detections of DDT and PCBs in human tissues: R. R. M. Sharpe, "Another DDT Connection," *Nature* 375 (1995): 538–39; W. J. Rogan et al., "Polychlorinated Biphenyls (PCBs) and Dichlorodiphenyl Dichloroethene (DDE) in Human Milk: Effects on Growth, Morbidity, and Duration of Lactation," *AJPH* 77 (1987): 1294–97.

6 DDT can remain in soil for several decades: J. B. Diamond and R. B. Owen, "Long-Term Residue of DDT Compounds in Forest Soils in Maine," *Environmental Pollution* 92 (1996): 227–30.

6–7 archival film clips appear in "Rachel Carson's *Silent Spring*," documentary film by Peace River Films, aired on PBS, *The American Experience*, 8 Feb. 1993.

7 old magazine ads for DDT are reprinted in E. P. Russell III, "'Speaking of Annihilation': Mobilizing for War against Human and Insect Enemies, 1914–1945," *Journal of American History* 82 (1996): 1505–29; and in J.

Curtis et al., *After* Silent Spring: *The Unsolved Problem of Pesticide Use in the United States* (New York: NRDC, 1993), 2.

7 DDT for polio control: T. R. Dunlap, *DDT: Scientists, Citizens and Public Policy* (Princeton, N.J.: Princeton Univ. Press, 1981), 65.

7 DDT in paint: This ad, for Sherwin-Williams, appeared in 1946. See E. C. Helfrick as told to M. Riddle, "Mass Murder Introduces Sherwin-Williams' 'Pestroy,'" *Sales Management*, 15 Oct. 1946, pp. 60–64. See also E. P. Russell III, *The Nature of War: Pest Control, Chemical Warfare, and American Culture, 1914–1962* (in preparation).

7 DDT in blankets: DDT was also incorporated into starch finishes. See T. F. West and G. W. Campbell, *DDT and Newer Persistent Pesticides* (New York: Chemical Publishing Co., 1952), 163–74. In addressing the question of whether the routine use of DDT in textiles could pose threats to human health, these authors reached the following conclusion: "Extensive investigations have been carried out and it would appear that DDT is as safe as many chemicals at present in everyday use, and probably a good deal safer than many" (173).

7 quotes from fellow baby boomers: Jean Powers of Dover, Mass., and John Gephart of Ithaca, N.Y.

7 "the harmless aspect of the familiar": R. Carson, *Silent Spring* (Boston: Houghton Mifflin, 1962), 20.

7–8 "It is not my contention . . .": Ibid., 12.

8 Carson on future generations: Ibid., 13.

8 "killer of killers," "the atomic bomb of the insect world": J. Warton. *Before* Silent Spring: *Pesticides and Public Health in Pre-DDT America* (Princeton, N.J.: Princeton Univ. Press, 1974), 248–55.

8 failure of DDT: Carson, *Silent Spring*, 20–23, 58, 103, 107–9, 112, 113, 120–22, 125, 143–44, 206–7, 225, 267–73; T. R. Dunlap, *DDT: Scientists, Citizens and Public Policy* (Princeton, N.J.: Princeton Univ. Press, 1981), 63–97.

8 DDT in breast milk: E. P. Laug et al., "Occurrence of DDT in Human Fat and Milk," *AMA Archives of Industrial Hygiene and Occupational Medicine* 3 (1951): 245–46.

9 DDT's ongoing presence: USDA, *Pesticide Data Program, Annual Summary Calendar Year 1994* (Washington, D.C: USDA, Agricultural Marketing Service, 1994), 13; R. G. Harper et al., "Organochlorine Pesticide Contamination in Neotropical Migrant Passerines," *Archives of Environmental Contamination and Toxicology* 31 (1996): 386–90; ATSDR, "DDT, DDE, and DDD" (fact sheet) (Atlanta: ATSDR, 1995); R. G. Lewis et al., "Evaluation of Methods for Monitoring the Potential Exposure of Small Children to Pesticides in the Residential Environment," *Archives of Environmental Contamination and Toxicology* 26 (1996): 37–46; W. H. Smith et al., "Trace Organochlorine Contamination of the Forest Floor of the White Mountain National Forest, New Hampshire," *Environmental Science and Technology* 27 (1993): 2244–46; EPA, *Deposition of Air Pollutants to the Great Lakes: First Report to Congress*, EPA-453/R-93-055 (Washington, D.C.: EPA, 1994).

9 export of DDT and other banned pesticides: In 1992, 600,000 lbs. of DDT were shipped out of U.S. ports. Some analysts suspect this cargo may represent a transshipment—cargo imported and then exported again. Poor labeling of pesticide exports make careful tracking very difficult. J. Raloff, "The Pesticide Shuffle," *Science News* 149 (1996): 174–75; Foundation for the Advancement of Science and Education, *Exporting Risk: Pesticide Exports from U.S. Ports* (Los Angeles: Foundation for the Advancement of Science and Education, 1996); J. Wargo, *Our Children's Toxic Legacy: How Science and Law Fail to Protect Us from Pesticides* (New Haven, Conn.: Yale Univ. Press, 1996), 163–64; D. J. Hanson, "Administration Seeks Tighter Curbs on Exports of Unregistered Pesticides," *Chemical and Engineering News*, 14 Feb. 1994, 16–18; Monica Moore, Pesticide Action Network, personal communication.

9 uses of lindane: M. Moses, *Designer Poisons: How to Protect Your Health and Home from Toxic Pesticides* (San Francisco: Pesticide Education Center, 1995); EPA, *Suspended, Cancelled and Restricted Pesticides*, 20T-1002 (Washington, D.C.: EPA, 1990); Curtis, *After* Silent Spring.

9–10 aldrin and dieldrin: J. B. Barnett and K. E. Rodgers, "Pesticides," in J. H. Dean et al. (eds.), *Immunotoxicology and Immunopharmacology*, 2nd ed. (New York: Raven Press, 1994), 191–211; R. Spear, "Recognized and Possible Exposures to Pesticides," in W. J. Hayes and E. R. Laws Jr. (eds.), *Handbook of Pesticide Toxicology*, vol. 1. (New York: Academic Press, 1991), 245–46; EPA, 1990, *Suspended*; Carson, *Silent Spring*, 26.

10 chlordane and heptachlor: Spear, "Possible Exposures," 245; P. F. Infante et al., "Blood Dyscrasias and Childhood Tumors and Exposure to Chlordane and Heptachlor," *Scandinavian Journal of Work Environment and Health* 4 (1978): 137–50.

10 pesticides in baby food: Dunlap, *DDT*, 68.

10 women with breast cancer have higher levels of DDE and PCBs in their tumors: M. Wasserman, "Organochlorine Compounds in Neoplastic and Adjacent Apparently Normal Breast Tissue," *Bulletin of Environmental Contamination and Toxicology* 15 (1976): 478–84.

10–11 the Finnish study: H. Mussalo-Rauhamaa et al., "Occurrence of beta-Hexachlorocyclohexane in Breast Cancer Patients," *Cancer* 66 (1990): 2124–28. Lindane is the gamma isomer of hexachlorocyclohexane.

11 the Connecticut study: F. Falck Jr. et al., "Pesticides and Polychlorinated Biphenyl Residues in Human Breast Lipids and Their Relation to Breast Cancer," *AEH* 47 (1992): 143–46.

11 the New York City study: M. S. Wolff et al., "Blood Levels of Organochlorine Residues and Risk of Breast Cancer," *JNCI* 85 (1993): 648–52; D. J. Hunter and K. T. Kelsey, "Pesticide Residues and Breast Cancer: The Harvest of a Silent Spring?" *JNCI* 85 (1993): 598–99; M. P. Longnecker and S. J. London, "Re: Blood Levels of Organochlorine Residues and Risk of Breast Cancer" (letter and response by M. S. Wolff), *JNCI* 85 (1993): 1696–97.

11 the Québec study: É. Dewailly et al., "High Organochlorine Body Burden in Women with Estrogen Receptor–Positive Breast Cancer," *JNCI* 86 (1994): 232–34. Increasing incidence of receptor-positive breast cancer is

largely responsible for the increase in breast cancer rates that occurred between the mid-1970s and the mid-1980s. See A. G. Glassand and R. N. Hoover, "Rising Incidence of Breast Cancer: Relationship to State and Receptor Status," *JNCI* 82 (1990): 693–96.

12–13 the California study: N. Krieger et al., "Breast Cancer and Serum Organochlorines: A Prospective Study among White, Black and Asian Women," *JNCI* 86 (1994): 589–99; B. MacMahon, "Pesticide Residues and Breast Cancer?" *JNCI* 86 (1994): 572–73; S. S. Sternberg, "Re: DDT and Breast Cancer" (and responses by the authors), *JNCI* 86 (1994): 1094–96; J. E. Brody, "Strong Evidence in a Cancer Debate," *New York Times*, 20 Apr. 1994, p. C-11; D. A. Savitz, "Re: Breast Cancer and Serum Organochlorines: A Prospective Study among White, Black, and Asian Women," *JNCI* 86 (1994): 1255. Questions about the tubes' red caps have been raised by Dr. Devra Lee Davis.

13 breast cancer among women born between 1947 and 1958: D. L. Davis et al., "Decreasing Cardiovascular Disease and Increasing Cancer among Whites in the United States from 1973 through 1987: Good News and Bad News," *JAMA* 271 (1994): 431–37.

13 pesticide use since *Silent Spring*: Pesticide use doubled between 1964 and 1982, as measured by weight of active pesticidal ingredients. See Wargo, *Toxic Legacy*, 132.

13 failure to pursue research on cancer's environmental connections: See, for example, M. S. Wolff, "Pesticides—How Research Has Succeeded and Failed in Informing Policy: DDT and the Link with Breast Cancer," *EHP* 103, suppl. 6 (1995): 87–91.

TWO *silence*

15–16 Carson's concern about pesticide debates: L. J. Lear, "Rachel Carson's *Silent Spring*," *Environmental History Review* 17 (1993): 23–48. See also Lear's definitive biography, *Rachel Carson: Witness for Nature* (New York: Holt, 1997).

16 letter from Duxbury: T. T. Williams, "The Spirit of Rachel Carson," *Audubon* 94 (1992): 104–7; P. Brooks, *The House of Life: Rachel Carson at Work* (Boston: Houghton Mifflin, 1989), 229–35.

16 "Knowing what I do . . .": Carson's letter to Freeman, June 28, 1958, reprinted in M. Freeman (ed.), *Always, Rachel: The Letters of Rachel Carson and Dorothy Freeman* (Boston: Beacon, 1995), 259.

16–17 Iroquois County: Rachel Carson, *Silent Spring* (Boston: Houghton Mifflin, 1962), 91–100.

17 refusal of scientists to send Carson information: Dr. Linda Lear, personal communication.

17 threat of defunding: Carson, *Silent Spring*, 94–95.

17 "The other day . . .": Carson's letter to Freeman, 27 June 1962, reprinted in Freeman, *Always, Rachel*, 408.

17 Carson's speech to the Press Club is quoted in Brooks, *House of Life*, 302–4.

20 20 years of life lost: Dr. Devra Lee Davis, personal communication.

21 Carson's cancer diagnosis and physical ailments: Carson's letters to Freeman,

1960–1964, in Freeman, *Always, Rachel*; Brooks, *House of Life*; Dr. Linda Lear, personal communication.

21 Carson's relief at finishing *Silent Spring*: Carson's letter to Freeman, 6 Jan. 1962, in Freeman, *Always, Rachel*, 391.

21 two quotes from letters to Freeman: 3 Nov. 1963, and 9 Jan. 1964, ibid., 490, 515. See also letters dated 6 Jan. 1962; 2 Mar. 1963; and 25 Apr. 1963.

24 Carson's letters to Freeman that speak openly: 3 Jan. 1961; 23 Mar. 1961; 25 Mar. 1961; and 18 Sept. 1963, ibid., 326, 364, 365–66, 469.

24 letters that speak elliptically: 17 Jan. 1961; 15 Feb. 1961; 25 Oct. 1962; 25 Dec. 1962; and 2 Jan. 1964, ibid., 331, 346, 414, 420, 508.

24 Freeman's reference to Carson's mastectomy: Freeman's letter to Carson, 30 Apr. 1960, ibid., 305.

24–25 their entreaties and admissions: See, for example, Freeman's letter to Carson, 6 Mar. 1963, ibid., 441.

25 the darker story: Freeman's letters to Carson, 4 and 17 Mar. 1961, ibid., 356, 363.

25 confessions and recantations: Carson's letters to Freeman, 23 Jan. 1962; 26 Mar. 1962; 10 Apr. 1962; 14 Feb. 1963; 18 Feb. 1963; 2 Mar. 1963; 14 Jan. 1964, ibid., 395, 399, 404, 434–37, 439–40, 516.

25 Carson's prohibition of discussions about her health: M. Spock, "Rachel Carson: A Portrait," *Rachel Carson Council News* 82 (1994): 1–4; Dr. Linda Lear, George Washington University, personal communication.

25 quotes instructing Dorothy: Carson's letters to Freeman, 1 Apr. and 20 May 1962, in Freeman, *Always, Rachel*, 401, 405.

25–26 photographs and old film clips: Beinecke Library archives, Yale University; "Rachel Carson's *Silent Spring*," documentary film by Peace River Films, aired on PBS, *The American Experience*, 8 Feb. 1993.

27 farmers and housewives with cancer: Carson, *Silent Spring*, 227–30.

27 first line of evidence: Ibid., 219–20.

27–28 second and third lines of evidence: Ibid., 221.

28 "whatever seeds of malignancy . . .": Ibid., 226.

28 death certificates and children's cancers: Ibid., 221–22.

28 animals with cancer: Ibid., 222–39.

28–29 cellular mechanisms of carcinogenesis: Ibid., 231–35.

28 effect on sex hormones: Ibid., 235–37.

28 effect on metabolism: Ibid., 231–32.

28–29 Carson's prediction: Ibid., 232–33.

29 interspecies differences in susceptibility: H. C. Pitot III and Y. P. Dragan, "Chemical Carcinogens," in D. Klaassen (ed.), *Casarett and Doull's Toxicology: The Basic Science of Poisons*, 5th ed. (New York: McGraw-Hill, 1996), 248–49; NRC, *Animals as Sentinels of Environmental Health Hazards* (Washington, D.C.: National Academy Press, 1991).

29 uncontrolled human experiment: A lack of unexposed controls makes human studies difficult but not impossible. Theoretically, all that is required for such studies are measurable differences in exposure levels among segments of the human population. For example, all of us are believed to carry detectable levels of dioxin in our tissues. The question of whether dioxin

contributes to human cancers can be addressed by studies that compare cancer incidence rates among those heavily, moderately, and lightly exposed. All other things being equal, a positive trend would indicate a dose-response relationship, which is considered strong evidence by cancer researchers. The wider the spread in exposure levels, the more likely the relationship—if indeed one exists—will reveal itself. As such, researchers interested in conducting human studies often look for "natural experiments" where an unfortunate event—such as a toxic spill of some sort—has exposed an identifiable sector of the population to a heavy dose of the substance in question. Disease rates among this group can then be compared to those of the general population whose exposures to this substance may be common and ongoing but are occurring at much lower levels.

THREE *time*

Unless otherwise stated in the following notes, all statistics cited in this chapter on U.S. cancer incidence and mortality rates come from the National Cancer Institute's SEER Program Registry: L. A. G. Reis et al. (eds.), *SEER Cancer Statistics Review 1973–1991: Tables and Graphs,* NIH Pub. 94-2989 (Bethesda, Md.: NCI, 1994). Statistics on cancer incidence and mortality in Illinois come from H. L. Howe and M. Lehnherr, *Incidence in Illinois by County, 1986–1990,* Epidemiological Report, ser. 92, no. 4 (Springfield, Ill.: IDPH, 1992). Statistics on cancer incidence and mortality in Massachusetts come from S. Gershman, *Cancer Incidence in Massachusetts, 1982–1990* (Boston: MDPH, Massachusetts Cancer Registry, 1993).

32 number of cancer diagnoses in 1995: ACS, *Cancer Facts and Figures—1995* (Atlanta: ACS, 1995).

35–36 uncertainties in data ascertainment: H. Menck and C. Smart (eds.), *Central Cancer Registries: Design, Management, and Use* (Chur, Switzerland: Harwood Academic Press, 1994); O. M. Jensen et al. (eds.), *Cancer Registration: Principles and Methods,* IARC Scientific Publication 95 (Lyon, France: IARC, 1991).

35 *Cancer Registry News* is a publication of the Massachusetts Cancer Registry.

36 percentage of recent upsurge in breast cancer attributable to earlier detection: R. N. Proctor, *Cancer Wars: How Politics Shapes What We Know and Don't Know about Cancer* (New York: Basic Books, 1995), 251; J. M. Liff, "Does Increased Detection Account for the Rising Incidence of Breast Cancer?" *AJPH* 81 (1991): 462–65.

36 rise in breast cancer that predates mammography: E. J. Feuer and L.-M. Wun, "How Much of the Recent Rise in Breast Cancer Incidence Can Be Explained by Increases in Mammography Utilization?" *AJE* 136 (1992): 1423–36; J. R. Harris, "Breast Cancer," *NEJM* 327 (1992): 319–28.

37–38 For an excellent history of cancer registration in the United States, see E. R. Greenberg et al., "Measurements of Cancer Incidence in the United States: Sources and Uses of Data," *JNCI* 68 (1982): 743–49. For an overview of the system of state registries, see USDHHS, *A National Pro-*

gram of Cancer Registries At-a-Glance, 1994– 1995 (Atlanta: CDC, 1995). For an accessible discussion of the problems with state cancer registries prior to the Cancer Registries Amendment Act of 1992, see J. H. Healey, "The Cancer Weapon America Needs Most," *Reader's Digest*, June 1992, 69–72.

37 data exchange in Illinois: H. L. Howe et al., *Effect of Interstate Data Exchange on Cancer Rates in Illinois, 1986–1990*, Epidemiological Report Series, 94:1 (Springfield, Ill.: IDPH, 1994).

37 percentage of populace sampled by SEER: NCI, *The Nation's Investment in Cancer Research: A Budget Proposal for Fiscal Year 1997–98* (Washington, D.C.: NCR, 1996).

37–38 SEER is not necessarily representative of the entire U.S. population in age and ethnic makeup. See C. Fry et al., "Representativeness of the Surveillance, Epidemiology and End Results Programs Data: Recent Trends in Cancer Mortality Rate," *JNCI* 84 (1992): 872–77.

38 mortality rates as more reliable: The renowned biostatistician Dr. John Bailar, for example, holds this view, which he discusses in J. C. Bailar III and E. M. Smith, "Progress against Cancer?" *NEJM* 314 (1986): 1226–32, and more recently at an evaluation of the National Cancer Program, as described in J. C. Bailar III, "Observations on Some Recent Trends in Cancer" (presentation at the President's Cancer Panel Meeting, NIH, Bethesda, Md., 22 Sept. 1993).

38–39 childhood cancers: L. L. Robison et al., "Assessment of Environmental and Genetic Factors in the Etiology of Childhood Cancers: The Children's Cancer Group Epidemiology Program," *EHP* 103, suppl. 6 (1995): 111–16; S. H. Zahm and S. S. Devesa, "Childhood Cancer: Overview of Incidence Trends and Environmental Carcinogens," *EHP* 103, suppl. 6 (1995): 177–84.

39 greater exposure of children: J. Wargo, *Our Children's Toxic Legacy: How Science and Law Fail to Protect Us from Pesticides* (New Haven, Conn.: Yale University Press, 1996); L. Mott et al., *Handle with Care: Children and Environmental Carcinogens* (New York: NRDC, 1994).

40 40 percent of Americans will contract cancer: NCI, *Nation's Investment*, 2, and E. J. Feuer, "Lifetime Probability of Cancer," *JNCI* 89 (1997): 279.

40 cancer as a leading cause of death: J. L. Cresanta, "Epidemiology of Cancer in the United States," *Primary Care* 19 (1992): 419–41.

40 increases in cancer incidence are seen in all age groups: S. S. Devesa et al., "Recent Cancer Trends in the United States," *JNCI* 87 (1995): 175–82.

41 decline in cancer mortality between 1991 and 1995: P. Cole and B. Rodu, "Declining Cancer Mortality in the United States," *Cancer* 78 (1996): 2045–48.

41 one-fourth are lung cancer deaths: T. Beardsley, "A War Not Won," *Scientific American*, Jan. 1994, 130–38.

41 lung cancer rates rising among American women: Ibid.

41 percentage of lung cancer deaths due to smoking: Cresanta, "Epidemiology of Cancer."

41 lung cancer among nonsmokers: S. S. Epstein, "Profiting from Cancer: Vested Interests and the Cancer Epidemic," *The Ecologist* 22 (1992): 233–39.

41 rise in testicular cancer: R. Bergstrom et al., "Increase in Testicular Cancer Incidence in Six European Countries: A Birth Cohort Phenomenon," *JNCI* 88 (1996): 727–33; NCI, *Cancer Rates and Risks*, 4th ed., NIH Pub. 96-691 (Bethesda, Md.: NCI, 1996), 194–96; G. R. Bunin et al., "Carcinogenesis," in M. Paul (ed.), *Occupational and Environmental Reproductive Hazards: A Guide for Clinicians* (Baltimore: Williams & Wilkins, 1993), 76–88.

41 Only part of the observed increase in brain cancer incidence among the elderly is attributable to changing diagnostic techniques. A. P. Polednak, "Time Trends in Incidence of Brain and Central Nervous System Cancers in Connecticut," *JNCI* 83 (1991): 1679–81.

41–42 decline of stomach cancer: D. G. Hoel et al., "Trends in Cancer Mortality in Fifteen Industrialized Countries, 1969–86," *JNCI* 84 (1992): 313–20.

42 decline of cervical cancer: "Lack of Test Tied to Cervical Cancer," *New York Times*, 7 June 1995, p. C-12.

42 decline in breast cancer mortality among whites: K. C. Chu, "Recent Trends in U.S. Breast Cancer Incidence, Survival, and Mortality Rates," *JNCI* 88 (1996): 1571–79; K. Smigel, "Breast Cancer Death Rates Decline for White Women," *JNCI* 87 (1995): 173.

42 quote by Philip Landrigan: P. J. Landrigan, "Commentary: Environmental Diseases—A Preventable Epidemic," *AJPH* 82 (1992): 941–43.

42 1995 assessment: Devesa, "Recent Cancer Trends."

42–43 quote by Landrigan: Landrigan, "Environmental Diseases."

43 quotes by Hueper and Conway: W. C. Hueper and W. D. Conway, *Chemical Carcinogenesis and Cancers* (Springfield, Ill.: Charles Thomas, 1964), 17, 158.

43 quotes about the ISCR: IDPH, *Cancer Incidence in Illinois by County, 1985–87* (Springfield, Ill.: IDPH, 1989).

45 report of the National Cancer Advisory Board: P. Calabresi et al., *Cancer at a Crossroads: A Report to the Congress for the Nation* (Bethesda, Md.: NCI, 1994), 6, 17, 21, B-6.

45 birth cohort study: D. L. Davis, "Decreasing Cardiovascular Disease and Increasing Cancer among Whites in the United States from 1973 through 1987: Good News and Bad News," *JAMA* 271 (1994): 431–37. These results have been replicated in Sweden, where researchers, making use of one of the world's oldest and most reliable cancer registries, have shown increasing cancer rates extending into the 1950s birth cohort: H-O Adami et al., "Increasing Cancer Risk in Younger Birth Cohorts in Sweden," *Lancet* 341 (1993): 773–77.

45 quote by Davis: personal communication.

47 melanoma accounts for 75 percent of deaths: NCI, *Cancer Rates and Risks*, 163.

48 incidence in whites is tenfold that in blacks: B. K. Armstrong and D. R. Dallas, "Cutaneous Malignant Melanoma," in D. Schottenfeld and J. F. Fraumeni Jr. (eds.), *Cancer Epidemiology and Prevention*, 2nd ed. (Oxford, England: Oxford University Press, 1996), 1282–1312.

48 location of melanoma in men and women: ACS, *Facts on Skin Cancer*, 88-400M-Rev.5/93-No.2049 (Atlanta: ACS, 1988).

48 association with UV radiation: R. Marks, "Prevention and Control of Melanoma: The Public Health Approach," *CA* 46 (1996): 199–216; H. K. Koh et al., "Etiology of Melanoma," *Cancer Treatment and Research* 65 (1993): 1–28.

48–49 rubber, plastics, electronics, and metal industries: P. J. Nelemans, "Melanoma and Occupation: Results of a Case-Control Study in the Netherlands," *British Journal of Industrial Medicine* 50 (1993): 642–46; N. E. L. Hall and K. D. Rosenman, "Cancer by Industry: Analysis of a Population-Based Cancer Registry with an Emphasis on Blue-Collar Workers," *American Journal of Industrial Medicine* 19 (1991): 145–59.

49 disappearing ozone, UV exposure, and melanoma rates: D. S. Rigel et al., "The Incidence of Malignant Melanoma in the United States: Issues as We Approach the 21st Century," *Journal of the American Academy of Dermatology* 34 (1996): 839–47; J. R. Herman et al., "UV-B Increases (1979–1992) from Decreases in Total Ozone," *Geophysical Research Letters* 23 (1996): 2117–20; J. D. Longstreth et al., "Effects of Increased Solar Radiation on Human Health," *Ambio* 24 (1995): 153–65; A. Leaf, "Loss of Stratospheric Ozone and Health Effects of Increased Ultraviolet Light," in E. Chivian et al. (eds.), *Critical Condition: Human Health and the Environment* (Cambridge, Mass.: MIT Press, 1993), 139–50; B. Coldiron, "Thinning of the Ozone Layer: Facts and Consequences," *Journal of American Academy of Dermatology* 27 (1992): 653–62;

49 number of fatal skin cancers due to ozone loss: N. Wright, *Environmental Sciences*, 4th ed. (Englewood Cliffs, N.J.: Prentice Hall, 1993), 379; A. A. Skolnick, "Is Ozone Loss to Blame for Melanoma Upsurge?" *JAMA* 265 (1991): 3218; EPA, *Ultraviolet Radiation and Melanoma: With a Special Focus on Assessing the Risks of Stratospheric Ozone Depletion*, EPA 400/1-87/001D (Washington, D.C.: EPA, Office of Air and Radiation, 1987).

49 Coco Chanel: R. M. Mackie, "Malignant Melanoma—The Story Unfolds," in T. Heller et al. (eds.), *Preventing Cancers* (Buckingham, England: Open University Press, 1992), 68–77.

49 quote from the *Journal of the American Academy of Dermatology*: Rigel, "The Incidence of Malignant Melanoma," 842.

49–50 Some chlorine-containing compounds, such as CFCs, that deplete the ozone layer have been banned from manufacture under the terms of the Montreal Protocol on Substances that Deplete the Ozone Layer. CFCs are not the only class of ozone depleters. The pesticide methyl bromide, for example, is thought to play a significant role.

51–52 AIDS and non-Hodgkin's lymphoma: L. K. Altman, "Lymphomas Are on the Rise in the U.S., and No One Knows Why," *New York Times*, 24 May 1994, p. C-3; P. Hartge et al., "Hodgkin's and non-Hodgkin's Lymphomas," in R. Doll et al. (eds.), *Trends in Cancer Incidence and Mortality*, Cancer Surveys 19/20 (Plainview, N.Y.: Cold Spring Harbor Laboratory Press, 1994), 423–53.

52 lymphomas and phenoxy herbicides: S. H. Zahm and A. Blair, "Pesticides and Non-Hodgkin's Lymphoma," *Cancer Research* 52 (1992 suppl.): 5485s–88s. Non-Hodgkin's lymphoma is also associated with exposure to organophosphate and organochlorine insecticides. See S. H. Zahm, "The Role of

Agricultural Pesticide Use in the Development of Non-Hodgkin's Lymphoma in Women," *AEH* 48 (1993): 253–58.

52 military history of phenoxy herbicides: D. E. Lilienfeld and M. A. Gallo, "2,4-D, 2,4,5-T, and 2,3,7,8-TCDD: An Overview," *Epidemiologic Reviews* 11 (1989): 28–58.

52 trade names: Shirley Briggs of the Rachel Carson Council in Chevy Chase, Md., has compiled detailed descriptions of the properties of some 700 pesticides, including a list of their trade names. See S. A. Briggs, *Basic Guide to Pesticides: Their Characteristics and Hazards* (Bristol, Penn.: Taylor & Francis, 1992).

52 evidence for an association: P. Hartge et al., "Hodgkin's and Non-Hodgkin's Lymphomas," in Doll, *Trends in Cancer Incidence*, 423–53; Institute of Medicine, *Veterans and Agent Orange: Health Effects of Herbicides Used in Vietnam* (Washington, D.C.: National Academy Press, 1994); D. D. Weisenburger, "Epidemiology of Non-Hodgkin's Lymphoma: Recent Findings Regarding an Emerging Epidemic," *Annals of Oncology* 1, suppl. 5 (1994): s19–s24; Zahm and Blair, "Pesticides and Non-Hodgkin's Lymphoma"; S. Zahm et al., "A Case-Control Study of Non-Hodgkin's Lymphoma and the Herbicide 2,4-Dichlorophenoxyacetic Acid (2,4-D) in Eastern Nebraska," *Epidemiology* 1 (1990): 349–56; L. Hardell et al., "Malignant Lymphoma and Exposure to Chemicals, Especially Organic Solvents, Chlorophenols and Phenoxy Acids: A Case-Control Study," *British Journal of Cancer* 43 (1981): 169–76.

52–53 quote by Zahm and Blair: Zahm and Blair, "Pesticides and Non-Hodgkin's Lymphoma," 5487s.

53 quote from study of Vietnam veterans: Institute of Medicine, *Veterans and Agent Orange*, 6.

53 lymphoma in dogs: H. M. Hayes et al., "Case-Control Study of Canine Malignant Lymphoma: Positive Association with Dog Owner's Use of 2,4-Dichlorophenoxyacetic Acid Herbicides," *JNCI* 83 (1991): 1226–31.

53 percentage of households using herbicides for lawn care: Zahm and Blair, "Pesticides and Non-Hodgkin's Lymphoma," 5487s.

54 multiple myeloma, blacks, and the elderly: J. Higginson et al., "Multiple Myeloma and Macroglobulimenia," in J. Higginson et al. (eds.), *Human Cancer: Epidemiology and Environmental Causes* (Cambridge, England: Cambridge Univ. Press, 1992), 465–75; NCI, "What You Need to Know about Multiple Myeloma," NIH Pub. 93-1575 (Bethesda, Md.: NCI, 1992).

54 Ascertainment of multiple myeloma mortality, as provided by death certificate data, is also thought to be particularly accurate. See L. J. Herrington et al., "Multiple Myeloma," in Schottenfeld and Fraumeni, *Cancer Epidemiology*; B. D. Goldstein, "Is Exposure to Benzene a Cause of Multiple Myeloma?" in D. L. Davis and D. Hoel (eds.), *Trends in Cancer Mortality in Industrialized Countries*, Annals, vol. 609 (New York: New York Academy of Sciences, 1990), 225–30.

54 association with radiation exposure: Higginson, "Macroglobulimenia," 467; NCI, "What You Need"; L. Tomatis, *Cancer: Causes, Occurrence and Control*, IARC Scientific Pub. 100 (Lyon, France: IARC, 1990), 159.

55 association with chemicals: Herrinton, "Multiple Myeloma"; S. H. Zahm et

al., "Pesticides and Multiple Myeloma in Men and Women in Nebraska," in H. H. McDuffie et al. (eds.), *Supplement to Agricultural Health and Safety: Workplace, Environment, Sustainability* (Saskatoon, Saskatchewan, Canada: Univ. of Saskatchewan Press, 1995), 75–81; S. H. Zahm and A. Blair, "Cancer among Migrants and Seasonal Farmworkers: An Epidemiological Review and Research Agenda," *American Journal of Industrial Medicine* 24 (1993): 753–66; Institute of Medicine, *Veterans and Agent Orange*, 528; M. Eriksson and M. Karlsson, "Occupational and Other Environmental Factors and Multiple Myeloma: A Population Based Case-Control Study," *British Journal of Industrial Medicine* 49 (1992): 95–103.

55 multinational mortality trends: J. Schwartz, "Multinational Trends in Multiple Myeloma," in Davis and Hoel, *Trends in Cancer Mortality*, 215–24.

55 multiple myeloma and benzene: B. D. Goldstein, "Is Exposure to Benzene a Cause of Human Multiple Myeloma?" in Davis and Hoel, *Trends in Cancer Mortality*, 215–24.

55 For a comprehensive description of the health effects of benzene, including its role in leukemia, see U.S. Agency for Toxic Substances and Disease Registry, *Toxicological Profile for Benzene* (Atlanta: USDHHS, 1993).

56 quote from the ATSDR: ATSDR, *Case Studies in Environmental Medicine: Benzene Toxicity* (Atlanta: USDHHS, 1992), 7.

56 leukemia in southeastern Massachusetts: M. S. Morris and R. S. Knorr, "Adult Leukemia and Proximity-Based Surrogates for Exposure to Pilgrim Plant's Nuclear Emissions," *AEH* 51 (1996): 266–74; M. S. Morris and R. S. Knorr, *Southeastern Massachusetts Health Study Final Report: Investigation of Leukemia Incidence in 22 Massachusetts Communities, 1978–86* (Boston: MDPH, 1990); L. Tye, "Screening Sought in Cancer Link to Pilgrim," *Boston Globe*, 19 Sept. 1989, pp. 21, 25; R. W. Clapp et al., "Leukaemia Near Massachusetts Nuclear Power Plant," *Lancet* 1987: 1324–25.

56 quote about leukemia risk: MDPH, *Southeastern Massachusetts*, 2.

FOUR *space*

58 history and environmental problems of Normandale: T. L. Aldous, "Community Dreads Threat of Disease," *PDT*, 14 Sept. 1991, pp. A-2, A-12.

59 half of cancers occur in industrialized countries: D. L. Davis et al., "International Trends in Cancer Mortality in France, West Germany, Italy, Japan, England and Wales and the U.S.A.," *Lancet* 336 (1990): 474–81.

59 geography of breast cancer: J. L. Kelsey and P. L. Horn-Ross, "Breast Cancer: Magnitude of the Problem and Descriptive Epidemiology," *Epidemiologic Reviews* 15 (1993): 7–16; P. Pisani, "Breast Cancer: Geographic Variations and Risk Factors, *Journal of Environmental Pathology, Toxicology and Oncology* 11 (1992): 313–16.

59 replication of rising rates in Europe and Japan: D. G. Hoel et al., "Trends in Cancer Mortality in Fifteen Industrialized Countries, 1969–86," *JNCI* 84 (1992): 313–20; D. L. Davis and D. Hoel (eds.), *Trends in Cancer Mortality in Industrialized Countries*, Annals, vol. 609 (New York: New York Academy of Sciences, 1990), 5–48.

59 brain cancer among the elderly: A. Ahlbom, "Some Notes on Brain Tumor

Epidemiology," in Davis and Hoel, *Trends in Cancer Mortality*, 179–85; D. L. Davis et al., "Is Brain Cancer Mortality Increasing in Industrialized Countries?" in ibid., 191–204.

60 Every few years IARC publishes a monograph entitled *Cancer Incidence in Five Continents*.

60 role of the WHO: Davis and Hoel, *Trends in Cancer Mortality*, 1–2; L. Tomatis, *Cancer: Causes, Occurrence and Control*, IARC Scientific Pub. 100 (Lyon, France: IARC, 1990), 21.

60 80 percent of cancer attributable to the environment: For a detailed discussion of the historical origins of this statistic, see R. N. Proctor, *Cancer Wars: How Politics Shapes What We Know and Don't Know about Cancer* (New York: Basic Books, 1995), 54–74; and J. T. Patterson, *The Dread Disease: Cancer and Modern American Culture* (Cambridge, Mass.: Harvard Univ. Press, 1987), 284–85.

60 what *environment* means in this context: See, for example, J. F. Fraumeni Jr., "Epidemiologic Approaches to Cancer Etiology," *Annual Review of Public Health* 3 (1982): 85–100.

61 quote from IARC: Tomatis, *Cancer: Causes*, 48.

61–62 migrants and changing cancer risks: E. V. Kliewar and K. R. Smith, "Breast Cancer Mortality among Immigrants in Australia and Canada," *JNCI* 87 (1995): 1154–61; N. Angier, "Woman's Move Can Change Her Risk of Breast Cancer," *New York Times*, 2 Aug. 1995, p. A-17; H. Shimizu et al., "Cancers of the Prostate and Breast among Japanese and White Immigrants in Los Angeles County," *British Journal of Cancer* 63 (1991): 963–66; Tomatis, *Cancer: Causes*; D. B. Thomas and M. R. Karagas, "Cancer in First and Second Generation Americans," *Cancer Research* 47 (1987): 5771–76.

62 cancer in Normandale: Aldous, "Community Dreads Threat."

62 quote from Normandale resident: Ibid.

63 cancer death maps: L. W. Pickle et al., *Atlas of U.S. Cancer Mortality among Whites: 1950–1980*, (NIH) 87-2900 (Washington, D.C.: GPO, 1987); L. W. Pickle et al., *Atlas of U.S. Cancer Mortality among Nonwhites: 1950–1980*, (NIH) 90-1582 (Washington, D.C.: GPO, 1990).

63 nonrandom distribution of cancer deaths: L. W. Pickle et al., "The New United States Cancer Atlas," *Recent Results in Cancer Research* 14 (1989): 196–207.

63 The appearance of "clusters" in geographic maps depends on color patterns and scale. See, for example, S. Lewandowsky et al., "Perception of Clusters in Statistical Maps," *Applied Cognitive Psychology* 7 (1993): 533–51.

63–64 1988 study: C. S. Stokes and K. D. Brace, "Agricultural Chemical Use and Cancer Mortality in Selected Rural Counties in the U.S.A.," *Journal of Rural Studies* 4 (1988): 239–47.

63–64 Public Data Access maps: B. A. Goldman, *The Truth about Where You Live: An Atlas for Action on Toxins and Mortality* (New York: Random House, 1991).

64 geographic patterns of specific cancers: In addition to the two cancer atlases cited above, see also S. S. Devesa, "Recent Cancer Patterns among Men

and Women in the United States: Clues for Occupational Research," *Journal of Occupational Medicine* 36 (1994): 832–41; and S. H. Zahm et al., "Pesticides and Multiple Myeloma in Men and Women in Nebraska," in H. H. McDuffie et al. (eds.), *Supplement to Agricultural Health and Safety: Workplace, Environment, Sustainability* (Saskatoon, Saskatchewan, Canada: Univ. of Saskatchewan Press, 1995), 75–81; J. L. Kelsey and P. L. Horn-Ross, "Breast Cancer: Magnitude of the Problem and Descriptive Epidemiology," *Epidemiologic Reviews* 15 (1993): 7–16; Pickle, "New United States."

64–65 workplace carcinogens: P. F. Infante, "Cancer and Blue-Collar Workers: Who Cares?" *New Solutions* (winter 1995): 52–57; J. Randal, "Occupation as a Carcinogen: Federal Researcher Suggests Change in Cancer Registries," *JNCI* 86 (1994): 1748–50; J. Landrigan, "Cancer Research in the Workplace" (presentation at the President's Cancer Panel meeting, NIH, Bethesda, Md., 22 Sept. 1993).

65 high cancer rates in farmers: S. H. Zahm and A. Blair, "Cancer among Migrant and Seasonal Farmers," *American Journal of Industrial Medicine* 24 (1993): 753–66; A. Blair et al., "Clues to Cancer Etiology from Studies of Farmers," *Scandinavian Journal of Work Environment and Health* 18 (1992): 209–15; D. L. Davis et al., "Agricultural Exposures and Cancer Trends in Developed Countries," *EHP* 100 (1992): 39–44. Excesses in non-Hodgkin's lymphoma and brain cancer did not always attain statistical significance.

65 other occupations with high cancer rates: G. Tornling et al., "Mortality and Cancer Incidence in Stockholm Firefighters," *American Journal of Industrial Medicine* 25 (1994): 219–28; E. L. Hall and K. D. Rosenman, "Cancer by Industry: Analysis of a Population-Based Cancer Registry with an Emphasis on Blue-Collar Workers," *American Journal of Industrial Medicine* 19 (1991): 145–59; Davis and Hoel, *Trends in Cancer Mortality*, x.

65 professional jobs with high cancer rates: E. A. Holly, "Intraocular Melanoma Linked to Occupations and Chemical Exposure," *Epidemiology* 7 (1996): 55–61; B. B. Arnetz et al., "Mortality among Petrochemical Science and Engineering Employees," *AEH* 46 (1991): 237–48.

65 cancer in dentists, dental assistants, and chemotherapy nurses: L. M. Pottern et al., "Occupational Cancer among Women: A Conference Overview," *Journal of Occupational Medicine* 36 (1994): 809–13.

65–66 children's cancers related to parental exposures: L. M. O'Leary et al., "Parental Exposures and Risk of Childhood Cancer: A Review," *American Journal of Industrial Medicine* 20 (1991): 17–35.

66 gender imbalance: W. J. Nicholson and D. L. Davis, "Analysis of Changes in the Ratios of Male-to-Female Cancer Mortality: A Hypothesis-Generating Exercise," in Davis and Hoel, *Trends in Cancer Mortality*, 290–99.

66 cancer among women hairdressers, women autoworkers, and women in traditionally female-held jobs: Pottern, "Occupational Cancer"; J. M. Stellman, "Where Women Work and the Hazards They May Face on the Job," *Journal of Occupational Medicine* 36 (1994): 814–25.

66–67 Cancer-causing properties of vinyl chloride: ATSDR, *Case Studies in Environmental Medicine: Vinyl Chloride Toxicity* (Atlanta: ATSDR, 1990).

67 a 1977 study: L. Chiazze et al., "Mortality among Employees of PVC Fabricators," *Journal of Occupational Medicine* 19 (1977): 623–28. See also P. F. Infante and J. Pesak, "A Historical Perspective of Some Occupationally Related Diseases of Women," *Journal of Occupational Medicine* 36 (1994): 826–31.

67 exposure of general populace to vinyl chloride: ATSDR, *Vinyl Chloride Toxicity*; IEPA, "Vinyl Chemical Information Sheet," IEPA/ENV/87-001-11 (Springfield, Ill.: IEPA, 1987); J. Gilbert et al., "Identification by Gas Chromatography–Mass Spectrometry of Vinyl Chloride Oligomers and Other Low-Molecular-Weight Components in Poly(Vinyl Chloride) Resins for Food Packaging Applications," *Journal of Chromatography* 237 (1982): 249–61.

67 quotes from the ATSDR: ATSDR, *Vinyl Chloride Toxicity*, 4.

68 lack of follow-up studies on vinyl chloride and women: Infante and Pesak, "Historical Perspective."

68 interactive effects of vinyl chloride and ethanol: ATSDR, *Vinyl Chloride Toxicity.*

68–69 two health studies in Normandale: IDPH, "Incidence of Cancer in Pekin (Tazewell County), Illinois" (Springfield, Ill.: IDPH, 1991); G. Poquette, "Normandale Cancer Study" (memorandum) (Tremont, Ill.: Tazewell County Health Department, 5 Mar. 1992.

69 headline: T. L. Aldous, "Study: Area Cancer Rates Normal," *PDT*, 19 Dec. 1991, pp. A-2, A-12.

69 National Research Council study: NRC, *Environmental Epidemiology: Public Health and Hazardous Wastes* (Washington, D.C.: National Academy Press, 1991).

69 ATSDR report: ATSDR, *ATSDR Biennial Report to Congress 1991 and 1992* (executive statement) (Atlanta: ATSDR, 1992). Quotes are from pp. 2, 3, and 7.

69–70 Congress enacted legislation on hazardous waste sites in 1980 when it passed the Comprehensive Environmental Response, Compensation and Liability Act (CERCLA), which is generally known as Superfund. The goals of the bill are to inventory hazardous waste sites, to establish priorities for cleanup based on relative danger, to contain dangerous releases, and ultimately to remediate through elimination of unsafe sites. The nomenclature surrounding hazardous waste is confusing. Waste sites appearing on the National Priorities List are called Superfund sites, whereas waste sites inventoried under the program but not on the NPL are commonly referred to as CERCLA sites.

70 number of people living near Superfund sites: NRC, *Environmental Epidemiology*, 2.

70 750 million tons: J. Griffith and W. B. Riggan, "Cancer Mortality in U.S. Counties with Hazardous Waste Sites and Ground Water Pollution," *AEH* 44 (1989): 69–74.

70 quote from the NRC: NRC, *Environmental Epidemiology*, 19.

71 cancer in New Jersey: G. R. Najem et al., "Female Reproductive Organs and Breast Cancer Mortality in New Jersey Counties and the Relationship with Certain Environmental Variables," *Preventive Medicine* 14 (1985): 620–35; G. R. Najem et al., "Clusters of Cancer Mortality in New Jersey Municipalities, with Special Reference to Chemical Toxic Waste Disposal Sites and Per Capita Income," *International Journal of Epidemiology* 14 (1985): 528–37; G. R. Najem et al., "Gastrointestinal Cancer Mortality in New Jersey Counties and the Relationship to Environmental Variables," *International Journal of Epidemiology* 12 (1983): 276–89.

71 cancer in counties with hazardous waste sites and groundwater contamination: Griffith and Riggan, "Cancer Mortality in U.S. Counties." See also R. Hoover and J. F. Fraumeni Jr., "Cancer Mortality in U.S. Counties with Chemical Industries," *Environmental Research* 9 (1975): 196–207.

71–72 Public Data Access: Goldman, *Where You Live*, 116.

72 Ecological fallacy can also refer to the mistake of applying group attributes to individuals. For example, in a famous study of suicide and religion in nineteenth-century Europe, researchers found that suicide rates rose as the proportion of Protestants living in a given region increased. The obvious conclusion—that Protestants are more likely to kill themselves than Catholics—does not necessarily follow, however. It is entirely possible that all the suicides in Protestant-dominated areas had occurred among Catholics: perhaps people become increasingly vulnerable to suicide as they become an increasingly isolated minority. This latter explanation turned out not to be the case, but an ecological study of groups could not distinguish between these two mutually exclusive conclusions. For a lively discussion of this study, see J. Esteve et al., *Descriptive Epidemiology: Statistical Methods in Cancer Research*, vol. IV (Lyon, France: IARC Scientific Pub. No. 128, 1994), 150–54.

73 quote from Peter Infante: P. F. Infante and G. K. Pohl, "Living in a Chemical World: Actions and Reactions to Industrial Carcinogens," *Teratogenesis, Carcinogenesis and Mutagenesis* 8 (1988): 225–49.

73–74 For an introduction to epidemiological methods and, from a variety of viewpoints, their limitations, see K. J. Rothman and C. Poole, "Causation and Causal Inference," in D. Schottenfeld and J. F. Fraumeni Jr. (eds.), *Cancer Epidemiology and Prevention*, 2nd ed. (Oxford, England: Oxford Univ. Press, 1996), 3–10; N. Krieger, "Epidemiology and the Web of Causation: Has Anyone Seen the Spider?" *Social Science and Medicine* 39 (1994): 887–903; D. Trichopoulos and E. Petridou, "Epidemiologic Studies and Cancer Etiology in Humans," *Medicine, Exercise, Nutrition, and Health* 3 (1994): 206–25; S. Wing, "Limits of Epidemiology," *Medicine and Global Survival* 1 (1994): 74–86; M. S. Legator and S. F. Strawn (eds.), *Chemical Alert! Community Action Handbook* (Austin: Texas Univ. Press, 1993).

74 Mary Wolff's study: M. S. Wolff et al., "Blood Levels of Organochlorine Residues and Risk of Breast Cancer," *JNCI* 85 (1993): 648–52.

74–78 the vexations of cluster studies: See the supplemental issue of the *AJE* 132 (1990), which contains the proceedings of the National Conference on Clustering of Health Events, held in Atlanta, Ga., 16–17 Feb. 1989.

See also G. Taubes, "Epidemiology Faces Its Limits," *Science* 269 (1995): 164–69; Legator and Strawn, *Chemical Alert*; CDC, "Guidelines for Investigating Clusters of Health Events," *Morbidity and Mortality Weekly Report* 39/RR-11 (1990): 1–23; and K. J. Rothman, "Clustering of Disease," *AJPH* 77 (1987): 13–15.

75–76 Power and significance can be described in several ways. One of the most commonly used measures in epidemiology is the confidence interval—a computed range with a given probability (95 percent) that the true value of the variable lies within it. Further explanations of epidemiological statistics for laypersons can be found in M. J. Scott and B. L. Harper, "Lots of Information: What to Do with It. Statistics for Nonstatisticians," in Legator and Strawn, *Chemical Alert*.

76 eight to twenty times higher: R. R. Neutra, "Counterpoint from a Cluster Buster," *AJE* 132 (1990): 1–8.

76–77 TCE: ATSDR, *Case Studies in Environmental Medicine: Trichloroethylene Toxicity* (Atlanta, Ga.: ATSDR, 1992).

77–78 the eleven blue men: B. Roueche, *Eleven Blue Men and Other Narratives of Medical Detection* (Boston: Little, Brown, 1954). The significance of this case study for cancer clusters is discussed by Neutra, "Counterpoint."

78 two excesses attained statistical significance: This reanalysis was conducted by Richard Clapp, an epidemiologist at Boston University and the former director of the Massachusetts Cancer Registry.

79 Long Island: E. L. Lewis-Michl et al., "Breast Cancer Risk and Residence Near Industry or Traffic in Nassau and Suffolk Counties, Long Island, New York," *AEH* 51 (1996): 255–65; J. Melius et al., "Residence Near Industries and High Traffic Areas and the Risk of Breast Cancer on Long Island" (Albany: New York State Dept. of Health, 1994); D. J. Schemo, "Long Island Breast Cancer Is Possibly Linked to Chemical Sites," *New York Times*, 13 Apr. 1994, pp. A-1, B-6.

79 after five years of study: New York State Department of Health, Department of Community and Preventative Medicine at the State University of New York in Stony Brook, Nassau County Department of Health, and Suffolk County Department of Health Services, *The Long Island Breast Cancer Study*, Reports 1–3 (1988–1990).

79–80 CDC: G. Kolata, "Long Island Breast Cancer Called Explainable by U.S.," *New York Times*, 19 Dec. 1992, p. A-9.

80 quote by Joan Swirsky: J. Swirsky, "Who's Minding the Store? Breast Cancer Study Begins," *Merrick Life*, 15 Sept. 1994.

81 history of the Upper Cape: S. Rolbein, *The Enemy Within: The Struggle to Clean Up Cape Cod's Military Superfund Site* (Orleans, Mass.: Association for the Preservation of Cape Cod, 1995).

81 cancer rates in the Upper Cape: MDPH, *Cancer Incidence in Massachusetts, 1982–90* (Boston: MDPH, 1993).

81 1991 study: A. Aschengrau and D. M. Ozonoff, *Upper Cape Cancer Incidence Study. Final Report* (Boston: Mass. Depts. of Public Health and Environmental Protection, 1991).

81–82 Silent Spring Institute: J. G. Brody et al., "Mapping Out a Search for Environmental Causes of Breast Cancer, *Public Health Reports* 111 (1996): 495–507; "Cape Cod Breast Cancer and Environment Study Overview" (Newton, Mass.: Silent Spring Institute, July 12, 1995).

82 quote from 1991 study: Aschengrau and Ozonoff, *Upper Cape*, ix.

83 Cape Cod water pipes: A. Aschengrau et al., "Cancer Risk and Tetrachloroethylene-Contaminated Drinking Water in Massachusetts," *AEH* 48 (1993): 284–92; T. Webster and H. S. Brown, "Exposure to Tetrachloroethylene via Contaminated Drinking Water Pipes in Massachusetts: A Predictive Model," *AEH* 48 (1993): 293–97.

84 dry-cleaners: N. S. Weiss, "Cancer in Relation to Occupational Exposure to Perchloroethylene," *Cancer Causes and Control* 6 (1995): 257–66.

84 1983 study: C. D. Larsen et al., "Tetrachloroethylene Leached from Lined Asbestos-Cement Pipe into Drinking Water," *Journal of the American Water Works Association* 75 (1983): 184–88.

84 quote from 1993 study: Aschengrau, "Tetrachloroethylene-Contaminated," 291.

84–85 Normandale: T. L. Aldous, "State to Probe Cancer in Normandale," *PDT*, 4 Oct. 1991, p. A-2; T. L. Aldous, "Study: No Cancer Cluster," *PDT*, 6 Mar. 1992, pp. A-1, A-12.

85 newspaper investigation of death certificates: Ibid.

86 quote from Normandale widower: Aldous, "Area Cancer Rates Normal."

FIVE *war*

88–89 World War II in *Silent Spring*: R. Carson, *Silent Spring* (Boston: Houghton Mifflin, 1962). (See especially Chapters 2 and 3.)

89 All life was caught in the crossfire: Ibid., 8.

89–90 graphs of chemical production: From International Trade Commission, Washington, D.C.

91 CFCs and stratospheric chlorine: S. Solomon, "Progress towards a Quantitative Understanding of Antarctic Ozone," *Nature* 347 (1990): 347–54.

92 inactive substances can shed or off-gas: J. Gilbert et al., "Identification by Gas Chromatography–Mass Spectrometry of Vinyl Chloride Oligomers and Other Low-Molecular-Weight Components in Poly (Vinyl Chloride) Resins for Food Packaging Applications," *Journal of Chromatography* 237 (1982): 249–61.

92 the worst of both worlds: Some critics contend that exposure to estrogen-mimicking industrial chemicals is unlikely to play a significant role in breast cancer because our exposure levels to these substances are far less than the exposure to naturally occurring estrogens in food crops. See, for example, S. H. Safe, "Environmental and Dietary Estrogens and Human Health: Is There a Problem?" *EHP* 103 (1995): 346–51. This argument ignores the fact that plant estrogens are readily broken down and excreted by the human body, while synthetic chemicals with estrogenic properties can be stored in fatty tissue for years. Furthermore, plant-based estrogens appear to have a protective effect against breast cancer. See J. Barrett, "Phytoestrogens: Friends or Foes?" *EHP* 104 (1996): 478–82.

More broadly, some researchers have argued that public concern about synthetic, environmental carcinogens is misplaced because the majority of carcinogens to which we are exposed are "natural" and include pest-repelling chemicals manufactured by food crops themselves. See, for example, B. N. Ames et al., "The Causes and Prevention of Cancer," *Proceedings of the National Academy of Science* 92 (1995): 5258–65; and NRC, *Carcinogens and Anticarcinogens in the Human Diet* (Washington, D.C.: National Academy Press, 1996). Again, this kind of accounting is incomplete. Unlike natural chemicals, daily exposures to very tiny amounts of synthetic chemicals are cumulative. As Devra Davis has pointed out, natural carcinogens can often be dismantled by human enzymes before they cause harm or, in the case of many fruits and vegetables, are often accompanied by equally potent anticarcinogens. As John Wargo has pointed out, it is hardly prudent to avoid regulating synthetic carcinogens just because we also have exposures to natural ones. If anything, an awareness of our exposures to unavoidable natural carcinogens should generate greater urgency toward eliminating the avoidable synthetic ones. Moreover, natural carcinogens in foodstuffs present only one route of exposure. Unlike their synthetic counterparts, plant-generated chemicals do not spill into waterways, pollute groundwater, contaminate sport fish, waft up from dump sites, or drift into other continents. Presumably, natural carcinogens have not skyrocketed in production over the past half century. They cannot explain the coincident rise in cancer incidence rates. See J. Wargo, *Our Children's Toxic Legacy: How Science and Law Fail to Protect Us from Pesticides* (New Haven, Conn.: Yale Univ. Press, 1996), 127; the exchange of letters between Bruce Ames and his critics in the Dec. 1990 through Feb. 1991 issues of *Science*, vols. 250 and 251; and W. Linjinsky, "Environmental Cancer Risks—Real and Unreal" (editorial), *Environmental Research* 50 (1989): 207–9.

92 collect in tissues high in fat: J. D. Sherman, *Chemical Exposure and Disease: Diagnostic and Investigative Techniques* (Princeton, N.J.: Princeton Scientific Publishing, 1994); L. S. Welch, "Organic Solvents," in M. Paul (ed.), *Occupational and Environmental Reproductive Hazards: A Guide for Clinicians* (Baltimore: Williams & Wilkins, 1993), 267–79.

93 chloroform: ATSDR, *Toxicological Profile for Chloroform* (Atlanta: ATSDR, 1993); IEPA, "Chloroform: Chemical Information Sheet" (Springfield, Ill.: IEPA, Office of Chemical Safety, 1990).

94 DDT in World War II: E. P. Russell III, "'Speaking of Annihilation': Mobilizing for War against Human and Insect Enemies, 1914–1945," *Journal of American History* 82 (1996): 1505–29; T. R. Dunlap, *DDT: Scientists, Citizens, and Public Policy* (Princeton, N.J.: Princeton Univ. Press, 1981), 61–62; J. Whorton, *Before* Silent Spring: *Pesticides and Public Health in Pre-DDT America* (Princeton, N.J.: Princeton Univ. Press, 1974), 248–55.

94 Hitler's head: This ad appeared in the trade magazine *Soap and Sanitary Chemicals* in April 1944 and is reprinted in Russell, "'Speaking of Annihilation.'"

94 phenoxy herbicides: D. E. Lilienfeld and M. A. Gallo, "2,4-D, 2,4,5-T, and 2,3,7,8-TCDD: An Overview," *Epidemiologic Reviews* 11 (1989): 28–58.

94 parathion and other organophosphates: Sherman, *Chemical Exposure and Disease*, 24; H. W. Chambers, "Organophosphorous Compounds: An Overview," in J. E. Chambers and P. E. Levi (eds.), *Organophosphates: Chemistry, Fate, and Effects* (San Diego: Academic Press, 1992), 3–17.

94 mechanisms of action: L. J. Fuortes et al., "Cholinesterase-Inhibiting Insecticide Toxicity," *American Family Physician* 47 (1993): 1613–20; F. Matsumura, *Toxicology of Insecticides*, 2nd ed. (New York: Plenum, 1985), 111–202.

94 organophosphates as German nerve gas: Sherman, *Chemical Exposure and Disease*, 161; J. Borkin, *The Crime and Punishment of I. G. Farben* (New York: Harper & Row, 1978), 722–23.

94–95 phenoxy herbicides in war: P. F. Cecil, *Herbicidal Warfare: The Ranch Hand Project in Vietnam* (New York: Praeger, 1986); A. Ihde, *The Development of Modern Chemistry* (New York: Harper & Row, 1964), 722–23.

95 by 1960, 2,4-D accounted for half: Lilienfeld and Gallo, "2,4-D, 2,4,5-T."

95 for more on the rise of herbicide use in the United States, see NRC, *Pesticides in the Diets of Infants and Children* (Washington, D.C.: National Academy Press, 1993), 15.

95 graphs of pesticide use: W. J. Hayes Jr. and E. R. Laws (eds.), *Handbook of Pesticide Toxicology*, vol. 1, *General Principles* (New York: Academic Press, 1991), 22.

95 capturing 90 percent of the market: NRC, *Pesticides in the Diets*, 15.

95 pesticide use statistics: EPA, *Pesticide Industry Sales and Usage 1992–93 Market Estimates*, 733-K-94-001 (Washington, D.C.: EPA, 1994), 14, table 4; NRC, *Pesticides in the Diets*, 15.

95 current U.S. annual use: This figure includes nonconventional pesticide use, such as wood preservatives and disinfectants. Herbicides, fungicides, and insecticides alone totaled 1.25 billion pounds in 1995 (Jay Feldman, National Coalition against the Misuse of Pesticides, personal communication).

95 family pesticide use as an important route of exposure: M. Moses, *Designer Poisons: How to Protect Your Health and Home from Toxic Pesticides* (San Francisco: Pesticide Education Center, 1995).

95 82 percent of households use pesticides: R. W. Whitmore et al., *The National Home and Garden Pesticide Survey*, vol. 1, *Executive Summary: Results and Recommendations*, RT1/5100/17-01F (Washington, D.C.: EPA, 1992). See also the excellent review by S. H. Zahm and A. Blair, "Carcinogenic Risk from Pesticides," in General Motors Cancer Research Fund, *1992 Accomplishments in Cancer Research* (Philadelphia: J. B. Lippincott, 1993), 266–78.

95 families in Missouri: J. R. Davis et al., "Family Pesticide Use in the Home, Garden, Orchard, and Yard," *Archives of Environmental Contamination and Toxicology* 22 (1992): 260–66.

95 persistence of pesticidal residues indoors: Moses, *Designer Poisons*, 25–30.

95–96 pesticides in carpet fibers and house dust: R. G. Lewis et al., "Evaluation of Methods for Monitoring the Potential Exposure of Small Children to Pesticides in the Residential Environment," *Archives of Environmental Contamination and Toxicology* 26 (1994): 37–46; M. Moses et al., "Environmental Equity and Pesticide Exposure," *Toxicology and Industrial Health* 9 (1993): 913–59.

96 Los Angeles study: R. A. Lowengart et al., "Childhood Leukemia and Parents' Occupational and Home Exposures," *JNCI* 79 (1987): 39–46.

96 Denver study: J. K. Leiss and D. A. Savitz, "Home Pesticide Use and Childhood Cancer: A Case-Control Study," *AJPH* 85 (1995): 249–52.

96 brain tumors in children and household pesticides: J. R. Davis et al., "Family Pesticide Use and Childhood Brain Cancer," *Archives of Environmental Contamination and Toxicology* 24 (1993): 87–92.

96 lindane-containing lice shampoo: According to the National Pediculosis Association, six million Americans contract lice each year. Lindane, an organochlorine, is banned for agricultural use in several countries and is tightly restricted in the United States. Nevertheless, some lice shampoos still contain lindane, which the USDHHS classifies as a substance "which may reasonably be anticipated to be a carcinogen" (USDHHS, *Seventh Annual Report on Carcinogens*, Summary [Research Triangle Park, N.C.: USDHHS, 1994], 241–44). The use of lindane-based lice shampoo has also been linked to the development of cancer-related blood disorders in humans. See A. E. Rauch et al., "Lindane (Kwell)-Induced Aplastic Anemia," *Archives of Internal Medicine* 150 (1990): 2393–95.

96 rise in brain cancers in children: L. A. G. Ries et al., *SEER Cancer Statistics Review, 1971–1991: Tables and Graphs*, NIH Pub. 94-2789 (Bethesda, Md.: NCI, 1994), 428.

96 rise of petrochemicals: R. F. Sawyer, "Trends in Auto Emissions and Gasoline Composition," *EHP* 101, suppl. 6 (1993): 5–12; Ihde, *Modern Chemistry*.

96 Germany's artificial fertilizer: Ihde, *Modern Chemistry*, 680–81.

96–97 chlorine gas and chlorinated solvents: International Programme on Chemical Safety, WHO, "Chlorine and Hydrogen Chloride," *Environmental Health Criteria* 21 (1982): 54–60; Dr. Edmund Russell III, personal communication.

97 after the war ended: A. Thackary et al., *Chemistry in America, 1876–1976* (Dordrecht, Netherlands: Reidel, 1985).

97 by the 1930s: Ihde, *Modern Chemistry*.

97 the all-out assaults of World War II: Ibid.

97 fear of national leaders: Dr. Edmund Russell III, personal communication.

97 quote by Aaron Ihde: Ihde, *Modern Chemistry*, 674.

97–99 transformation from a carbohydrate-based economy to a petrochemical-based one: D. Morris and I. Ahmed, *The Carbohydrate Economy: Making Chemicals and Industrial Materials from Plant Matter* (Washington, D.C.: Institute for Local Self-Reliance, 1992). For an entertaining history of plant-derived plastics and their replacement by petrochemical plastics, see S. Fenichell, *Plastic: The Making of a Synthetic Century* (New York: HarperBusiness, 1996).

98 production, use, and carcinogenicity of formaldehyde: USDHHS, *Seventh Annual Report*, 214–19.

98 formaldehyde in foam insulation: IDPH, "Urea Formaldehyde Foam Insulation" (pamphlet) (Springfield, Ill.: IDPH, 1992).

98 formaldehyde as an indoor air pollutant: M. C. Marbury and R. A. Krieger, "Formaldehyde," in J. M. Samet and J. D. Spengler (eds.), *Indoor Air Pol-*

lution: A Health Perspective (Baltimore: Johns Hopkins Univ. Press, 1991), 223–51.

98 routes of exposure to formaldehyde: USDHHS and U.S. Labor Department, "Formaldehyde: Evidence of Carcinogenicity," *Joint NIOSH/OSHA Current Intelligence Bulletin* 34 (1980).

98 soybeans as a formaldehyde predecessor: Morris and Ahmed, *Carbohydrate Economy.*

99 other oil-based plants: Ibid.

99 synthetic cutting fluids in machine shops: Y. T. Fan, "N-Nitrosodiethanolamine in Synthetic Cutting Fluids: A Part-per-Hundred Impurity," *Science* 196 (1977): 70–71.

99 contaminants in cutting fluids: USDHHS, *Seventh Annual Report*, 282.

99 quote from cutting-fluid study: Fan, "N-Nitrosodiethanolamine," 71.

99 percentage of chemicals tested: NRC, *Toxicity Testing: Strategies to Determine Needs and Priorities* (Washington, D.C.: National Academy Press, 1984). Updated estimate obtained from Dr. James Huff, Environmental Carcinogenesis Program, National Institute for Environmental Health Sciences, Jan. 1997.

99–100 review of new chemicals under TSCA: L. Ember, "Pollution Prevention: Study Says Chemical Industry Lags," *Chemical and Engineering News*, 20 Mar. 1995, p. 6.

100 FFDCA and FIFRA: For a thoughtful discussion of the loopholes and shortcomings of both of these laws, see Wargo, *Our Children's Toxic Legacy;* and GAO, *Food Safety: Changes Needed to Minimize Unsafe Chemicals in Food*, Report to the Chairman, Human Resources and Intergovernmental Relations Subcommittee, Committee on Government Operations, House of Representatives, GAO/RCED-94-192, Sept. 1994.

100 issuing everyone a driver's license: D. Ozonoff, "Taking the Handle off the Chlorine Pump" (presentation at the public health forum, "Environmental and Occupational Health Problems Posed by Chlorinated Organic Chemicals," Boston Univ. School of Public Health, 5 Oct. 1993).

100 NRC report: NRC, *Toxicity Testing.*

100 history of right-to-know laws: B. A. Goldman, "Is TRI Useful in the Environmental Justice Movement?" (presentation to the Toxics Release Inventory Data Use Conference, Boston, Mass., 6 Dec. 1994), reprinted in *EPA Proceedings: Toxics Release Inventory (TRI) Data Use Conference, Building TRI and Pollution Prevention Partnerships,* EPA/749-R-95-001 (Washington, D.C.: EPA, 1995), 133–37; and Paul Orum, Working Group on Community-Right-to-Know, personal communication.

101 deficiencies of TRI: Goldman, "Is TRI Useful"; Working Group on Community Right-to-Know, "Environmental Groups Blast EPA for Toxics Reporting Loophole," press release, Washington D.C., July 25, 1994; Paul Orum, personal communication.

101 phantom reductions: Working Group on Community Right-to-Know, "New Toxics Data Show Little Progress in Source Reduction," press release, Washington, D.C., 27 Mar. 1995.

101–102 declines in releases not tethered to decline in production: Inform, Inc., *Toxics Watch 1995* (New York: Inform, Inc., 1995). Since TRI's inception, there has been no overall decline in toxic waste generation by the 74,000 plants that, as of 1995, had to report to TRI. As the EPA itself admits, the total amount of toxic chemical waste generated by industry has actually increased, not declined, since the first TRI report was released. Nevertheless, *reported* releases dropped about 35 percent between 1988 and 1992. Some analysts attribute this apparent discrepancy to better containment. "So while industry may be improving its management of toxic chemical waste, clearly there are still many opportunities for preventing pollution by reducing the use of toxic chemicals," concluded EPA administrator Carol Browner in 1992 (EPA, *1992 Toxics Release Inventory: Public Data Release*, 745-R-94-001 [Washington, D.C.: EPA, 1994]).

102 impact of the TRI report: J. H. Cushman, "Efficient Pollution Rule under Attack," *New York Times*, 28 June 1995, p. A-16; K. Schneider, "For Communities, Knowledge of Polluters Is Power," *New York Times*, 24 Mar. 1991, p. A-5.

102 quotes from chemical industry representatives: Reprinted in *Working Notes on Community-Right-to-Know* (Washington, D.C.: Working Group on Community Right-to-Know, May–June 1995), 3.

103 the most recent TRI: EPA, *1994 Toxics Release Inventory: Public Data Release*, 745-R-96-002 (Washington, D.C.: EPA, 1996).

104 a hydrologist's description: L. Hoberg et al., *Groundwater in the Peoria Region*, Cooperative Research Bulletin 39 (Urbana: ISGWS, 1950), 53.

104–105 history of Pekin: *Pekin, Illinois, Sesquicentennial (1824–1974): A History* (Pekin, Ill.: Pekin Chamber of Commerce, 1974).

105 one of Pekin's distilleries: Midwest Grain Products, *1994 Annual Report*.

105 "Cats" in action: P. A. Letourneau (ed.), *Caterpillar Military Tractors*, vol. 1. (Minneapolis: Iconografix, 1994).

105 sugar-beet fields and starch explosion: *Pekin, Illinois, Sesquicentennial*, 68.

106 pollution from Powerton: "Pekin Edison Plant Named Worst Pollutor," *Bloomington Daily Pantagraph*, 10 Aug. 1974; J. Simpson, "Conservationist Blasts Pekin Energy Plant," *Bloomington Daily Pantagraph*, 30 July 1971.

106 Keystone: E. Hopkins, "Keystone Plans Costly Cleanup," *PJS*, 3 July 1993, p. A-1; E. Hopkins, "Region Awash in Toxic Chemicals: Study," *PJS*, 25 July 1993, p. A-2.

107 quote from toxicologist and newspaper's conclusion: E. Hopkins, "Emissions List Ranks Region 13th," *PJS*, 19 Mar. 1995, pp. A-1, A-22.

107 statistics on toxic emissions in the Pekin-Peoria: From TRI. See also Hopkins, "Region Awash."

107 Captan: EPA, *Suspended, Cancelled, and Restricted Pesticides*, 20T-1002 (Washington, D.C.: EPA, 1990).

107 documents: From the Right-to-Know Network's copies of EPA's TRI, PCS, and FINDS databases. These searches were conducted by Kathy Grandfield on 1 Jan. 1995. Additional data for Tazewell County were provided by Joe Goodner, TRI coordinator at the IEPA in Springfield.

108 Tazewell doubled the amount of hazardous waste: IEPA, *Summary of Annual Reports on Hazardous Waste in Illinois for 1991 and 1992: Generation, Treatment, Storage, Disposal, and Recovery*, IEPA/BOL/94-155 (Springfield, Ill.: IEPA, 1994), 61.

108 received four times more waste than it produced: IEPA, *Illinois Nonhazardous Special Waste Annual Report for 1991* (Springfield, Ill.: IEPA, 1993), table K.

108 the spill report: The report is part of the Tazewell County, Illinois, Area Report taken from the Right-to-Know Network's copy of EPA's ERNS database.

109 toxicity and production of methyl chloride: National Institute for Occupational Safety and Health, *NIOSH Current Intelligence Bulletin*, 43, NIOSH Pub. 84-117 (Cinninnati: NIOSH, 1984).

109 estrogenicity of postwar chemicals: D. M. Klotz et al., "Identification of Environmental Chemicals with Estrogenic Activity Using a Combination of *In Vitro* Assays," *EHP* 104 (1996): 1084–89; "Masculinity at Risk" (editorial), *Nature* 375 (1995): 522; R. M. Sharpe, "Another DDT Connection," *Nature* 375 (1995): 538–39; "Male Reproductive Health and Environmental Oestrogens" (editorial), *Lancet* 345 (1995): 933–35; Institute for Environment and Health, *Environmental Oestrogens: Consequences to Human Health and Wildlife* (Leicester, England: Univ. of Leicester, 1995); J. Raloff, "Beyond Estrogens: Why Unmasking Hormone-Mimicking Pollutants Proves So Challenging," *Science News* 148 (1995): 44–46.

109–110 DDE: W. R. Kelce et al., "Persistent DDT Metabolite *p,p*'-DDE is a Potent Androgen Receptor Antagonist," *Nature* 375 (1995): 581–85. More recently, researchers have discovered estrogenic properties in two other DDT metabolites, both isomers of DDD, previously thought to be nonestrogenic. See Klotz, "Identification of Environmental Chemicals."

109–110 persistence of DDE: Dr. Mary Wolff, Mt. Sinai School of Medicine, personal communication.

110 concern focuses on reproduction and wildlife: The most readable summary is T. Colborn et al., *Our Stolen Future: Are We Threatening Our Fertility, Intelligence, and Survival?—A Scientific Detective Story* (New York: Dutton, 1996).

110 long-running conversations about cancer: D. L. Davis and H. L. Bradlow, "Can Environmental Estrogens Cause Breast Cancer?" *Scientific American*, Oct. 1995, 166–72; D. L. Houghton and L. Ritter, "Organochlorine Residues and Risk of Breast Cancer," *Journal of American College of Toxicology* 14 (1995): 71–89; T. Key and G. Reeves, "Organochlorines in the Environment and Breast Cancer," *British Medical Journal* 308 (1994): 1520–21; D. L. Davis et al., "Medical Hypothesis: Xenoestrogens as Preventable Causes of Breast Cancer," *EHP* 101 (1993): 372–77; U.S. House of Representatives, *Health Effects of Estrogenic Pesticides: Hearing before the Subcommittee on Health and the Environment*, Cong., sess., 21 Oct. 1993; R. Coosen and F. L. van Velsen, "Effects of the ß-Isomer of Hexachlorocyclohexane on Estrogen-Sensitive Human Mammary Tumor Cells," *Toxicology and Applied Pharmacology* 101 (1989): 310–18; J. A. Nelson, "Effects of Dichlorodiphenyltrichloroethane (DDT) Analogs and Polychlorinated Biphenyl (PCB) Mixtures on 17ß-[^{3}H]estradiol Binding to Rat Uterine Receptor," *Biochemical Pharmacology* 23 (1974): 447–51; R. M. Welch et al., "Estrogenic Action of DDT and Its Analogs," *Toxicology and Applied Phar-*

macology 14 (1969): 358–67; C. Huggins and N. C. Yang, "Induction and Extinction of Mammary Cancer," *Science* 137 (1962): 257–62.

110 Carson's mention of endocrine disruption: R. Carson, *Silent Spring* (Boston: Houghton Mifflin, 1962), 212, 235–37.

110–111 Soto and Sonnenschein's discovery: A. M. Soto et al., "*p*-Nonylphenol: An Estrogenic Xenobiotic Released from 'Modified' Polystyrene," *EHP* 92 (1991): 167–73.

111 estrogenic activity in other substances: A. M. Soto et al., "The Pesticides Endosulfan, Toxaphene, and Dieldrin Have Estrogenic Effects on Human Estrogen-Sensitive Cells," *EHP* 102 (1994): 380–383.

111 chemicals identified as estrogenic: B. Hileman, "Concerns Broaden over Chlorine and Chlorinated Hydrocarbons," *Chemical and Engineering News*, 19 Apr. 1993, 11–20.

111 many plasticizers and surfactants are estrogenic: J. A. Brotons et al., "Xenoestrogens Released from Lacquer Coatings in Food Cans," *EHP* 103 (1995): 608–12; J. Raloff, "Additional Sources of Dietary Estrogens," *Science News* 147 (1995): 341; R. White et al., "Environmentally Persistent Alkylphenolic Compounds Are Estrogenic," *Endocrinology* 135 (1994): 175–82; A. V. Krishnan et al., "Bisphenol-A: An Estrogenic Substance Is Released from Polycarbonate Flasks during Autoclaving," *Endocrinology* 132 (1993): 2279–86

111 APEOs in New Jersey drinking water: L. B. Clark et al., "Determination of Alkylphenol Ethoxylates and Their Acetic Acid Derivatives in Drinking Water by Particle Beam Liquid Chromotography/Mass Spectrometry," *International Journal of Environmental Analytic Chemistry* 47 (1992): 167–80.

111 English researchers found APEOs stimulate breast cancer cells and feminize fish: White, "Alkylphenolic Compounds."

111–112 source of estrogens in sewage: J. Kaiser, "Scientists Angle for Answers," *Science* 274 (1996): 1837–38; L. C. Folmar et al., "Vitellogenin Induction and Reduced Serum Testosterone Concentrations in Feral Male Carp (*Cyprinus carpio*) Captured Near a Major Metropolitan Sewage Treatment Plant," *EHP* 104 (1996): 1096–1101; C. E. Purdom et al., "Estrogenic Effects of Effluents from Sewage Treatment Works," *Chemistry and Ecology* 8 (1994): 275–85; S. Jobling and J. P. Sumpter, "Detergent Components in Sewage Effluent Are Weakly Estrogenic to Fish: An *In Vitro* Study Using Rainbow Trout (*Oncorhynchus mykiss*) Hypatocytes," *Aquatic Toxicology* 27 (1993): 361–72.

111 phthalates: S. Jobling et al., "A Variety of Environmentally Persistent Chemicals, Including Some Phthalate Plasticizers, Are Weakly Estrogenic," EHP 103 (1995): 582–87; J. Raloff, "Newest Estrogen Mimics the Commonest?" *Science News* 148 (1995): 47.

111 production, use, and toxicity of DEHP: IEPA, "Di(2-ethylhexyl)phthalate: Chemical Information Sheet," IEPA/ENV/93-006 (Springfield, Ill.: IEPA, 1993).

111 DEHP in blood bags: S. D. Pearson and L. A. Trissel, "Leaching of Diethylhexl Phthalate from Polyvinyl Chloride Containers by Selected

Drugs and Formulation Components," *American Journal of Hospital Pharmacology* 50 (1993): 1405–9; R. J. Jaeger and R. J. Rubin, "Migration of a Phthalate Ester Plasticizer from Polyvinyl Chloride Blood Bags in Stored Human Blood and Its Localization in Human Tissues," *NEJM* 287 (1972): 1114–18.

112 half of endocrine disrupters are organochlorines: Hileman, "Concerns Broaden."

113–114 creation and use of organochlorines: International Programme of Chemical Safety, WHO, "Chlorine and Hydrogen Chloride."

115 1993 resolution: American Public Health Association Resolution 9304, "Recognizing and Addressing the Environmental and Occupational Health Problems Posed by Chlorinated Organic Chemicals," *AJPH* 84 (1994): 514–15.

115 quote from the IJC: International Joint Commission, *Sixth Biennial Report on Great Lakes Water Quality* (Washington, D.C., and Ottawa, Ontario: International Joint Commission, 1992), 5.

115 In addition, Canada's Ontario Task Force on the Primary Prevention of Cancer has recommended the phaseout of all industrial feedstocks that contain chlorine, as well as all persistent toxic substances known or suspected to be carcinogens. See *Recommendations for the Primary Prevention of Cancer* (Ottawa, Ontario: Ministry of Health, Mar. 1995).

115–116 quote from David Ozonoff: "On the Need to Ban Organochlorines" (presentation to the Massachusetts Breast Cancer Coalition Conference, "Breast Cancer and Environment: Our Health at Risk," Boston, Mass., 28 Oct. 1996).

116 For more information on alternatives to solvents, contact the Silicon Valley Toxics Coalition, 760 N. First St., in San Jose, CA 96112, (408) 287-6707, or the Toxics Use Reduction Institute, One University Ave., Univ. of Massachusetts, Lowell, MA 01854, (508) 434-3050.

116 formaldehyde for embalming: Toxics Use Reduction Institute, *Formaldehyde Use Reduction in Mortuaries*, Technical Report 24 (Lowell, Mass.: TURI, 1994).

116–117 alternatives to dry-cleaning: H. Black, "A Cleaner Bill of Health," *EHP* 104 (1996): 488–90; S. B. Williams et al., "Fabric Compatibility and Cleaning Effectiveness of Dry Cleaning with Carbon Dioxide," *Los Alamos National Laboratory Report*, LA-UR-96-822 (Los Alamos, N.M.: Los Alamos National Laboratory, 1996).

117 statistics on perchloroethylene: U.S. Agency for Toxic Substances and Disease Registry, *Toxicological Profile for Tetrachloroethylene*, TP-92118 (Atlanta: ATSDR, 1993); USDHHS, *Seventh Annual Report on Carcinogens* (Rockville, Md.: USDHHS, 1994), 375; EPA, *Chemical Summary for Perchloroethylene*, 749-F-94-020a (Washington, D.C.: EPA Office of Pollution Prevention and Toxics, 1994).

117 dry-cleaning in New York: D. Wallace et al., *Upstairs, Downstairs: Perchloroethylene in the Air in Apartments Above New York City Dry Cleaners* (New York: Consumers Union, 1995); M. Green, *Clothed in Controversy: The Risk to New Yorkers from Dry Cleaning Emissions and What Can Be Done About It* (New York: Office of the Public Advocate for the City of New York, 1994).

SIX *animals*

119–120 additive effect of estrogen mimics: A. M. Soto et al., "The E-SCREEN Assay as a Tool to Identify Estrogens: An Update on Estrogenic Environmental Pollutants," *EHP* 103, suppl. 7 (1995): 113–22.

120 estrogenic activity in a variety of pesticides: A. M. Soto et al., "The Pesticides Endosulfan, Toxaphene, and Dieldrin Have Estrogenic Effects on Human Estrogen-Sensitive Cells," *EHP* 102 (1994): 380–83.

120 history of toxaphene: R. L. Metcalf, "An Increasing Public Concern," in D. Pimentel and H. Lehman (eds.), *The Pesticide Question: Environment, Economics, and Ethics* (New York: Routledge, 1993), 426–30; H. P. Hynes, *The Recurring Silent Spring* (New York: Pergamon, 1989), 156.

120 Carson on toxaphene: R. Carson, *Silent Spring* (Boston: Houghton Mifflin, 1962), 141–42.

120 marine animals and toxaphene: J. Paasivirta et al., "Chloroterpenes and Other Organochlorines in Baltic, Finnish, and Arctic Wildlife," *Chemosphere* 22 (1991): 47–55.

120 within the range of concentrations found in salmon: Soto, "Endosulphan, Toxaphene."

121 immortality of cancer cell lines: G. B. Dermer, *The Immortal Cell: Why Cancer Research Fails* (Garden City Park, N.Y.: Avery, 1994).

121 names of breast cancer cells lines: A. Leibovitz, "Cell Lines from Human Breast," in R. J. Hay et al. (eds.), *Atlas of Human Tumor Cell Lines* (New York: Academic Press, 1994), 161–84; Dr. Carlos Sonnenschein, personal communication.

121 the coin of the realm: J. Ricci, "One Nun's Living Legacy," *Detroit Free Press*, 30 Sept. 1984, pp. F-1, F-4.

121–122 origins of MCF-7: H. D. Soule, "A Human Cell Line from a Pleural Effusion Derived from a Breast Carcinoma," *JNCI* 51 (1973): 1409–16.

122 The Immaculate Heart of Mary was the setting for the Greenpeace Chlorine-Free Great Lakes Conference, 4–6 Dec. 1992.

122 old newspaper clipping: Ricci, "Living Legacy."

123 concordance: Soto, "E-SCREEN Assay."

123 effects of endosulfan: Soto, "Endosulfan, Toxaphene."

123 the standard yardstick: NRC, *Science and Judgement in Risk Assessment* (Washington, D.C.: National Academy Press, 1994), 56–67.

123 strengths and limitations of epidemiology: USDHHS, *Seventh Annual Report on Carcinogens*, Summary (Research Triangle Park, N.C.: USDHHS, 1994), 4–5.

123 advantages of animal assays: NRC, *Science and Judgement.*

124 one-third of carcinogens discovered in animals: J. Huff, "Chemicals Causally Associated with Cancers in Humans and in Laboratory Animals: A Perfect Concordance," in M. P. Waalkes and J. M. Ward (eds.), *Carcinogenesis* (New York: Raven Press, 1994), 25–37.

124 1918 study: K. Yamagiwa and K. Ishikawa, "Experimental Study of the Pathogenesis of Carcinoma," *Journal of Cancer Research* 3 (1918): 1–29.

124 by the 1930s: S. S. Epstein, *The Politics of Cancer* (San Francisco: Sierra Club Books, 1979), 56–57.

124 1938 dog studies: W. C. Hueper et al., "Experimental Production of Bladder Tumors in Dogs by Administration of beta-Naphthylamine," *Journal of Industrial Hygiene and Toxicology* 20 (1938): 46–84.

124 coincident rise of synthetic dyes and bladder cancer among textile workers: E. K. Weisburger, "General Principles of Chemical Carcinogenesis," in Waalkes and Ward, *Carcinogenesis*, 1–23; NIOSH, *Special Occupational Hazard Review for Benzidene-Based Dyes*, DHEW (NIOSH) Pub. 80-109 (Cincinnati: NIOSH, 1980).

124 International Labor Organization: L. Tomatis, *Cancer: Causes, Occurrence and Control*, IARC Scientific Pub. 100 (Lyon, France: IARC, 1990), 129.

124 bladder cancer among workers in rubber and metal industries: P. Vineis and S. Di Prima, "Cutting Oils and Bladder Cancer," *Scandinavian Journal of Work Environment and Health* 9 (1983): 449–50; R. R. Monson and K. Nakano, "Mortality among Rubber Workers: I. White Male Union Employees in Akron, Ohio," *AJE* 103 (1976): 284–96; P. Cole et al., "Occupation and Cancer of the Lower Urinary Tract," *Cancer* 29 (1972): 1250–60.

124–125 The science historian Robert Proctor provides an excellent overview of Wilhelm Hueper's struggles in *Cancer Wars: How Politics Shapes What We Know and Don't Know about Cancer* (New York: Basic Books, 1995), 36–48.

125 number of carcinogens identified: Huff, "Chemicals Causally Associated."

125 quote from IARC: IARC, *IARC Monographs on the Evaluation of Carcinogenic Risks to Humans*, suppl. 7 (Lyon, France: IARC, 1987), 17–34. In this context, "sufficient evidence" means that cancer develops in two or more species of animals or during two or more independent studies of any one species.

125 U.S. system of classification: NRC, *Science and Judgement*, 58–60.

125 K. G. Thigpen, "Biennial Report on Carcinogens: A Work in Progress," *EHP* 103 (1995): 806–7.

126 quote from the carcinogen report: USDHHS, *Seventh Annual Report*, 1.

126 benzene: Ibid., 34–37.

126–127 PCBs: Ibid., 324–28; ATSDR, *ATSDR Case Studies in Environmental Medicine: Polychlorinated Biphenyl (PCB) Toxicity* (Atlanta: ATSDR, 1990); NIOSH, *Occupational Exposure to Polychlorinated Biphenyls (PCBs)*, DHEW (NIOSH) Pub. 77-225, (Cincinnati: NIOSH, 1977).

127 a program of intense surveillance and assessment: "Naming Carcinogens," *EHP* 103 (1995): 657–58.

127 schools in Cape Cod: I. Rosen, "More Chemicals Found at Bourne School," *Boston Globe*, 9 Sept. 1995, p. 14.

127 quote from a 1980 review: NIOSH, *Special Occupational Hazard Review for Benzidene-based Dyes*, DHEW (NIOSH) Pub. 80-109 (Cincinnati: NIOSH, 1980) 1–2.

128 quote from a 1994 report: USDHHS, *Seventh Annual Report*, 37–39.

128 quote from a 1996 report: ATSDR, "Benzidine Fact Sheet," CAS 92-87-5 (Atlanta: ATSDR, 1996).

128–129 description of animal assays: R. A. Griesemer and S. L. Eustis, "Gender Differences in Animal Bioassays for Carcinogenicity," *Journal of Occupational Medicine* 36 (1994): 855–59.

129 transpecies extrapolation: Dr. Ross Hume Hall, personal communication.

129 five of the top ten sites: J. Huff et al., "Chemicals Associated with Site-Specific Neoplasia in 1394 Long-Term Carcinogenesis Experiments in Laboratory Rodents," *EHP* 93 (1991): 247–70.

129 description of breast development: C. W. Daniel and G. B. Silverstein, "Postnatal Development of the Rodent Mammary Gland," in M. C. Neville and C. W. Daniel (eds.), *The Mammary Gland: Development, Regulation, and Function* (New York: Plenum, 1987), 3–36; J. Russo and I. H. Russo, "Development of the Human Mammary Gland," in ibid., 67–93; S. Z. Haslam, "Role of Sex Steroid Hormones in Normal Mammary Gland Function," in ibid., 499–533.

129–130 breast cancer in rodents and humans: S. Nandi et al., "Hormones and Mammary Carcinogenesis in Mice, Rats, and Humans: A Unifying Hypothesis," *Proceedings of the National Academy of Science* 92 (1995): 3650–57.

130 agreement between species is high: Huff, "Site-Specific Neoplasia," 247–70.

130 lessons from molecular biology: Dr. Donald Malins, Pacific Northwest Research Foundation, personal communication.

130–131 high-dose animal assays: J. Sherman, *Chemical Exposure and Disease: Diagnostic and Investigative Techniques*, (Princeton, N.J.: Princeton Scientific Publishing, 1994), 27–28. For a discussion of the challenges to this system, see G. W. Lucier, "Mechanism-based Toxicology in Cancer Risk Assessment: Implications for Research, Regulation, and Legislation," *EHP* 104 (1996): 84–87; and J. Marx, "Animal Carcinogen Testing Challenged," *Science* 250 (1990): 743–45.

131 estimates of carcinogens in commerce: V. A. Fung et al., "The Carcinogenesis Bioassay in Perspective: Application in Identifying Human Cancer Hazards," *EHP* 103 (1995): 680–83.

131 estimates of carcinogens identified and regulated: The USDHHS's *Seventh Annual Report on Carcinogens* contains 166 individual listings.

131–132 St. Lawrence whales: C. H. Farnsworth, "Scientists Are Puzzled over the Deaths of Whales in the St. Lawrence," *New York Times*, 22 Aug. 1995, p. C-12; D. E. Sargent and W. Hoek, "An Update of the Status of White Whales *Delphinapterus leucas* in the St. Lawrence Estuary, Canada," in J. Prescott and M. Gauquelin (eds.), *Proceedings of the International Forum for the Future of the Beluga* (Sillery, Québec: Presses de l'Université du Québec, 1990), 59–74.

132 bladder cancer in whales and smelter workers: P. Béland, "About Carcinogens and Tumors," *Canadian Journal of Fisheries and Aquatic Sciences* 45 (1988): 1855–56; D. Martineau et al., "Transitional Cell Carcinoma of the Urinary Bladder in a Beluga Whale (*Delphinapterus leucas*)," *Journal of Wildlife Diseases* 22 (1985): 289–94.

132–133 1988 study: D. Martineau et al., "Pathology of Stranded Beluga Whales (*Delphinapterus leucas*) from the St. Lawrence Estuary, Québec, Canada," *Journal of Comparative Pathology* 98 (1988): 287–311.

133 1994 autopsy reports: S. de Guise et al., "Tumors in St. Lawerence Beluga Whales," *Veterinary Pathology* 31 (1994): 444–49.

133 cancers so far identified in the St. Lawrence beluga: D. Martineau, "Intesti-

nal Adenocarcinomas in Two Beluga Whales (*Delphinapterus leucas*) from the Estuary of the St. Lawrence River," *Canadian Veterinary Journal* 36 (1995): 563–65; C. Cirard et al., "Adenocarcinoma of the Salivary Gland in a Beluga Whale (*Delphinapterus leucas*)," *Journal of Veterinary Diagnostic Investigation* 3 (1991): 264–65.

133 beluga reproductive problems: S. de Guise et al., "Possible Mechanisms of Action of Environmental Contaminants on St. Lawrence Beluga Whales (*Delphinapterus leucas*)," *EHP* 103, suppl. 4 (1995): 73–77; A. Motluk, "Deadlier Than the Harpoon?" *New Scientist*, 1 July 1995, 12–13; D. Martineau et al., "Levels of Organochlorine Chemicals in Tissues of Beluga Whales (*Delphinapterus leucas*) from the St. Lawrence Estuary, Québec, Canada," *Archives of Environmental Contamination and Toxicology* 16 (1987): 137–47; R. Masse et al., "Concentrations and Chromatographic Profile of DDT Metabolites and Polychlorobiphenyl (PCB) Residues in Stranded Beluga Whales (*Delphinapterus leucas*) from the St. Lawrence Estuary, Canada," *Archives of Environmental Contamination and Toxicology* 15 (1986): 567–79.

133 airborne deposition of chlordane and toxaphene: D. Muir, "Levels and Possible Effects of PCBs and Other Organochlorine Contaminants and St. Lawrence Belugas," in Prescott and Gauquelin, *Future of the Beluga*, 219–23.

134 eels, whales, and Mirex: P. Béland et al., "Toxic Compounds and Health and Reproductive Effects in St. Lawrence Beluga Whales," *Journal of Great Lakes Research* 19 (1993): 766–75; T. Colborn, *Great Lakes, Great Legacy?* (Washington, D.C.: Conservation Foundation, 1990), 140.

134 life history of eels: R. Carson, *Under the Sea Wind* (New York: Penguin Books, 1941), 209–72. Comparison to willow leaves, p. 265.

134 quote from Ezra Pound: "Portrait d'une Femme," *Personae* (New York: New Directions, 1926).

135 benzo[a]pyrene and St. Lawrence belugas: P. Béland, "The Beluga Whales of the St. Lawrence River," *Scientific American*, May 1996, 74–81.

135–136 chemistry and carcinogenicity of benzo[a]pyrene: H. C. Pitot and Y. P. Dragan, "Chemical Carcinogenesis," in C. D. Klaassen, *Casarett and Doull's Toxicology: The Basic Science of Poisons*, 5th ed. (New York: McGraw-Hill, 1996), 202–12; USDHHS, *Seventh Annual Report*, 328–35.

136 mechanism of action: M. E. Hahn and J. J. Stegeman, "The Role of Biotransformation in the Toxicity of Marine Pollutants," in Prescott and Gauquelin, *Future of the Beluga*, 185–98.

136 whales and DNA adducts: D. Martineau et al., "Pathology and Toxicology of Beluga Whales from the St. Lawrence Estuary, Québec, Canada: Past, Present and Future," *The Science of the Total Environment* 154 (1994): 201–15; L. R. Shugart and C. Theodorakis, "Environmental Toxicology: Probing the Underlying Mechanisms," *EHP* 102, suppl. 12 (1994): 13–17; L. R. Shugart et al., "Detection and Quantitation of Benzo[a]pyrene-DNA Adducts in Brain and Liver Tissues of Beluga Whales (*Delphinapterus leucas*) from the St. Lawrence and Mackenzie Estuaries," in Prescott and Gauquelin, *Future of the Beluga*, 219–23.

138 bladder cancer and occupation: F. Barbone et al., "Occupation and Bladder

Cancer in Prodenone (North-East Italy): A Case-Control Study," *International Journal of Epidemiology* 23 (1994): 58–65; M. McCredie, "Bladder and Kidney Cancers," in R. Doll et al. (eds.), *Trends in Cancer Incidence and Mortality* (Plainview, N.Y.: Cold Spring Harbor Laboratory Press, 1994), 343–68; H. Anton-Culver et al., "Occupation and Bladder Cancer Risk," *AJE* 136 (1992): 89–94; T. Skov et al., "Risk for Cancer of the Urinary Bladder among Hairdressers in the Nordic Countries," *American Journal of Industrial Medicine* 17 (1990): 217–23; D. T. Silverman et al., "Occupational Risks of Bladder Cancer in the United States: I. White Men," *JNCI* 81 (1989): 1472–80; D. T. Silverman et al., "Occupational Risks of Bladder Cancer in the United States: II. Nonwhite Men," *JNCI* 81 (1989): 1480–83.

138 in one English factory: P. Vineis, "Epidemiology of Cancer from Exposure to Arylamines," *EHP* 102, suppl. 6 (1994): 7–10.

138 Clinton County: L. D. Budnick et al., "Cancer and Birth Defects Near the Drake Superfund Site, Pennsylvania," *AEH* 39 (1984): 409–13.

138 association with chemical manufacturing plants: W. J. Blot and J. F. Fraumeni Jr., "Geographic Epidemiology of Cancer in the United States," in D. Schottenfeld and J. F. Fraumeni Jr. (eds.), *Cancer Epidemiology and Prevention*, 1st ed. (Philadelphia: Saunders, 1982), 189–90.

138 Massachusetts water pipes: A. Aschengrau et al., "Cancer Risk and Tetrachloroethylene-Contaminated Drinking Water in Massachusetts," *AEH* 48 (1993): 284–92.

138 Taiwan study: B. J. Pan et al., "Excess Cancer Mortality among Children and Adolescents in Residential Districts Polluted by Petrochemical Manufacturing Plants in Taiwan," *JTEH* 43 (1994): 117–29.

138 bladder cancer in young women: J. M. Piper et al., "Bladder Cancer in Young Women," *AJE* 123 (1986): 1033–42.

138 dogs with bladder cancer: L. T. Glickman et al., "Epidemiologic Study of Insecticide Exposures, Obesity, and Risk of Bladder Cancer in Household Dogs," *JTEH* 28 (1989): 407–14; H. M. Hayes, "Bladder Cancer in Pet Dogs: A Sentinel for Environmental Cancer?" *AJE* 114 (1981): 229–33.

139 quote by Leone Pippard: L. Pippard, "Ailing Whales, Water and Marine Management Systems: An Urgency for Fresh, New Approaches," in Prescott and Gauquelin, *Future of the Beluga*, 14–15.

139 Clyde Dawe's discovery: J. C. Harshbarger, "Introduction to Session on Pathology and Epizootiology," *EHP* 90 (1991): 5.

140 Registry of Tumors in Lower Animals: J. C. Harshbarger, "Role of the Registry of Tumors in Lower Animals in the Study of Environmental Carcinogenesis in Aquatic Animals," *Annals of the New York Academy of Sciences* 298 (1977): 280–89; J. C. Harshbarger, "The Registry of Tumors in Lower Animals," in *Neoplasia and Related Disorders in Invertebrates and Lower Vertebrate Animals*, NCI Monograph 31 (1969), xi–xvi.

140 cancer in lower animals linked to environmental contamination: Dr. John Harshbarger, Registry of Tumors in Lower Animals, personal communiciation.

140 tumors associated with contaminated sediments: M. J. Moore and M. S.

Myers, "Pathobiology of Chemical-Associated Neoplasia in Fish," in D. C. Malins and G. K. Ostrander (eds.), *Aquatic Toxicology: Molecular, Biochemical and Cellular Perspectives* (Boca Raton, Fla.: Lewis, 1994), 327–86.

140 laboratory experiments with contaminated sediments: J. C. Harshbarger and J. B. Clark, "Epizootiology of Neoplasms in Bony Fish of North America," *Science of the Total Environment* 94 (1990): 1–32; Dr. William Hawkins, Gulf Coast Research Laboratory, personal communication.

140–141 fish echoing human trends: Harshbarger and Clark, "Epizootiology of Neoplasms."

141 liver tumor epizootics coincide with chemical production: Harshbarger, personal communication.

141 number of liver cancer epizootics: Ibid.

141 skin cancer in fish: Harshbarger and Clark, "Epizootiology of Neoplasms."

141 other fish epizootics: Ibid.

141 the registry's international survey: J. C. Harshbarger, "Neoplasms in Wild Fish from the Marine Ecosystem Emphasizing Environmental Interactions," in J. A. Couch and J. W. Fournie (eds.), *Pathobiology of Marine and Estuarine Organisms* (Boca Raton, Fla.: CRC Press, 1993).

141 quote from William Hawkins: Personal communication.

141 ecotoxicology: G. McMahon, "The Genetics of Human Cancer: Implications for Ecotoxicology," *EHP* 102, suppl. 12 (1994): 75–80; J. J. Stegeman and J. J. Lech, "Cytochrome P-450 Monooxygenase Systems in Aquatic Species: Carcinogen Metabolism and Biomarkers for Carcinogen and Pollutant Exposure," *EHP* 90 (1991): 101–9.

141–142 dosimetry of DNA adducts: McMahon, "Genetics of Human Cancer"; J. E. Stein et al., "Molecular Epizootiology: Assessment of Exposure to Genotoxic Compounds in Teleosts," *EHP* 102, suppl. 12 (1994): 19–23.

142 quote by George Bailey: "OSU Develops Living Environmental Warning System," press release, Oregon State University, 1 Sept. 1992.

142 aquarium studies: Harshbarger, personal communication.

142 lobsters: This phenomenon may not decrease cancer risk to human consumers. See C. B. Cooper et al., "Risks of Consumption of Contaminated Seafood: The Quincy Bay Case Study," *EHP* 90 (1991): 133–40.

142 animal sentinels: T. Colborn et. al., *Our Stolen Future: Are We Threatening Our Fertility, Intelligence, and Survival?—A Scientific Detective Story* (New York: Dutton, 1996); G. A. Le Blanc, "Are Environmental Sentinels Signaling?" *EHP* 103 (1995): 888–90; T. Colborn, "The Wildlife/Human Connection: Modernizing Risk Decisions," *EHP* 102, suppl. 12 (1994): 55–59; L. J. Guillette Jr., "Endocrine-Disrupting Environmental Contaminants and Reproduction: Lessons from the Study of Wildlife," in D. R. Popkin and L. J. Peddle (eds.), *Women's Health Today: Perspectives on Current Research and Clinical Practice* (Pearl River, N. Y.: Parthenon, 1994), 201–7; NRC, *Animals as Sentinels of Environmental Health Hazards* (Washington, D.C.: National Academy Press, 1991); F. L. Rose and J. C. Harshbarger, "Neoplastic and Possibly Related Skin Lesions in Neotenic Tiger Salamanders from a Sewage Lagoon," *Science* 196 (1977): 315–17.

142 animal viruses: J. A. Couch and J. C. Harshbarger, "Effects of Carcinogenic

Agents on Aquatic Animals: An Environmental and Experimental Overview," *Environmental Carcinogenesis Reviews* 3 (1985): 63–105.

142–143 turtles: L. J. Guillette Jr. et al., "Organization versus Activation: The Role of Endocrine-Disrupting Contaminants (EDCs) during Embryonic Development in Wildlife," *EHP* 103, suppl. 7 (1995): 156–64.

143 red-eared sliders: J. M. Bergeron et al., "PCBs as Environmental Estrogens: Turtle Sex Determination as a Biomarker of Environmental Contamination," *EHP* 102 (1994): 780–81.

143 Deer Island: E. Kales and D. Kales, *All about the Boston Harbor Islands* (Boston: Marlborough House, 1976); Water Resources Authority of Massachusetts, "History of Deer Island" (pamphlet) (Boston: n.d.).

143–144 flounder of Deer Island Flats: M. J. Moore and J. J. Stegeman, "Hepatic Neoplasms in Winter Flounder *Pleuronectes americanus* from Boston Harbor, Massachusetts, U.S.A.," *Diseases of Aquatic Organisms* 20 (1994): 33–48; R. A. Murchelano and R. E. Wolke, "Neoplasms and Nonneoplastic Liver Lesions in Winter Flounder, *Pseudopleuronectes americanus*, from Boston Harbor, Massachusetts," *EHP* 90 (1991): 17–26.

144 the Connecticut experiments: G. R. Gardner et al., "Carcinogenicity of Black Rock Harbor Sediment to the Eastern Oyster and Trophic Transfer of Black Rock Harbor Carcinogens from the Blue Mussel to the Winter Flounder," *EHP* 90 (1991): 53–66.

144–147 my description of fish with cancer in the pilgrimage passage is based in part on sworn testimony presented at the hearing on fish cancer epidemics before the Subcommittee on Fisheries and Wildlife Conservation and the Environment of the Committee on Merchant Marine and Fisheries, U.S. House of Representatives, 98th Congress: *The Causes of Reported Epidemics of Cancer in Fish and the Relationship between These Occurrences and Environmental Quality and Human Health*, serial no. 98-40 (Washington, D.C.: GPO, 21 Sept. 1983).

145 quotation by Leone Pippard: Pippard, "Ailing Whales."

145 clams in northern Maine: R. J. Van Beneden, "Molecular Analysis of Bivalve Tumors: Models for Environmental/Genetic Interactions," *EHP* 102, suppl. 12 (1994): 81–83; R. J. Van Beneden, "Implications for the Presence of Transforming Genes in Gonadal Tumors in Two Bivalve Mollusk Species," *Cancer Research* 53 (1993): 2976–79; G. R. Gardner, "Germinomas and Teratoid Siphon Anomalies in Softshell Clams, *Mya arenaria*, Environmentally Exposed to Herbicides," *EHP* 90 (1991): 43–51. Military dogs in Vietnam exposed to these same herbicides in the form of Agent Orange suffered from usually high rates of testicular cancer. See H. M. Hayes et al., "Excess of Seminomas Observed in Vietnam Service U.S. Military Working Dogs," *JNCI* 82 (1990): 1042–46.

145 elevated cancer rates: W. B. Riggan et al., *U.S. Cancer Mortality Rates and Trends, 1950–1979*, vol. 4, *Maps*, EPA/600/1-83/015e (Research Triangle Park, N.C.: EPA, Health Effects Research Laboratory, 1987).

145 quote from Rebecca Van Beneden: Personal communication.

145 English sole in Seattle: D. C. Malins and S. J. Gunselman, "Fourier-Transform Infrared Spectroscopy and Gas Chromatography–Mass Spec-

trometry Reveal a Remarkable Degree of Structural Damage in the DNA of Wild Fish Exposed to Toxic Chemicals," *Proceedings of the National Academy of Science* 91 (1994): 13038–41; M. S. Myers et al., "Relationships between Hepatic Neoplasms and Related Lesions and Exposure to Toxic Chemicals in Marine Fish from the U.S. West Coast," *EHP* 90 (1991): 7–15.

145–146 mummichogs in Virginia: W. K. Vogelbein, "Hepatic Neoplasms in the Mummichog *Fundulus heteroclitus* from a Creosote-Contaminated Site," *Cancer Research* 50 (1990): 5978–86.

146 catfish in Ohio: P. C. Baumann and J. C. Harshbarger, "Decline in Liver Neoplasms in Wild Brown Bullhead Catfish after Coking Plant Closes and Environmental PAHs Plummet," *EHP* 103 (1995): 168–70.

146 Fox River in Illinois: J. A. Couch and J. C. Harshbarger, "Effects of Carcinogenic Agents on Aquatic Animals: An Environmental and Experimental Overview," *Environmental Carcinogenesis Reviews* 3 (1985): 63–105; E. R. Brown et al., "Frequency of Fish Tumors Found in a Polluted Watershed as Compared to Nonpolluted Canadian Waters, *Canada Research* 33 (1973): 189–98.

146–147 earth mounds on Buffalo Rock: D. C. McGill, *Michael Heizer: Effigy Tumuli, The Reemergence of Ancient Mound Building* (New York: Abrams, 1990). Susan Post of the Illinois Natural History Survey informs me that hikers are no longer allowed to walk on the mounds, which are eroding.

SEVEN *earth*

149–150 Illinois now has approximately 80,000 farms. See IFB, *Farm and Food Facts* (Bloomington, Ill.: IFB, 1994), 4, 19.

150 changes in land ownership and disappearance of farm animals: J. Bender, *Future Harvest: Pesticide-Free Farming* (Lincoln: Univ. of Nebraska Press, 1994), 2.

150 decline in crop diversity: IDENR, *The Changing Illinois Environment: Critical Trends, Summary Report*, IDENR/RE-EA-94/05 (Springfield, Ill.: IDENR, 1994), 54–55.

151 postwar changes in agricultural economy: A. Rosenfeld et al., *Agrichemicals in America: Farmers' Reliance on Pesticides and Fertilizers, A Study of Trends over the Last 25 Years* (Washington, D.C.: Public Voice for Food and Health Policy, 1993); Dr. David Pimentel, Cornell University, personal communication.

151 crop rotation: Bender, *Future Harvest.*

151 alfalfa: Ibid.

151 insecticides as ecological narcotics: P. Debach and D. Rosen, *Biological Control by Natural Enemies*, 2nd ed. (Cambridge, England: Cambridge Univ. Press, 1991), 27–28.

152 pests resistant in 1950: Ibid., 27.

152 pests resistant in 1960: R. Carson, *Silent Spring* (Boston: Houghton Mifflin, 1962), 265. "The Rumblings of an Avalanche" is the title of Chapter 16.

152 pests resistant in 1990: C. A. Edwards, "The Impact of Pesticides on the Environment," in D. Pimentel and H. Lehman (eds.), *The Pesticide Question: Environment, Economics, and Ethics* (New York: Routledge, 1993),

13–46; G. P. Georghiou, "Overview of Insecticide Resistance," in M. B. Green et al. (eds.), *Managing Resistance to Agrochemicals* (Washington, D.C.: American Chemical Society, 1990), 18–41.

152 herbicide-resistant weeds: Edwards, "Impact of Pesticides."

152 quote from a recent study: S. B. Powles and J. A. M. Holtum, *Herbicide Resistance in Plants: Biology and Biochemistry* (Boca Raton, Fla.: Lewis, 1994), 2.

153 In addition to certain insect species, bats also serve an important role as a natural enemy of agricultural pests. A dietary mainstay of the big brown bat in the midwestern United States, for example, is the adult cucumber beetle. Mammalogist John Whittaker of Indiana State University estimates that an average-sized colony of these bats—which are found throughout North America—can consume approximately 38,000 such beetles in a single season. Such a consumption rate, Whittaker estimates, means that 18 *million* cucumber beetle larvae will not be produced in the following year. These larvae are known to farmers as corn rootworm, a nasty pest responsible for much of the insecticide use in Corn Belt agriculture. The gradual disappearance of the big brown bat throughout large segments of its former range is a matter of grave concern. Whether its numbers are declining because of habitat loss, pesticide poisoning, or some other yet-to-be identified cause is not known. See J. O. Whittaker, "Bats, Beetles, and Bugs: More Brown Bats Mean Less Agricultural Pests," *Bats* 11 (1993). See also J. O. Whittaker and P. Clem, "Food for the Evening Bat *Nycticeius Humeralis* from Indiana," *American Midland Naturalist* 127 (1992): 211–14.

153 parasitic wasps: R. Wiedenmann, "Using Natural Enemies for Pest Control," *INHS* 340 (1996): 5; Debach and Rosen, *Biological Control*, 52–58.

153 resurgence and secondary pests in *Silent Spring*: Carson, *Silent Spring*, 245–75.

153–154 unforeseen consequences of chemical pest control: D. Pimentel et al., "Assessment of Environmental and Economic Impacts of Pesticide Use," in Pimentel and Lehman, *Pesticide Question*, 54–55.

154 weeds as secondary pests: This phenomenon is known as "weed composition shifting." See NRC, *Ecologically Based Pest Management: New Solutions for a New Century* (Washington, D.C.: National Academy Press, 1996).

154 doubling of crop loss due to insect damage: D. Pimentel et al., "Assessment of Environmental and Economic Impacts of Pesticide Use," in Pimentel and Lehman, *Pesticide Question*, 47–84.

155 insect loss in corn: D. Pimentel et al., "Environmental and Economic Effects of Reducing Pesticide Use," *BioScience* 41 (1992): 402–9.

155 pesticide use in specific regions of the country: Rosenfeld, *Agrichemicals in America*.

155 reasons for declining use of insecticides: Ibid.; D. Pimentel and L. Levitan, "Pesticides: Amounts Applied and Amounts Reaching Pests," *BioScience* 36 (1986): 86–91.

155 ranking of corn: D. Pimentel et al., "Environmental and Economic Impacts of Reducing U.S. Agricultural Pesticide Use," in Pimentel and Lehman, *Pesticide Question*, 225.

156 natural history of soybeans: American Soybean Association, *Soy Stats: A*

Reference Guide to Important Soybean Facts and Figures (St. Louis: American Soybean Association, 1994); S. L. Post, "Miracle Bean," *The Nature of Illinois* (fall 1993): 1, 3; Illinois Soybean Association, *Soybeans: The Gold That Grows* (pamphlet) (Bloomington, Ill.: Illinois Soybean Association, n.d.).

157 natural history of corn: *The Nature of Corn* (pamphlet) (Springfield: Illinois State Board of Education, 1996).

158 average yields of corn and beans: IFB, *Farm and Food Facts*, 4.

159 Illinois weeds: IDENR, *The Changing Illinois Environment: Critical Trends*, vol. 3, IDENR/RE-EA-94/05 (Springfield, Ill.: IDENR, 1994), 84; R. L. Zimdahl, *Fundamentals of Weed Science* (San Diego: Academic Press, 1993); M. J. Chrispeels and D. Sadava, *Plants, Food, and People* (San Francisco: Freeman, 1977), 163–64.

159 density of seedbank: F. Forcella et al., "Weed Seedbanks of the U.S. Corn Belt: Magnitude, Variation, Emergence, and Application," *Weed Science* 40 (1992): 636–44.

159 history of cultivation: F. Knobloch, *The Culture of Wilderness: Agriculture as Colonization in the American West* (Chapel Hill: Univ. of North Carolina Press, 1996).

159 direction of weed control research: D. D. Buhler et al., "Integrated Weed Management Techniques to Reduce Herbicide Inputs in Soybeans," *Agronomy Journal* 84 (1992): 973–78.

159 genetic engineering of crops: "Herbicide-Tolerant Crops," *The Gene Exchange*, June 1994, pp. 6–8.

159–160 herbicides applied to corn and beans: IASS, *Agricultural Fertilizer and Chemical Usage: Corn—1993* (Springfield, Ill.: IDA, 1994); IASS, *Agricultural Fertilizer and Chemical Usage: Soybeans—1993* (Springfield, Ill.: IDA, 1994).

160 names of herbicides: Some trade names represent blends of two or more active ingredients; not all may be used in Illinois. See *Farm Chemicals Handbook '96* (Willoughby, Ohio: Meister, 1996); L. P. Gianessi and J. E. Anderson, *Pesticide Use in Illinois Crop Production* (Washington, D.C.: National Center for Food and Agricultural Policy, 1995); S. A. Briggs, *Basic Guide to Pesticides: Their Characteristics and Hazards* (Washington, D.C.: Taylor & Francis, 1992).

160 poisoning mechanisms of herbicides: A. Cobb, *Herbicides and Plant Physiology* (New York: Chapman & Hall, 1992).

160 2,4-D in Illinois: IASS, *Agricultural Fertilizer and Chemical Usage: Corn—1993* and *Agricultural Fertilizer and Chemical Usage: Soybeans—1993.*

160 atrazine as one of the top two: The other one is alachlor (Dr. Penelope Fenner-Crisp, EPA Office of Pesticide Programs, personal communication).

160 uses of triazine herbicides: "EPA Begins Special Review of Triazine Pesticides" (EPA press release, 10 Nov. 1994).

160 poisoning mechanism of triazines: J. W. Gronwald, "Resistance to Photosystem II Inhibiting Enzymes," in S. B. Powles and J. A. M. Holtum (eds.), *Herbicide Resistance in Plants: Biology and Biochemistry* (Boca Raton, Fla.: Lewis, 1994), 27–60; M. D. Devine et al., *Physiology of Herbicide Action* (Englewood Cliffs, N.J.: Prentice Hall, 1993), 113–40.

161 triazines in ground and surface water: B. Hileman, "Concerns Broaden over Chlorine and Chlorinated Hydrocarbons," *Chemical and Engineering News* (19 Apr. 1993): 11–20.
161 quote from the EPA: EPA, *The Triazine Herbicides, Atrazine, Simazine, and Cyanazine: Position Document 1, Initiation of Special Review*, OPP-30000-60, FRL-4919-5 (Washington, D.C.: EPA, 1994), 49.
161 triazines in raindrops: Ibid., 32–33.
161 effect of triazines on prairie plants: Ibid., 50.
161 triazines as carcinogens: Ibid., 1; "Triazine–Human Breast Cancer Possible Link Noted by EPA," *Pesticide and Toxic Chemical News* 23 (16 Nov. 1994): 3–4.
161 atrazine's restrictions in other countries: Hileman, "Concerns Broaden."
161 atrazine's use on corn: EPA, *Triazine Herbicides*, 52; IASS, *Corn—1993*.
161–162 uses of simazine: EPA, "EPA Begins Special Review."
162 quote by the EPA: EPA, *Triazine Herbicides*, 7.
162 demise of cyanazine: "Cyanazine Pesticide Voluntarily Cancelled and Uses Phased Out" (EPA press release, 2 Aug. 1995).
162 use of cyanazine in Illinois corn: IASS, *Corn—1993*.
162 atrazine and breast cancer in rats: EPA, *Triazine Herbicides*, 8–9; Hileman, "Concerns Broaden"; J. C. Eldridge, "A Hypothesis for Mammary Tumorigenesis in Female Sprague-Dawley Rats Exposed to Chlorotriazine Herbicides," *Journal of American College of Toxicology* 9 (1990): 650.
162 atrazine and menstrual cycling in rats: J. C. Eldridge, "Acute and Chronic Effects of Oral Chlorotriazine Administration on Rat Estrous Cycling Activity," *Biology of Reproduction* 44, suppl. 1 (1991): 133.
162 atrazine and hamster ovaries: C. Taets and A. L. Rayburn, "The Clastogenic Potential of Herbicides Found in Illinois Groundwater III," *Proceedings of the Sixth Annual Conference on Agricultural Chemicals in Illinois Groundwater* (Carbondale: Illinois Groundwater Consortium, 1996), 219–26; D. P. Biradar and A. L. Rayburn, "Flow Cytogenetic Analysis of Whole Cell Clastogenicity of Herbicides Found in Groundwater," *Archives of Environmental Contamination and Toxicology* 28 (1995): 13–17.
162 quote from author: "Further Study Needed to Weigh Risks Linked to Herbicide, Scientist Says" (Univ. of Illinois, Urbana-Champaign, press release, 2 Aug. 1995).
162 triazines' effect on sex hormones: S. H. Zahm and A. Blair, "Carcinogenic Risks from Pesticides," in General Motors Research Fund, *1992 Accomplishments in Cancer Research* (Philadelphia: J. B. Lippincott, 1993), 266–78.
162 triazines and ovarian cancer in Italy: A. Donna et al., "Triazine Herbicides and Ovarian Epithelial Neoplasms," *Scandinavian Journal of Work Environment and Health* 15 (1989): 47–53; A. Donna et al., "Ovarian Mesothelial Tumors and Herbicides: A Case-Control Study," *Carcinogenesis* 5 (1984): 941–42.
162–163 special review: EPA, *Triazine Herbicides*.
163 quote by EPA: Ibid., 17.
163–164 Carson on tolerances: Carson, *Silent Spring*, 181–84.
164 quote by Carson: Ibid., 183.

164 number of tolerances: C. M. Benbrook, *Pest Management at a Crossroads* (Yonkers, N.Y.: Consumers Union, 1996), 69. For a cogent summary of the follies of food tolerances, see also J. Wargo, *Our Children's Toxic Legacy: How Science and Law Fail to Protect Us from Pesticides* (New Haven, Conn.: Yale Univ. Press, 1996); and C. Osteen, "Pesticide Use Trends and Issues in the United States," in Pimentel and Lehman, *Pesticide Question*, 309–36.

164 effects on children: NRC, *Pesticides in the Diets of Infants and Children* (Washington, D.C.: National Academy Press, 1993), 8.

165 pesticides in baby food: R. Wiles and K. Davies, *Pesticides in Baby Food* (Washington, D.C.: Environmental Working Group, 1995).

165 deficient enforcement: GAO, *Food Safety: Changes Needed to Minimize Unsafe Chemicals in Food*, Report to the Chairman, Human Resources and Intergovernmental Relations Subcommittee, Committee on Government Operations, House of Representatives, GAO/RCED-94-192 (Washington D.C.: GAO, 1994).

165 residues in 35 percent of food: Edwards, "Impact of Pesticides."

165 review of FDA monitoring data: S. Elderkin et al., *Forbidden Fruit: Illegal Pesticides in the U.S. Food Supply* (Washington, D.C.: Environmental Working Group, 1995).

165 quote by FDA: "F.D.A. Is Accused of Ignoring Illegal Use of Pesticides on Produce," *New York Times*, 17 Feb. 1995, p. A-25.

165–166 fraction of food shipments tested by FDA: Elderkin, *Forbidden Fruit*, 10–11.

166 sensitivity of compliance testing: Fenner-Crisp, personal communication.

166 quote by the USDA: USDA, *Pesticide Data Program, Annual Summary Calendar Year 1993* (Washington, D.C.: USDA, Agricultural Marketing Service, 1993), vii.

166 banning pesticides not coordinated with revoking tolerances: GAO, *Pesticides: Reducing Exposure to Residues of Canceled Pesticides*, Report to the Chairman, Environment, Energy, and Natural Resources Subcommittee, Committee on Government Operations, House of Representatives, GAO/RCED-95-23 (Washington, D.C.: GAO, 1994), 2, 4–5.

166 estimate of EPA official: Ibid., 6.

166–167 Food Quality Protection Act: The word *safe* is defined as a "reasonable certainty that no harm will result from aggregate exposure to the pesticide chemical residue, including all anticipated dietary exposures for which there is reliable information" (Benbrook, *Pest Management*, 108). J. H. Cushman Jr., "Pesticide Bill Advances in House without Rancor or Opponents," *New York Times*, 8 July 1996, pp. A-1, A-20; B. Riley, "Congress Enacts a Faustian Safety Bargain," *Journal of Pesticide Reform* 16, no. 3 (1996): 12–13; Science and Environmental Health Policy Project, *Beyond Delaney: Preventing Exposure to Hazardous Pesticides* (Washington, D.C.: Physicians for Social Responsibility and Environment Working Group, 1995).

167–168 pesticides in fish: GAO, *Pesticides: Reducing Exposure*, 4–5, 17–27, 35–37.

168 pesticides in mothers' milk: This topic is discussed in detail in Chapter 11.

169 chlorinated insecticides in animal products: NRC, *Animals as Sentinels of Environmental Health Hazards* (Washington, D.C.: National Academy Press, 1991); R. Spear, "Recognized and Possible Exposure to Pesticides," in W. J. Hayes Jr. and E. R. Laws Jr. (eds.), *Handbook of Pesticide Toxicology*, vol. 1 (New York: Academic Press, 1991), 245–74; J. A. Pennington and E. L. Gunderson, "History of the Food and Drug Administration's Total Diet Study—1961 to 1987," *Journal of the Association of Official Analytical Chemists* 70 (1987): 772–82.

169 average DDT intake: D. V. Reed et al., "Chemical Contaminants Monitoring: The FDA Pesticides Monitoring Program," *Journal of the Association of Official Analytical Chemists* 70 (1987): 591–95.

169 average dieldrin intake: Pennington and Gunderson, "Total Diet Study."

169 Carson's remark: Carson, *Silent Spring*, 173–84.

169 Iowa soybean farmers: J. Feldman, "Risk Assessment: A Community Perspective," *EHP* 103, suppl. 6 (1995): 153–58.

170 how-to manual: Bender, *Future Harvest.*

170 1989 report: NRC, *Alternative Agriculture* (Washington, D.C.: National Academy Press, 1989). See also the 1996 NRC report, *Ecologically Based Pest Management*.

170 subsequent studies: Benbrook, *Pest Management;* D. Pimentel, "Environmental and Economic Effects of Reducing Pesticide Use," *BioScience* 41 (1991): 402–9.

170 polls of farmers and consumers: Rosenfeld, *Agrichemicals in America*, 1, 9; C. E. Sachs, "Growing Public Concerns over Pesticides in Food and Water," in Pimentel and Lehman, *Pesticide Question.*

170 one in three shoppers: "Shoppers Seeking Out Organic Produce," *Wall Street Journal*, 23 Sept. 1994, p. B-1.

170 sale of organic food: M. Burros, "Developing a Taste for Organic Milk," *New York Times*, 30 Oct. 1996, p. C-16.

170 supporting ecological methods of pest control: The pesticide industry spends more on print advertising than the federal government does in supporting research into alternative agriculture (Benbrook, *Pest Management*).

170–171 true costs of pesticides: H. Wade et al., *The Interagency Study of the Impact of Pesticide Use on Ground Water in North Carolina* (Raleigh: North Carolina Pesticide Board, 1997); D. Pimentel, "Environmental and Economic Costs of Pesticide Use, *BioScience* 42 (1992): 750–60. See also Benbrook et al. (*Pest Management*), who point out that efforts to regulate pesticides siphon billion of dollars of public funds to feed a fractured, overwhelmed research and regulatory apparatus.

EIGHT *air*

174 Paracelsus: M. P. Hall, *The Secret Teachings of All Ages* (Los Angeles: Philosophical Research Society, 1988), CVII–CVIII.

175 DDT and PCBs in Hubbard Brook: W. H. Smith et al., "Trace Organochlorine Contamination of the Forest Floor of the White Mountain National Forest, New Hampshire," *Environmental Science and Technology* 27 (1993): 2244–46; "DDT and PCBs, Long Banned in the U.S., Found in

Remote Forest, Suggesting Global Distribution via the Atmosphere" (Yale Univ. press release, 14 Dec. 1993).

175–176 rain-fed bogs: R. A. Rapaport et al., "'New' DDT Inputs to North America: Atmospheric Deposition," *Chemosphere* 14 (1985): 1167–73.

176 DDT in Mexico: L. López-Carrillo, "Is DDT Use a Public Health Problem in Mexico?" *EHP* 104 (1996): 584–88.

176 a lake in England: G. Sanders et al., "Historical Inputs of Polychlorinated Biphenyls and Other Organochlorines to a Dated Lacustrine Sediment Core in Rural England," *Environmental Science and Technology* 26 (1992): 1815–21.

176 the world's trees: S. L. Simonich and R. A. Hites, "Global Distribution of Persistent Organochlorine Compounds," *Science* 269 (1995): 1851–54.

176–177 global distillation: J. Raloff, "The Pesticide Shuffle," *Science News* 149: 174–75; B. G. Loganathan and K. Kannon, "Global Organochlorine Contamination Trends: An Overview," *Ambio* 23 (1994): 187–91; F. Wania and D. Mackay, "Global Fractionation and Cold Condensation of Low Volatility Organochlorine Compounds in Polar Regions," *Ambio* 22 (1993): 10–18.

177 Lake Baikal: J. R. Kucklick et al., "Organochlorines in the Water and Biota of Lake Baikal, Siberia," *Environmental Science and Technology* 28 (1994): 31–37.

177 quote about Lake Laberge: K. A. Kidd et al., "High Concentrations of Toxaphene in Fishes from a Subarctic Lake," *Science* 269 (1995): 240–42. See also J. Raloff, "Fishy Clues to a Toxaphene Puzzle," *Science News* 148 (1995): 38–39.

178 PCBs in the Great Lakes: EPA, *Deposition of Air Pollutants to the Great Waters*, EPA-453/R-93-055 (Washington, D.C.: EPA, 1994), x, 2.

178 quote from the EPA: Ibid., 71.

178 sport fishers in Wisconsin: B. J. Fiore et al., "Sport Fish Consumption and Body Burden Levels of Chlorinated Hydrocarbons: A Study of Wisconsin Anglers," *AEH* 44 (1989): 82–88.

179 meal of freshwater fish: The Great Lakes basin, for example, accumulates toxics from throughout the hemisphere. Researchers have calculated that a single meal of Great Lakes trout or salmon contains a dose of PCBs equivalent to drinking five liters of Great Lakes water every day for more than 200 years. See J. A. Foran et al., "Sport Fish Consumption Advisories and Projected Cancer Risks in the Great Lakes Basin," *AJPH* 79 (1989): 322–25. See also M. A. Hovinga et al., "Environmental Exposure and Lifestyle Predictors of Lead, Cadmium, PCB, and DDT Levels in Great Lakes Fish Eaters," *AEH* 48 (1993): 98–104; and J. J. Black and P. C. Baumann, "Carcinogens and Cancers in Freshwater Fish," *EHP* 90 (1991): 27–33.

179 percentage of pesticides reaching target pests: D. Pimentel and L. Levitan, "Pesticides: Amounts Applied and Amounts Reaching Pests," *BioScience* 36 (1986): 86–91.

179 emissions into air: EPA, *1992 Toxics Release Inventory*, EPA 745-R-94-001 (Washington, D.C.: EPA, 1994), 3, 79; D. A. Sheiman et al., *A Who's Who of American Toxic Air Polluters: A Guide to More than 1500 Factories in 46 States Emitting Cancer-Causing Chemicals* (New York: NRDC, 1989).

179 airborne carcinogens: A. Pintér et al., "Mutagenicity of Emission and Immission Samples around Industrial Areas," in H. Vainio et al. (eds.), *Complex Mixtures and Cancer Risk*, IARC Scientific Pub. 104 (Lyon, France: IARC, 1990), 269–76.

179 number of cities failing to meet standards: T. Wagner, *In Our Backyard: A Guide to Understanding Pollution and Its Effects* (New York: Van Nostrand Reinhold, 1994), 78.

179 number of Americans breathing illegal air: E. Friebele, "The Attack of Asthma," *EHP* 104 (1996): 22–25.

180 actual contribution elusive: G. Pershagen, "Air Pollution and Cancer," in Vainio, *Complex Mixtures*, 240–51.

180 epidemiological dilemma: C. M. Shy and R. J. Struba, "Air and Water Pollution," in D. Schottenfeld and J. F. Fraumeni Jr. (eds.), *Cancer Epidemiology and Prevention* (Philadelphia: Saunders, 1982), 346. For an updated discussion, see C. M. Shy, "Air Pollution," in D. Schottenfeld and J. F. Fraumeni Jr. (eds.), *Cancer Epidemiology and Prevention*, 2nd ed. (Oxford, England: Oxford Univ. Press, 1996), 406–17.

180 fluidity of air: F. E. Speizer and J. M. Samet, "Air Pollution and Lung Cancer," in J. M. Samet (ed.), *Epidemiology of Lung Cancer* (New York: Marcel Dekker, 1994), 131–50; Wagner, *In Our Backyard*, 80; K. Hemminki, "Measurement and Monitoring of Individual Exposures," in L. Tomatis (ed.), *Air Pollution and Human Cancer* (New York: Springer-Verlag, 1990), 35–47.

180 transmutational quality of air: L. Fishbein, "Sources, Nature, and Levels of Air Pollutants," in Tomatis, *Air Pollution*, 9–34; L. Lewtas, "Experimental Evidence for Carcinogenicity of Air Pollutants," in ibid., 49–61.

180–181 ozone: K. Breslin, "The Impact of Ozone," *EHP* 103 (1995): 660–64; G. J. Jakab et al., "The Effects of Ozone on Immune Function," *EHP* 103, suppl. 2 (1995): 77–89.

181 survival rate of lung cancer: "Lung Cancer State at Diagnosis," *JNCI* 87 (1995): 1662.

182 guilt and blame: "Lung Cancer: Dying in Disgrace?" *Harvard Health Letter* 20 (1995): 4–6.

182 primacy of tobacco: J. M. Samet (ed.), *Epidemiology of Lung Cancer* (New York: Marcel Dekker, 1994).

182 "cigarette science": "Cigarette Science at Johns Hopkins," *Rachel's Environment and Health Weekly*, no. 464, 19 Oct. 1995.

182 more to the story than cigarettes: "Dying in Disgrace?"

182 lung cancer among nonsmokers: R. C. Brownson et al., "Lung Cancer in Nonsmoking Women: Histology and Survival Patterns," *Cancer* 75 (1995): 29–33.

182 deaths due to environmental tobacco smoke: T. Reynolds, "EPA Finds Passive Smoking Causes Lung Cancer," *JNCI* 85 (1993): 179–80.

182 unavoidable and interactive effects of air pollution: K. Hemminki and G. Pershagen, "Cancer Risk of Air Pollution: Epidemiological Evidence," *EHP* 102 (1994): 187–92.

182 adenocarcinoma: "Dying in Disgrace?"; Brownson, "Nonsmoking Women."

182 quote from *Harvard Health Letter*: "Dying in Disgrace?"

182–183 ecologic studies: These are reviewed in Hemminki and Pershagen, "Cancer Risk of Air Pollution"; and in G. Pershagen and L. Simonato, "Epidemiological Evidence on Air Pollution and Cancer," in Tomatis, *Air Pollution*.

183 chemical plants, paper mills, and petroleum industries: W. J. Blot, and J. F. Fraumeni Jr., "Geographic Patterns of Lung Cancer: Industrial Correlations," *AJE* 103 (1976): 539–50.

183 one recent cohort study: D. W. Dockery, "An Association between Air Pollution and Mortality in Six U.S. Cities," *NEJM* 329 (1993): 1753–59.

183 Swedish chimney sweeps: P. Gustavsson et al., "Excess Mortality among Swedish Chimney Sweeps," *British Journal of Industrial Medicine* 44 (1987): 738–43.

183 case-control studies: These are reviewed in D. Trichopoulos and E. Petridou, "Epidemiologic Studies and Cancer Etiology in Humans," *Medicine, Exercise, Nutrition, and Health* 3 (1993): 206–25.

183 study of Italian men: F. Barbone et al., "Air Pollution and Lung Cancer in Trieste, Italy," *AJE* 141 (1995): 1161–69.

183 follow-up study: A. Biggeri et al., "Air Pollution and Lung Cancer in Trieste, Italy: Spatial Analysis of Risk as a Function of Distance from Sources," *EHP* 104 (1996): 750–54.

183 Chinese studies: A. Hricko, "Environmental Problems behind the Great Wall," *EHP* 102 (1994): 154–59; L. Tomatis, "Air Pollution and Cancer: A New and Old Problem," Tomatis, *Air Pollution*, 1–7.

184 bladder cancer and diesel exhaust: D. Trichopoulos and E. Petridou, "Epidemiologic Studies and Cancer Etiology in Humans," *Medicine, Exercise, Nutrition, and Health* 3 (1994): 206–25.

184 breast cancer and air pollution: J. M. Melius et al., *Residence near Industries and High Traffic Areas and the Risk of Breast Cancer on Long Island* (Albany: New York State Dept. of Health, 1994).

184 benzo[a]pyrene and breast cancer: J. J. Morris and E. Seifter, "The Role of Aromatic Hydrocarbons in the Genesis of Breast Cancer," *Medical Hypotheses* 38 (1992): 177–84.

184 nitrogen dioxide and lung tumors: K. A. Fackelmann, "Air Pollution Boosts Cancer Spread," *Science News* 137 (1990): 221; A. Richters, "Effects of Nitrogen Oxide and Ozone on Blood-Borne Cancer Cell Colonization of the Lungs," *JTEH* 25 (1988): 383–90.

184 spread of cancer to lungs: E. Ruoslahti, "How Cancer Spreads," *Scientific American*, Sept. 1996, pp. 72–77.

184 quote by Richters: Fackelman, "Air Pollution Boosts," 221.

185–188 Quotations are from, in the order of presentation, the following sources: Biggeri, "Air Pollution and Lung Cancer," 750; Pershagen and Simonato, "Epidemiological Evidence on Air Pollution and Cancer," 67; ibid., 69; C. W. Sweet and S. J. Vermette, *Toxic Volatile Organic Chemicals in Urban Air in Illinois*, HWRIC RR-057 (Champaign: Hazardous Waste Research and Information Center, 1991), 1; Hemminki and Pershagen, "Cancer Risk of Air Pollution," 191; and Lewtas, "Experimental Evidence," 58.

186–187 the miasma theory: S. N. Tesh, *Hidden Arguments: Political Ideology and Disease Prevention Policy* (New Brunswick, N.J.: Rutgers Univ. Press, 1988), 8, 25–32.

187 rise in asthma rates: Friebele, "Attack of Asthma"; "40% Rise Reported in Asthma and Asthma Deaths," *New York Times*, 7 Jan. 1995, p. A-10.

187 rise in asthma mortality: CDC, "Asthma—United States, 1982–1992," *Morbidity and Mortality Weekly Report*, 6 Jan. 1995, 952–55.

188 asthma's link to air pollution: Friebele, "Attack of Asthma"; and CDC, "Asthma—United States." See also the collection of papers presented at the National Urban Air Toxics Research Center's 1994 conference, "Asthma as an Air Toxics End Point," and published in *EHP* 103, suppl. 6 (1995): 209–71.

NINE *water*

189–190 photograph of mussel gatherers: L. M. Talkington, *The Illinois River: Working for Our State*, Misc. Pub. 128 (Champaign: ISWS, 1991), 11.

190 demise of button factories: Ibid., 10–11.

190 demise of diving ducks: H. B. Mills, *Man's Effect on the Fish and Wildlife of the Illinois River*, Biological Notes 57 (Urbana: INHS, 1966).

190 demise of scaups and fingernail clams: F. C. Bellrose et al., *Waterfowl Populations and the Changing Environment of the Illinois River Valley*, Bulletin 32 (Urbana: INHS, 1979); Mills, *Man's Effect.*

190 quotes on bird identification: From D. and L. Stokes, *Stokes Field Guide to Birds, Eastern Region* (Boston: Little, Brown, 1996); National Geographic Society, *Field Guide to the Birds of North America*, 2nd ed. (Washington, D.C.: National Geographic Society, 1987); C. S. Robbins et al., *Birds of North America* (New York: Golden Press, 1966).

190 demise of dabbling ducks and aquatic plants: E. Hopkins, "Pollution Keeps Preying on Plants in Illinois River," *PJS*, 25 July 1993, p. A-2; Mills, *Man's Effect.*

190–191 fishing industry on the Illinois River: P. Ross and R. Sparks, *Identification of Toxic Substances in the Upper Illinois River*, Report 283 (Urbana: INHS, 1989).

191 geology and ecology of the Illinois River: M. Runkle, "Plight of the Illinois: A River in Transition," *Illinois Audubon* 236 (1991): 2–7; Talkington, *Illinois River;* Bellrose, *Waterfowl Populations;* Doug Blodgett, Illinois Natural History Survey, personal communication.

191 S&S Canal: Talkington, *Illinois River;* Bellrose, *Waterfowl Populations.*

191 photograph of Illinois fish: Mills, *Man's Effect.*

191–192 improvement after 1972: IDENR, *The Changing Illinois Environment: Critical Trends*, summary report, ILENR/RE-EA-94/05 (SR) 20M (Springfield, Ill.: IDENR, 1994), 16–17.

192 fish advisories: *Illinois 1994 Fishing Information* (Springfield, Ill.: IDC, 1994).

192 impact of barges and tugs: T. A. Butts and D. B. Shackleford, *Impacts of Commercial Navigation on Water Quality in the Illinois River Channel*, Research Report 122 (Champaign: ISWS, 1992); R. M. Sparks, "River Watch: The Surveys Look after Illinois' Aquatic Resources," *The Nature of*

Illinois (winter 1992): 1–4; W. J. Tucker, *An Intensive Survey of the Illinois River and Its Tributaries: A Comparison Study of the 1967 and 1978 Stream Conditions* (Springfield, Ill.: IEPA, n.d.); Runkle, "Plight of the Illinois."

192 toxic spills: M. Demissie and L. Keefer, *Preliminary Evaluation of the Risk of Accidental Spills of Hazardous Materials in Illinois Waterways*, HWRIC RR-055 (Champaign: Hazardous Waste Research and Information Center, 1991); Talkington, *Illinois River*, 14; "The Illinois River: Its History, Its Uses, Its Problems," *Currents* 5 (Champaign: ISWS, Jan.–Feb. 1993), 1–12; Blodgett, personal communication.

192 routine industrial discharges: E. Hopkins, "New Rules, Industry Initiatives May Cut Toxic Dumping in River," *PJS*, 19 Mar. 1995, p. A-23.

193 disappearance of fish, amphibians, crayfish, mussels: IDENR, *The Changing Illinois Environment*, 19–22; J. H. Cushman, "Freshwater Mussels Facing Mass Extinction," *New York Times*, 3 Oct. 1995, pp. C-1, C-7.

193 poem by Robert Frost: "The Oven Bird," in E. C. Lathem (ed.), *The Poetry of Robert Frost* (New York: Henry Holt, 1969).

193–194 description of maximum contaminant levels: EPA, *Drinking Water Regulations and Health Advisories*, EPA 822-R-96-001 (Washington, D.C.: EPA, 1996).

194 examples of specific MCLs: Ibid.

194 MCLs not a health-based standard: EPA, *Drinking Water Standard Setting: Question and Answer Primer*, G-206 (Washington, D.C.: EPA, 1994), 12.

194 maximum-contaminant-level goals: EPA, *Drinking Water Regulations*, i.

194 deficiencies in regulating drinking-water contaminants: *Trouble on Tap: Arsenic, Radioactive Radon, and Trihalomethanes in Our Drinking Water* (New York: NRDC, 1995); E. D. Olsen, *Think Before You Drink: The Failure of the Nation's Drinking Water System to Protect Public Health* (New York: NRDC, 1993).

195 cyanazine in drinking water: EPA, "Cyanazine: Notice of Preliminary Determination to Terminate Special Review; Notice of Receipt of Requests for Voluntary Cancellation," *Federal Register* 61 (1 Mar. 1996): 8185–203. (See especially pp. 8191–95.)

195 quote from NRC: NRC, *Environmental Epidemiology: Public Health and Hazardous Wastes* (Washington D.C.: National Academy Press, 1991), 10.

195 averaging contaminant levels: D. A. Goolsby et al., *Occurrence and Transport of Agricultural Chemicals in the Mississippi River Basin, July–August 1993*, Circular 1120-C (Boulder, Colo.: U.S. Geological Survey, 1993).

195 elevation of herbicides during spring planting: A. G. Taylor and S. Cook, "Water Quality Update: The Results of Pesticide Monitoring in Illinois' Streams and Public Water Supplies" (paper presented at the 1995 Illinois Agricultural Pesticides Conference, Univ. of Illinois, Urbana, 4–5 Jan. 1995).

195 1995 study of herbicides in drinking water: B. Cohen et al., *Weed Killers by the Glass: A Citizens' Tap Water Monitoring Project in 29 Cities* (Washington, D.C.: Environmental Working Group, 1995).

196 Children have higher rates of cell proliferation, as well as less developed detoxifying mechanisms. For a thoughtful discussion of age-related sensitivities to carcinogens, see J. Wargo, *Our Children's Toxic Legacy: How Sci-*

ence and Law Fail to Protect Us from Pesticides (New Haven, Conn.: Yale Univ. Press, 1996), 191–99.

196 Safe Drinking Water Act: S. Terry, "Drinking Water Comes to a Boil," *New York Times Magazine*, 26 Sept. 1993, pp. 42–65.

196 monitoring for farm chemicals in Illinois: IEPA, *Water Quality in Illinois, 1990–1991* (Springfield, Ill.: IEPA, IEPA/WPC/92-224, 1993); A. G. Taylor, "Pesticides in Illinois Public Water Supplies: Complying with the New Federal Drinking Water Standards," in *1993 Illinois Agricultural Pesticides Conference: Summary of Presentations* (Urbana: Univ. of Illinois Cooperative Extension Service, 1993).

197 1996 amendments: "Right-to-Know Added to Drinking Water Law," *Working Notes on Community Right-to-Know*, July–Aug. 1996, p. 1; J. H. Cushman, "Environment Bill's Approval Now Likely after Panel's Vote," *New York Times*, 7 June 96, p. A-28.

197 inhalation and skin absorption of volatile organic compounds: B. Lévesque et al., "Evaluation of Dermal and Respiratory Chloroform Exposure in Humans," *EHP* 102 (1994): 1082–87; C. P. Weisel and W. J. Chen, "Exposure to Chlorination By-products from Hot Water Uses," *Risk Analysis* 14 (1994): 101–6.

197 danger to women and infants: C. W. Forrest and R. Olshansky, *Groundwater Protection by Local Government* (Springfield, Ill.: IDENR and IEPA, 1993), 16.

197–198 a 1996 study: C. P. Weisel and Wan-Kuen Jo, "Ingestion, Inhalation, and Dermal Exposures to Chloroform and Trichloroethene from Tap Water," *EHP* (104): 48–51.

198 route of exposure: Ibid.; N. I. Maxwell et al., "Trihalomethanes and Maximum Contaminant Levels: The Significance of Inhalation and Dermal Exposures to Chloroform in Household Water," *Regulatory Toxicology and Pharmacology* 14 (1991): 297–312.

198 quote by Weisel and Jo: Weisel and Jo, "Ingestion, Inhalation," 48.

198–199 Rockford, Illinois: J. E. Keller and S. W. Metcalf, *Exposure Study of Volatile Organic Compounds in Southeast Rockford*, Epidemiological Report Series 91:3 (Springfield, Ill.: IDPH, 1991). See also K. Mallin, "Investigation of a Bladder Cancer Cluster in Northwestern Illinois," *AJE* 132, suppl. 1 (1990): 96–106.

199 The 1918 survey form: IEPA, *Pilot Groundwater Protection Program Needs Assessment for Pekin Public Water Supply Facility Number 1795040* (Springfield, Ill.: IEPA, Division of Public Water Supplies, 1992), appendix C.

199–200 lengthy report on Pekin's groundwater: IEPA, *Pilot Groundwater Protection.*

200 city's response: T. L. Aldous, "Committee Examines Aquifer Protection," *PDT*, 11 Dec. 1993, pp. A-1, A-12.

200 quote by mayor: Ibid.

200 a recent review by Kenneth Cantor: K. P. Cantor et al., "Water Pollution," in D. Schottenfeld and J. F. Fraumeni Jr. (eds.), *Cancer Epidemiology and Prevention*, 2nd ed. (Oxford, England: Oxford Univ. Press, 1996), 418–37.

200 most studies are ecological: Ibid.

201 ecologic studies of drinking water and cancer: L. D. Budnick et al., "Cancer

and Birth Defects near the Drake Superfund Site, Pennsylvania," *AEH* 39 (1984): 409–13; A. Aschengrau et al., "Cancer Risk and Tetrachloroethylene-Contaminated Drinking Water in Massachusetts," *AEH* 48 (1993): 284–92; J. Griffith et al., "Cancer Mortality in U.S. Counties with Hazardous Waste Sites and Ground Water Pollution," *AEH* 44 (1989): 69–74.

201 New Jersey study: J. Fagliano et al., "Drinking Water Contamination and the Incidence of Leukemia: An Ecologic Study," *AJPH* 80 (1990): 1209– 12.

201 Iowa study: Cited in Cantor, "Water Pollution."

201 Woburn study: S. W. Lagakos et al., "An Analysis of Contaminated Well Water and Health Effects in Woburn, Massachusetts," *Journal of the American Statistical Association* 395 (1986): 583–96. This community's struggle to receive recompense for their damages is documented in J. Harr, *A Civil Action* (New York: Random House, 1995).

201 North Carolina study: NRC, *Environmental Epidemiology*, 188; J. S. Osborne et al., "Epidemiologic Analysis of a Reported Cancer Cluster in a Small Rural Population," *AJE* 132, suppl. 1 (1990): 87–95.

201 Chinese study: Cited in Cantor, "Water Pollution."

201 German study: W. Hoffmann et al., "Radium-226–Contaminated Drinking Water: Hypothesis on an Exposure Pathway in a Population with Elevated Childhood Leukemias," *EHP* 101, suppl. 3 (1993): 113–15.

201–202 Finnish study: P. Lampi et al., "Cancer Incidence following Chlorophenol Exposure in a Community in Southern Finland," *AEH* 47 (1992): 167–75.

202 history of water chlorination: R. D. Morris et al., "Chlorination, Chlorination By-products, and Cancer: A Meta-analysis," *AJPH* 82 (1992): 955–63; S. Zierler, "Bladder Cancer in Massachusetts Related to Chlorinated and Chloraminated Drinking Water: A Case-Control Study, *AEH* 43 (1988): 195–200; R. L. Jolley et al. (eds.), *Water Chlorination: Chemistry, Environmental Impact, and Health Effects*, vol. 5 (Chelsea, Mich.: Lewis, 1985).

202 link between water chlorination and bladder and rectal cancers: Cantor, "Water Pollution."

202 quote by Kenneth Cantor: K. Cantor, "Water Chlorination, Mutagenicity, and Cancer Epidemiology" (editorial), *AJPH* 84 (1994): 1211–13.

202–203 disinfection by-products and trihalomethanes: Morris, "Chlorination." Trihalomethanes were first identified in drinking water in the 1970s. Only a few years ago, researchers discovered another group of volatile disinfection by-products: the haloacetic acids. There are undoubtedly others. Comprehensive knowledge about all of the chemical offspring of water chlorination is still lacking. Only about half (by weight) of the chlorinated by-products found in drinking water have thus far been identified (Ronnie Levin, EPA, personal communication).

203 regulation of trihalomethanes: Olsen, *Think Before You Drink*.

203 EPA chart: EPA, *National Primary Drinking Water Standards*, EPA 810-F-94-001A (Washington, D.C.: EPA, Office of Water, 1994).

203–204 Studies of water chlorination and bladder/rectal cancers are summarized in Cantor, "Water Pollution."

204 Kenneth Cantor's study: K. P. Cantor et al., "Bladder Cancer, Drinking

Water Source, and Tap Water Consumption: A Case-Control Study," *JNCI* 79 (1987): 1269–79.

204 quote by Cantor: Cantor, "Water Pollution," 427.

204–205 alternatives to chlorination: Cantor, "Water Pollution"; B. A. Cohen and E. D. Olsen, *Victorian Water Treatment Enters the 21st Century: Public Health Threats from Water Utilities' Ancient Treatment and Distribution Systems* (New York: NRDC, 1994); Terry, "Drinking Water." None of these alternatives alone provides a perfect solution. Moving chlorination to the end of the process, for example, decreases its killing time and so may increase the numbers of microoganisms in finished drinking water. There is no technological substitute for watershed protection.

206 New Jersey court examiner: Cantor, "Water Chlorination, Mutagenicity."

206–207 Sankoty Aquifer: W. H. Walker et al., *Preliminary Report on the Groundwater Resources of the Havana Region in West-Central Illinois,* Cooperative Groundwater Report 3 (Urbana: ISGWS, 1965); L. Horberg et al., *Groundwater in the Peoria Region,* Cooperative Bulletin 39 (Urbana: ISGWS, 1950).

207 types of aquifers: Horberg, *Groundwater in the Peoria Region,* 16; "Surveying Groundwater," *The Nature of Illinois* (winter 1992): 9–12.

207 1989 survey: IEPA, *Illinois American Water Company, Pekin, Facility Number 1795040 Well Site Survey Report* (Springfield, Ill.: IEPA, 1989).

207 Creve Coeur advisory: S. L. Burch and D. J. Kelly, *Peoria-Pekin Regional Ground-Water Quality Assessment,* Research Report 124 (Champaign: ISWS, 1993).

207 contaminants in North Pekin wells: Ibid.

208 quote from 1993 assessment: Burch and Kelly, *Peoria-Pekin Regional,* 56.

208 recharge areas: IEPA, *A Primer Regarding Certain Provisions of the Illinois Groundwater Protection Act* (Springfield, Ill.: IEPA, 1988); ISGS, *Ground-Water Contamination: Problems and Remedial Action,* Environmental Geology Notes 81 (Champaign: ISGS, 1977).

208–209 difficulty of remediation: W. T. Piver, "Contamination and Restoration of Groundwater Aquifers," *EHP* 100 (1992): 237–47; IDPH, *Chlorinated Solvents in Drinking Water* (Springfield, Ill.: IDPH, n.d.); ISGS, *Ground-Water Contamination.*

209 Pekin's ordinance: City of Pekin Groundwater Protection Area Ordinance.

209 growing public awareness: D. Rheingold, "Pekin Readies Water Watch," *PDT,* 17 Jan. 1994, pp. A-1, A-12.

209 quote by gas station owner: Ibid.

209 observation by superintendent: Kevin W. Caveny, personal communication.

210 ongoing detections in Pekin's wells: Kevin W. Caveny, personal communication; Interagency Coordinating Committee on Groundwater, *Illinois Groundwater Protection Program,* vols. 1 and 2, *Biennial Technical Appendices Report* (Springfield, Ill.: IEPA, 1994). In total, as of 1994, there were nine contaminated public water wells in Tazewell County; four of these were in Pekin, three in East Peoria, and one each in Mackinaw, Marquette Heights, North Pekin, and South Pekin.

210 sooner or later: contaminants in groundwater typically move between one

inch and one foot each year. The contaminants we drink now may be those spilled onto the ground decades ago.

210 percentage of Americans drinking ground and surface water: Ronnie Levin, EPA, personal communication.

210 observation by Rachel Carson: R. Carson, *Silent Spring* (Boston: Houghton Mifflin, 1962), 42.

210 My imaginary description of subterranean Illinois is inspired by old Tazewell County drilling logs, as well as by geological background information provided in M. A. Marino and R. J. Shichts, *Groundwater Levels and Pumpage in the Peoria-Pekin Area, Illinois, 1890–1966* (Urbana: ISWS, 1969), and in Horberg, *Groundwater in the Peoria Region.*

TEN *fire*

212 Epigraph is from John Knoepfle, *Poems from the Sangamon* (Urbana: Univ. of Illinois Press, 1985).

213–214 the plan: T. L. Aldous, "Developer Proposes a Site for Burner," *PDT*, 22 July 1992, pp. A-1, A-12.

214 recycling competes with incinerators: K. Schneider, "Burning Trash for Energy: Is It an Endangered Species?" *New York Times*, 11 Oct. 1994, p. C-18.

214 Columbus incinerator: Ibid.; S. Powers, "From Trash Burner to Cash Burner," *Columbus Dispatch*, 4 Sept. 1994, p. B-6; S. Powers, "Board Votes to Close Trash Plant," *Columbus Dispatch*, 2 Nov. 1994, p. A-1.

214 Albany incinerator: K. Nelis and R. Pitlyk, "Snow, Then Soot: ANSWERS Fallout a Blizzard of Blackness," *Albany Times Union*, 11 Jan. 1994, p. B-1.

214 incinerators release dioxin: D. R. Zook and C. Rappe, "Environmental Sources, Distribution and Fate of Polychlorinated Dibenzodioxins, Dibenzofurans, and Related Organochlorines," in A. Schecter (ed.), *Dioxins and Health* (New York: Plenum, 1994), 79–113.

214 dioxin harmful in trace amounts: T. Webster and B. Commoner, "Overview: The Dioxin Debate," in Schecter, *Dioxins and Health*, 1–50.

214 draft reassessment: EPA, *Estimating Exposure to Dioxin-Like Compounds*, vols. 1–3, EPA/600/6-88/005Ca,b,c (Washington, D.C.: EPA, 1994); EPA, *Health Assessment Document for 2,3,7,8-Tetrachlorodibenzo-p-dioxin (TCDD) and Related Compounds*, vols. 1–3, EPA/600/BP-92/001a,b,c (Washington, D.C.: EPA, 1994).

215 waxing and waning of incinerator popularity: Schneider, "Burning Trash for Energy."

216 if 1,800 tons go in, 1,800 tons come out: Actually, the final mass exceeds the mass of the material the incinerator is stoked with. Because oxygen combines with fuel in the process of burning, total combustion emissions—ash plus smoke plus vapors—are somewhat heavier than the solid ingredients fed into the incinerator initially.

216 John Kirby's demonstration: T. L. Aldous, "Hearing Has Havana Humming," *PDT*, 23 Oct. 1993, pp. A-1, A-10.

216 toxicity of incinerator ash: P. Connett and E. Connett, "Municipal Waste Incineration: Wrong Question, Wrong Answer," *The Ecologist* 24 (1994):

14–20; K. Schneider, "In the Humble Ashes of a Lone Incinerator, the Makings of a Law," *New York Times*, 18 Mar. 1994, p. A-22.

216 18 boxcars becomes 10 truckloads: The 18:10 ratio was part of the Pekin incinerator proposal.

217 formation of fly ash: Connett and Connett, "Municipal Waste Incineration"; T. G. Brna and J. D. Kilgroe, "The Impact of Particulate Emissions Control on the Control of Other MWC Air Emissions," *Journal of Air and Waste Management Association* 40 (1990): 1324–29.

217 types of dioxins and furans: M. J. Devito and L. S. Birnbaum, "Toxicology of Dioxins and Related Compounds," in Schecter, *Dioxins and Health*, 139–62.

217 TCDD: H. C. Pitot III and Y. P. Dragan, "Chemical Carcinogenesis," in C. D. Klaassen (ed.), *Cassarett and Doull's Toxicology: The Basic Science of Poisons*, 5th ed. (New York: McGraw-Hill, 1996), 201–67; Devito and Birnbaum, "Toxicology of Dioxins."

218 origins of dioxin: B. Paigen and S. Lester, "Where Dioxin Comes From," in L. Gibbs et al. (eds.), *Dying from Dioxin: A Citizen's Guide to Reclaiming Our Health and Rebuilding Democracy* (Boston: South End Press, 1995).

218 PVC plastic: Polyvinyl chloride, 59 percent chlorine by weight, is the dominant source of organically bound chlorine in hospital waste, where it takes the form of IV-bags, gloves, bedpans, tubing, and packaging. Much of this waste is incinerated. Indeed, medical waste incineration was identified in the EPA's 1994 reassessment as the single largest known source of dioxin generation in the United States. The resulting irony—that institutions dedicated to the prevention and cure of disease are contributing to a public health menace—has not been lost on the medical community. In 1996, the American Public Health Association adopted a resolution calling for the phaseout of PVC products in health care institutions and urging medical suppliers to bring to the market appropriate, affordable replacements for chlorinated plastics. See J. Thorton et al., "Hospitals and Plastics: Dioxin Prevention and Medical Waste Incineration," *Public Health Reports* 3 (1996): 298–313; K. B. Wagener et al., "Polymer Substitutes for Medical Grade Polyvinyl Chloride," in A. E. S. Green (ed.), *Medical Waste Incineration and Pollution Prevention* (New York: Van Nostrand Reinhold, 1993), 155–69.

218 dioxin and forest fires: Zook and Rappe, "Environmental Sources." Contemporary forest fires may release more dioxin than those that burned during preindustrial times because human sources of dioxin have distributed this contaminant globally. Dioxin is easily adsorbed onto vegetation (Dr. Thomas Webster, Boston University, personal communication).

218 evidence from sediment cores: Webster and Commoner, "Dioxin Debate"; R. M. Smith et al., "Measurement of PCDFs and PCDDs in Air Samples and Lake Sediments at Several Locations in Upstate New York," *Chemosphere* 25 (1992): 95–98; J. M. Czucwa, "Polychlorinated dibenzo-*p*-dioxins and Dibenzofurans in Sediments from Siskiwit Lake, Isle Royale," *Science* 226 (1992): 568–69; J. M. Czucwa and R. A. Hites, "Environmental Fate of Combustion Generated Polychlorinated Dioxins and Furans," *Environmental Science and Technology* 18 (1984): 444–50.

218 higher body burdens in industrialized countries: A. Schecter et al., "Dioxins in U.S. Food and Estimated Daily Intake," *Chemosphere* 29 (1994): 2261–65; A. Schecter et al., "Chlorinated Dioxins and Dibenzofurans in Human Tissue from General Populations: A Selective Review," *EHP* 102, suppl. 1 (1994): 159–71.

218 mummies and frozen Eskimos: W. Ligon Jr. et al., "Chlorodibenzofurans and Chlorodibenzo-*p*-dioxin Levels in Chilean Mummies Dated to about 2800 before the Present," *Environmental Science and Technology* 23 (1989): 1286–90; A. Schecter et al., "Sources of Dioxin in the Environment: A Study of PCDDs and PCDFs in Ancient, Frozen Eskimo Tissue," *Chemosphere* 17 (1988): 627–31.

220 vote of the Havana city council: T. L. Aldous, "Siting Battle Begins," *PDT*, 27 Oct. 1993, pp. A-1, A-12.

220 excerpt from testimony: C. West-Williams, "State Board Decision on Hold for Now," *PDT*, 7 Apr. 1994, pp. A-1, A-12.

221 John Kirby in Pekin: K. Kaufman, "PEC to Present Its Incinerator Plan," *PDT*, 10 Jan. 1990, p. B-2; S. Brown, "Company Forsakes Pekin Incinerator," *PJS*, 31 Mar. 1990, pp. A-1, A-2; K. Kauffman, "PEC: Incinerator Is a Way to Expand. Environmentalists: Not This Way," *Pekin Times*, 11 Jan. 1990, p. B-2.

221–222 dioxin in food: A. Schecter, "Congener-Specific Levels of Dioxins and Dibenzofurans in U.S. Food and Estimated Daily Dioxin Toxic Equivalent Intake," *EHP* 102 (1994): 962–66; T. Webster and P. Connett, "Critical Factors in the Assessment of Food Chain Contamination by PCDD/PCDF from Incinerators," *Chemosphere* 18 (1989): 1123–29.

222 dioxin in cow's milk near incinerators: A. K. D. Liem et al., "Occurrence of Dioxin in Cow's Milk in the Vicinity of Municipal Waste Incinerators and a Metal Reclamation Plant in the Netherlands, *Chemosphere* 23 (1991): 1675–84; P. Connett and T. Webster, "An Estimatation of the Relative Human Exposure to 2,3,7,8-TCDD Emissions via Inhalation and Ingestion of Cow's Milk," *Chemosphere* 16 (1987): 2079–84.

222 dioxin in rivers, fish, soil, and crops: B. Paigen, "What Is Dioxin?" in Gibbs, *Dying from Dioxin*, 35–46.

222 foraging farm animals can also accumulate dioxin directly from soil: M. X. Petreas, "Biotransfer and Bioaccumulation of PCDD/PCDFs from Soil: Controlled Exposure Studies of Chickens," *Chemosphere* 23 (1991): 1731–41.

222 cadmium: R. A. Goyer, "Toxic Effects of Metals," in Klaassen, *Cassarett and Doull's Toxicology*, 691–736; ATSDR, *Case Studies in Environmental Medicine: Cadmium Toxicity* (Atlanta: ATSDR, 1990); H. A. Hattemer-Frey and C. C. Travis, "Assessing the Extent of Human Exposure through the Food Chain to Pollutants Emitted from Municipal Solid Waste Incinerators," in H. A. Hattemer-Frey and C. Travis (eds.), *Health Effects of Municipal Waste Incineration* (Boca Raton, Fla.: CRC Press, 1991), 84–101.

222 cadmium in batteries: D. Wartenberg, "Do Dead Batteries Cause Cancer?" *Air and Waste* 43 (1993): 880–81.

223 U.S. military exposed to dioxin: Institute of Medicine, *Veterans and Agent Orange* (Washington, D.C.: National Academy Press, 1994).

223 clues from animals: M. J. DeVito et al., "Comparisons of Estimated Human Body Burdens of Dioxinlike Chemicals and TCDD Body Burdens in Experimentally Exposed Animals," *EHP* 103 (1995): 820–31.

223 quote from James Huff: J. Huff, "Dioxins and Mammalian Carcinogenesis," in Schecter, *Dioxins and Health*, 389–407.

223 dioxin and liver cancer: A. M. Tritscher et al., "Dose-Response Relationships from Chronic Exposure to 2,3,7,8-Tetrachlorodibenzo-*p*-dioxin in a Rat Tumor Promotion Model: Quantification and Immunolocalization of CYP1A1 and CYP1A2 in the Liver," *Cancer Research* 52 (1992): 3436–42.

223 dioxin and lung cancer: G. W. Lucier et al., "Receptor Mechanisms and Dose-Response Models for the Effects of Dioxin," *EHP* 101 (1993): 36–44.

223–224 interspecies differences in sensitivity: DeVito, "Estimated Human Body Burdens"; Physicians for Social Responsibility and Environmental Defense Fund, *Putting the Lid on Dioxins* (Washington, D.C.: Physicians for Social Responsibility and Environmental Defense Fund, 1994), 3. Dioxin researcher Tom Webster points out that a particular sensitivity (high or low) to one of dioxin's effects does not imply an equivalent sensitivity at other endpoints. In other words, a species particularly sensitive to dioxin's carcinogenic powers at low doses may or may not be sensitive to its acutely toxic effects at high doses—and vice versa (Webster, personal communication).

224 human studies: H. Becher et al., "Cancer Mortality in German Male Workers Exposed to Phenoxy Herbicides and Dioxin," *Cancer Causes and Control* 7 (1996): 312–21; L. Hardell et al., "Cancer Epidemiology," in Schecter, *Dioxins and Health*, 525–47; M. A. Fingerhut et al., "Cancer Mortality in Workers Exposed to 2,3,7,8-Tetrachlorodibenzo-*p*-dioxin," *NEJM* 324 (1991): 212–18; A. Manz et al., "Cancer Mortality among Workers in a Chemical Plant Contaminated with Dioxin," *Lancet* 338 (1991): 959–64; A. Zober et al., "Thirty-four-year Mortality Follow-up of BASF Employees Exposed to 2,3,7,8-TCDD after the 1953 Accident," *International Archives of Occupational and Environmental Health* 62 (1990): 139–57.

224–225 Seveso, Italy: P. A. Bertazzi and A. di Domenico, "Chemical, Environmental, and Health Aspects of the Seveso, Italy, Accident," in Schecter, *Dioxins and Health*, 587–632; P. A. Bertazzi et al., "Cancer Incidence in a Population Accidentally Exposed to 2,3,7,8-tetrachlorodibenzo-para-dioxin," *Epidemiology* 4 (1993): 398–406; R. Stone "New Seveso Findings Point to Cancer," *Science* 261 (1993): 1383.

225 Havana's feasibility study: T. L. Aldous, "Study: Trash Burner a Boon," *PDT*, 20 May 1992, pp. A-1, A-12.

225 the first rebuttal: T. Webster, "Comments on 'A Feasibility Study of Operating a Waste-to-Energy Facility in Mason County Near Havana, Illinois'" (unpub. ms., 7 Oct. 1992, 4 pp.).

225 the second rebuttal: T. L. Aldous, "Farm Bureau Members Oppose New Incinerator," *PDT*, 24 July 1992, p. A-1; S. Iyengar, "Farm Bureau: SIU Study Skewed," *PDT*, 8 Oct. 1992, pp. A-1, A-12.

226 the popcorn threat: Dr. Dorothy Anderson, personal communication.

226 Fourth of July: K. McDermott, "Havana Incinerator Backers Hot about 'Devil Burns' Parade Float," *SSJR*, 1 July 1992, p. 1.

226 letter to the editor in Havana: A. Robertson, *Mason City Banner Times*, 10 June 1992, p. 11.

226 letter to the editor in Forrest: C. Kaisner, "Suddenly in Forrest, Greed Has Become No. 1 Attitude," *Bloomington Daily Pantagraph*, 6 Aug. 1994.

227 letter about Kirby's smoking habits: R. Hankins, letter to the editor, *Mason County Democrat*, 3 June 1992, p. 2.

227 endorsement of risk: "Editorial," *Fairbury Blade*, 20 July 1994, p. 2.

227 condemnation of risk: "Dioxin Findings Raise New Fears" (editorial), *Jacksonville Journal-Courier*, 15 Sept. 1994, p. 10.

227–228 P450 enzymes and Ah receptors: Webster and Commoner, "Dioxin Debate"; G. Lucier et al., "Receptor Model and Dose-Response Model for the Effects of Dioxin," *EHP* 101 (1993): 36–44; T. R. Sutter et al., "Targets for Dioxin: Genes for Plasminogen Activator Inhibitor-2 and Interleukin-1B," *Science* 254 (1991): 415–18.

228 antiestrogenic qualities of dioxin: L. Birnbaum, "Endocrine Effects of Prenatal Exposure to PCBs, Dioxins, and Other Xenobiotics: Implications for Policy and Future Research," *EHP* 102 (1994): 676–79.

228 substances that bind with the Ah receptor: "The Problem with Tallying Dioxin," *Science News* 146 (1994): 206.

228–229 Ah receptors knocked out in mice: R. Stone, "Dioxin Receptor Knocked Out," *Science* 268 (1995): 638–39; P. Fernandez-Salguero et al., "Immune System Impairment and Hepatic Fibrosis in Mice Lacking the Dioxin-Binding Ah Receptor," *Science* 268 (1995): 722–26.

229 dioxin's other shadowy habits: T. Colborn et al., *Our Stolen Future: Are We Threatening Our Fertility, Intelligence, and Survival?—A Scientific Detective Story* (New York: Dutton, 1996), 110–21; A. P. van Birgelen et al., "Synergistic Effect of 2,2',4,4',5,5'-Hexachlorobiphenyl and 2,3,7,8-Tetra-Chlorodibenzo-*p*-Dioxin on Hepatic Porphyrin Levels in the Rat," *EHP* 104 (1996): 550–57; N. I. Kerkvliet, "Immunotoxicology of Dioxins and Related Compounds," in Schecter, *Dioxins and Health*, 199–225; S. P. Porterfield, "Vulnerability of the Developing Brain to Thyroid Abnormalities: Environmental Insults to the Thyroid System," *EHP* 102, suppl. 2 (1994): 125–30; S. Rier et al., "Endometriosis in Rhesus Monkeys (*Macaca mulatta*) following Chronic Exposure to 2,3,7,8-Tetrachlorodibenzo-*p*-dioxin," *Fundamental and Applied Toxicology* 21 (1993): 433–41.

229–230 John Kirby's career: T. L. Aldous, "Kirby Sees Havana Opportunity, Opposition," *PDT*, 22 Oct. 1993, pp. A-1, A-12; A. Lindstrom, "Sherman Horse Track Sure Bet—Promoters," *SSJR*, 6 Jan. 1977; "John Kirby, Williamsville, Ponders Race for Senate," *SSJR*, 9 Oct. 1973; J. O'Dell, "Hens with Glasses a Barnyard Spectacle," *SSJR*, 27 Aug. 1973; K. Watson, "John Kirby Eyes Candidacy," *SSJR*, 8 Aug. 1968; K. Watson, "Page Names Kirby," *SSJR*, 7 Jan. 1963.

230 quotes by Kirby: Aldous, "Kirby sees Havana."

232 quote from Summit: B. M. Rubin, "Summit's Push for Incinerator Sparks Unusual Bunch of Foes," *Chicago Tribune*, 24 Oct. 1993, Southwest sec., p. 1.

233 number of U.S. incinerators: B. Paigen, "How to Be a Dioxin Detective," in Gibbs, *Dying from Dioxin*, 205–36.

233 fish of the Vermilion: IEPA, *Illinois Water Quality Report, 1992–93*, vol. 1, IEPA/WPC/94-160 (Springfield, Ill.: IEPA, 1994).
234 quote by John Kirby: J. Knauer, "Incinerator's Future Smoldering after 'No' Vote," *Fairbury Blade*, 16 Nov. 1994, pp. 1, 3.
234 appellate court decision: E. Hopkins, "Court Backs Pollution Board's Incinerator Ruling," *PJS*, 13 Sept. 1995, p. B-5.
234 repeal of retail rate law: R. B. Dold, "Clearing the Air," *Chicago Tribune*, 12 Jan. 1996, pp. 1–23.
234 Malignant mesothelioma is a cancer of the membranes surrounding the lungs. It is caused almost exclusively by exposure to asbestos fibers. Cigarette smoke interacts synergistically with asbestos in creating risk.

ELEVEN *our bodies, inscribed*

235–236 tree-ring analysis: R. Phipps and M. Bolin, "Tree Rings—Nature's Signposts to the Past," *Illinois Steward* (summer 1993): 18–21.
236 organochlorine residues: Anne Colston Wentz, testimony before the Subcommittee on Health and the Environment of the Committee on Energy and Commerce, U.S. House of Representatives, Health Effects of Estrogenic Pesticides Hearings, 21 Oct. 1993 (Washington, D.C.: GPO serial no. 103-87, 1994), 133.
236 contaminants found in a variety of human tissues: D. Holzman, "Banking on Tissues," *EHP* 104 (1996): 606–10; M. Moses et al., "Environmental Equity and Pesticide Exposure," *Toxicology and Industrial Health* 9 (1993): 913–59. (See especially pp. 922–26.)
236–237 organochlorines in umbilical cords: L. W. Kanja et al., "A Comparison of Organochlorine Pesticide Residues in Maternal Adipose Tissue, Maternal Blood, Cord Blood, and Human Milk from Mother/Infant Pairs," *Archives of Environmental Contamination and Toxicology* 22 (1994): 21–24; H. Autrup, "Transplacental Transfer of Genotoxins and Transplacental Carcinogenesis," *EHP* 101, suppl. 2 (1993): 33–38.
237 chlorpyrifos in urine: "Chlorpyrifos Metabolites in 82% of U.S. Population," *Pesticide and Toxic Chemical News*, 8 Nov. 1995, pp. 15–16.
237 PCBs in blood as a measure of body burden: Å. Gergman et al., "Selective Retention of Hydroxylated PCB Metabolites in Blood," *EHP* 102 (1994) 464–69; A. Schecter et al., "Polychlorinated Biphenyl Levels in the Tissues of Exposed and Nonexposed Humans," *EHP* 102, suppl. 1 (1994): 149–58. Mixtures of PCBs also vary by source and by geographic region. The mix of PCB types found in the tissues of someone exposed primarily through eating freshwater fish is likely very different from that found in someone exposed on the job, for example. These complications make determining the health effects of PCB exposure an arduous task. Adding to the difficulty is the fact that the 209 different PCB variants have differing physiological properties. Some appear to be estrogenic, others antiestrogenic. See H. A. Tilson et al., "Polychlorinated Biphenyls and the Developing Nervous System: Cross-Species Comparisons," *Neurotoxicology and Teratology* 12 (1990): 239–48.
237 body fat: L. Kohlmeier et al., "Adipose Tissue as a Medium for Epidemio-

logic Exposure Assessment," *EHP* 103, suppl. 3 (1995): 99–106; B. G. Loganathan et al., "Temporal Trends of Persistent Organochlorine Residues in Human Adipose Tissue from Japan, 1928–1985," *Environmental Pollution* 81 (1993): 31–39; L. López-Carillo et al., "Is DDT Use a Public Health Problem in Mexico?" *EHP* 104 (1996): 584–88.

237–238 breast milk: M. N. Bates et al., "Chlorinated Organic Contaminants in Breast Milk of New Zealand Women," *EHP* 102, suppl. 1 (1994): 211–17; P. Fürst et al., "Human Milk as a Bioindicator for Body Burden of PCDDs, PCDFs, Organochlorine Pesticides, and PCBs," *EHP* 102 (1994): 187–93; M. R. Sim and J. J. McNeil, "Monitoring Chemical Exposure Using Breast Milk: A Methodological Review," *AJE* 136 (1992): 1–11.

238 1951 discovery of breast milk contaminants: E. P. Laug et al., "Occurrence of DDT in Human Fat and Milk," *A.M.A. Archives of Industrial Hygiene and Occupational Medicine* 3 (1951): 245–46.

238 Carson on breast milk: R. Carson, *Silent Spring* (Boston: Houghton Mifflin, 1962), 23.

238 PCBs in breast milk: W. J. Rogan et al., "Polychlorinated Biphenyls (PCBs) and Dichlorodiphenyl Dichloroethene (DDE) in Human Milk: Effects of Maternal Factors and Previous Lactation," *AJPH* 76 (1986): 172–77.

238 one-quarter of breast milk illegal: Ibid.

238 relationship between cancer and carcinogens in breast milk: Several small-scale studies have indicated that levels of organochlorinated pesticides in breast fat are higher in breast cancer patients than in controls. Other studies do not find this pattern. See discussion in Chapter 1.

238 Prolonged breast-feeding reduces modestly the risk of breast cancer to the mother, although it is not clear whether this benefit is due to the excretion of breast carcinogens or to some other unidentifed factor. See L. A. Brinton, "Breastfeeding and Breast Cancer Risk," *Cancer Causes and Control* 6 (1995): 199–208; and P. A. Newcomb, "Lactation and a Reduced Risk of Premenopausal Breast Cancer," *NEJM* 330 (1994): 81–87. See also É. Dewailly et al., "Protective Effect of Breast Feeding on Breast Cancer and Body Burden of Carcinogenic Organochlorines" (letter), *JNCI* 86 (1994): 803.

238 North Carolina study: Rogan, "Polychlorinated Biphenyls (PCBs)."

238–239 some breast milk contaminants beginning to drop: K. Norén et al., "Methylsufonyl Metabolites of PCBs and DDE in Human Milk in Sweden, 1972–1992," *EHP* 104 (1996): 766–73; Fürst, "Human Milk as a Bioindicator"; A. Somogyi and H. Beck, "Nurturing and Breast-feeding: Exposure to Chemicals in Breast Milk," *EHP* 101 (1993): 45–52.

239 apoptosis: H. C. Pitot III and Y. P. Dragan, "Chemical Carcinogenesis," in C. D. Klaassen (ed.), *Casarett and Doull's Toxicology: The Basic Science of Poisons*, 5th ed. (New York: McGraw-Hill, 1996), 227.

239 local and distant control of mitosis: Dr. Thomas Webster, Boston University, personal communication.

240 process of carcinogenesis: J. Felton, "Mechanisms of Cancer Induction and Progression: Endogenous and Environmental Factors," in *Evaluating the National Cancer Program: An Ongoing Process*, Proceedings of the Presi-

dent's National Cancer Panel Meeting, 22 Sept. 1993 (Bethesda, Md.: NCI), 14–16.

240 purposefulness: Robert Millikan, personal communication. See also S. B. Nuland, *How We Die: Reflections on Life's Final Chapter* (New York: Random House, 1993), 202–21.

240–241 invasiveness and primitivism: R. A. Weinberg, "How Cancer Arises," *Scientific American*, Sept. 1996, 62–70; E. J. Mange and A. P. Mange, *Basic Human Genetics* (Sunderland, Mass.: Sinauer, 1994), 350.

241 Inheritance of flawed genes is thought to account for a small percentage of all U.S. cancers. See F. P. Perera, "Uncovering New Clues to Cancer Risk," *Scientific American*, May 1996, 54–62.

241 oncogenes and tumor suppressor genes: W. K. Cavenee and R. L. White, "The Genetic Basis of Cancer," *Scientific American*, Mar. 1995, 72–79; J. C. Barrett and M. D. Shelby, "Mechanisms of Human Carcinogens," in R. D'Amato et al. (eds.), *Relevance of Animal Studies for the Evaluation of Human Cancer Risk* (New York: Wiley-Liss, 1992), 415–34.

241 accelerator and brakes: J. P. Oliner, "The Role of p53 in Cancer Development," *Scientific American*, Sept.–Oct. 1994, pp. 16–25.

242 colon cancers involve both oncogene and tumor suppressor gene mutations: Cavanee and White, "Genetic Basis of Cancer"; B. Vogelstein et al., "Genetic Alterations during Colorectal-Tumor Development," *NEJM* 319 (1988): 525–32.

242 p53 involved in half of cancers: Oliner, "Role of p53."

242 nature of p53 damage indicates carcinogen responsible: Perera, "Uncovering New Clues."

242 p53 mutations in breast cancer: R. Millikan, "Studying Environmental Influences and Breast Cancer Risk: Suggestions for an Integrated Population-Based Approach," *Breast Cancer Research and Treatment* 35 (1995): 79–89; B. Newman et al., "The Carolina Breast Cancer Study: Integrating Population-Based Epidemiology and Molecular Biology," *Breast Cancer Research and Treatment* 35 (1995) 51–60; P. J. Biggs et al., "Does a Gentoxic Carcinogen Contribute to Human Breast Cancer? The Value of Mutational Spectra in Unravelling the Aetiology of Cancer," *Mutagenesis* 8 (1993): 275–83.

242 benzo[a]pyrene and DNA adducts: Perera, "Uncovering New Clues."

242 disruption of spindle fibers: Barrett and Shelby, "Mechanisms of Human Carcinogens."

242 significance of DNA repair genes: B. Proujan, "DNA Repair," *EHP* 104 (1996): 18–19.

242–243 three stages of carcinogenesis: Pitot and Dragan, "Chemical Carcinogenesis"; S. H. Yupsa and C. C. Harris, "Molecular and Cellular Basis of Chemical Carcinogenesis," in D. Schottenfeld and J. F. Fraumeni Jr., *Cancer Epidemiology and Prevention* (Philadelphia: Saunders, 1982), 23–43.

243 role of the immune system (dioxin, leukemias and lymphomas, Soviet Union and T cells): R. Repetto and S. S. Baliga, *Pesticides and the Immune System: The Public Health Risk* (Washington, D.C.: World Resources Institute, 1996).

243–244 signal transduction: Weinberg, "How Cancer Arises"; Pitot and Dragan, "Chemical Carcinogenesis"; Barrett and Shelby, "Mechanisms of Human Carcinogens."

244 cancer progressors: Barrett and Shelby, "Mechanisms of Human Carcinogens."

244 carcinogens do not fall into neat categories: Pitot and Dragan, "Chemical Carcinogenesis"; Barrett and Shelby, "Mechanisms of Human Carcinogens."

244 dioxin may interfere with apoptosis: J. M. Samet, "Dioxin and Cancer: The Never-Ending Story," *Cancer Causes and Control* 7 (1996): 302–4.

244 differing sensitivities and risks: Perera, "Uncovering New Clues."

244–245 DDT as a cancer accelerator: J. D. Scribner and N. K. Moffet, "DDT Acceleration of Mammary Gland Tumors Induced in the Male Sprague-Dawley Rat by 2-Acetamidophenanthrene," *Carcinogenesis* 2 (1981): 1235–39.

245 quote by Ross Hume Hall: Personal communication.

245 biological markers: Perera, "Uncovering New Clues"; S. Anderson et al., "Genetic and Molecular Ecotoxicology: A Research Framework," *EHP* 102, suppl. 12 (1994): 3–8; M. Eubanks, "Biological Markers: The Clues to Genetic Susceptibility," *EHP* 102 (1994): 50–56; S. Blakeslee, "Genes Tell Story Why Some Get Cancer While Others Don't," *New York Times*, 17 May 1994, p. C-3; M. A. Saleh et al. (eds.), *Biomarkers of Human Exposure to Pesticides* (Washington, D.C.: American Chemical Society, 1994); F. P. Perera, "DNA Adducts and Related Biomarkers in Populations Exposed to Environmental Carcinogens," *EHP* 98 (1992): 133–37.

245 correlations in lab animals: F. A. Béland and M. C. Poirier, "Significance of DNA Adduct Studies in Animal Models for Cancer Molecular Dosimetry and Risk Assessment," *EHP* 99 (1993): 5–10.

245–246 Polish studies: S. Øvrebø et al., "Biological Monitoring of Polycyclic Aromatic Hydrocarbon Exposure in a Highly Polluted Area of Poland," *EHP* 103 (1995): 838–43; F. P. Perera et al., "Molecular and Genetic Damage in Humans from Environmental Pollution in Poland," *Nature* 360 (1992): 256–58; K. Hemminki et al., "DNA Adducts in Humans Environmentally Exposed to Aromatic Compounds in an Industrial Area of Poland," *Carcinogenesis* 11 (1990): 1229–31.

246 quote by Frederica Perera: Perera, "Uncovering New Clues."

246 vinyl chloride and signal transduction proteins: P. W. Brandt-Rauf et al., "Mutant p21 Protein as BioMarker of Chemical Carcinogenesis in Humans," in M. Mendelsohn et al. (eds.), *Biomarkers and Occupational Health: Progress and Perspectives* (Washington, D.C.: Joseph Henry Press, 1995), 163–73.

246 alterations in enzymes: Eubanks, "Biological Markers."

247 Minnesota pesticide applicators: V. F. Garry et al., "Pesticide Applicators with Mixed Pesticide Exposure: G-banded Analysis and Possible Relationship to Non-Hodgkin's Lymphoma," *Cancer Epidemiology, Biomarkers, and Prevention* 5 (1996): 11–16. See also V. F. Garry, "Survey of Health and Use Characterization of Pesticide Appliers in Minnesota," *AEH* 49 (1994): 337–43; and S. Lipkowitz et al., "Interlocus V-J Recombination Measures

Genetic Instability in Agriculture Workers at Risk for Lymphoid Malignancies," *Proceedings of the National Academy of Science* 89 (1992): 5301–05.

247 free radicals: B. Halliwell and O. I. Aruoma, *DNA and Free Radicals* (New York: Ellis Horwood, 1993); D. C. Malins et al., "The Etiology of Breast Cancer: Characteristic Alterations in Hydroxyl Radical-Induced DNA Base Lesions during Oncogenesis with Potential for Evaluating Incidence Risk," *Cancer* 71 (1993): 3036–43; S. S. Thorgeirsson, "Endogenous DNA Damage and Breast Cancer," *Cancer* 71 (1993): 2897–99.

248 1896 discovery: H. Magdelenet and P. Pouillart, "Steroid Hormone Receptors in Breast Cancer," in P. J. Sheridan et al. (eds.), *Steroid Receptors and Disease: Cancer, Autoimmune, Bone, and Circulatory Disorders* (New York: Marcel Dekker, 1988), 436–65.

248 breast cancer risk and estrogen exposure: M. C. Pike et al., "Estrogens, Progestogens, Normal Breast Proliferation, and Breast Cancer Risk," *Epidemiologic Reviews* 15 (1993): 17–35.

248 focus on xenoestrogens: D. L. Davis and H. L. Bradlow, "Can Environmental Estrogens Cause Breast Cancer?" *Scientific American*, Oct. 1995, 166–72; D. L. Davis, "Medical Hypothesis: Xenoestrogens as Preventable Causes of Breast Cancer," *EHP* 101 (1993): 372–77.

248–249 life history of estrogen: Davis and Bradlow, "Can Environmental Estrogens"; P. Toniolo et al., "Reliability of Measurements of Total, Protein-Bound, and Unbound Estradiol in Serum," *Cancer Epidemiology, Biomarkers, and Prevention* 3 (1994): 47–50.

249 behavior of xenoestrogens: Davis and Bradlow, "Can Environmental Estrogens"; N. M. Brown and C. A. Lamartiniere, "Xenoestrogens Alter Mammary Gland Differentiation and Cell Proliferation in the Rat," *EHP* 103 (1995): 708–13.

249 researchers have assumed small role for xenoestrogens: D. L. Houghton and L. Ritter, "Organochlorine Residues and Risk of Breast Cancer," *Journal of the American College of Toxicology* 14 (1995): 71–89.

249 xenoestrogens are common: P. Common, "Environmental Estrogenic Agents Area of Concern," *JAMA* 271 (1994): 414–16.

250 xenoestrogens may interact synergistically: S. F. Arnold et al., "Synergistic Activation of Estrogen Receptor with Combinations of Chemicals," *Science* 272 (1996): 1489–92.

250 xenoestrogens may be more bioavailable: S. F. Arnold et al., "A Yeast Estrogen Screen for Examining the Relative Exposure of Cells to Natural and Xenoestrogens," *EHP* 104 (1996): 544–48.

250 indirect effects of xenoestrogens: Davis and Bradlow, "Can Environmental Estrogens."

250–251 xenoestrogens alter estrogen metabolism: Davis and Bradlow, "Can Environmental Estrogens"; H. L. Bradlow et al., "Effects of Pesticides on the Ratio of 16 alpha/2-Hydroxyestrone: A Biologic Marker of Breast Cancer Risk," *EHP* 103, suppl. 7 (1995): 147–50; N. T. Telang et al., "Induction by Estrogen Metabolite 16 alpha-Hydroxyestrone of Genotoxic Damage and Aberrant Proliferation in Mouse Mammary Epithelial Cells," *JNCI* 84 (1992): 634–38.

251 cancer among adoptees: T. I. A. Sørensen et al., "Genetic and Environmental Influences on Premature Death in Adult Adoptees," *NEJM* 318 (1988): 727–32.

252 1974 breast cancer blip: ACS, *Breast Cancer Facts and Figures 1996* (Atlanta: ACS, 1995), fig. 2.

TWELVE *ecological roots*

255 bladder tumor experiment: R. A. Weinberg, "A Molecular Basis of Cancer," *Scientific American*, Nov. 1983, pp. 126–42.

256 genetic changes involved in bladder cancer: I. Orlow et al., "Deletion of the p16 and p15 Genes in Human Bladder Tumors," *JNCI* 87 (1995): 1524–29; S. H. Kroft and R. Oyasu, "Urinary Bladder Cancer: Mechanisms of Development and Progression," *Laboratory Investigation* 71 (1994): 158–74; P. Lipponen and M. Eskelinen, "Expression of Epidermal Growth Factor Receptor in Bladder Cancer as Related to Established Prognostic Factors, Oncoprotein Expression and Long-Term Prognosis," *British Journal of Cancer* 69 (1994): 1120–25.

256 aromatic amines and DNA adducts: D. Lin et al., "Analysis of 4-Aminobiphenyl-DNA Adducts in Human Urinary Bladder and Lung by Alkaline Hydrolysis and Negative Ion Gas Chromatography–Mass Spectrometry," *EHP* 102, suppl. 6 (1994): 11–16; P. L. Skipper and S. R. Tannenbaum, "Molecular Dosimetry of Aromatic Amines in Human Populations," *EHP* 102, suppl. 6 (1994): 17–21; S. M. Cohen and L. B. Ellwein, *EHP* 101, suppl. 5 (1994): 111–14.

256–257 slow and fast acetylators: P. Vineis and G. Ronco, "Interindividual Variation in Carcinogen Metabolism and Bladder Cancer Risk," *EHP* 98 (1992): 95–99.

256–257 One researcher offers the following reflection on the bladder cancer situation in England: "The continued use of known carcinogenic substances in British industry for many years after their identification, the wide range of industries with a known or suspected increased risk of bladder cancer, and our ignorance of the carcinogenic potential of many materials used in current manufacturing should be a cause for continuing concern" (R. R. Hall, "Superficial Bladder Cancer," *British Medical Journal* 308 [1994]: 910–13).

257 rise in bladder cancer incidence and its link to cigarette smoking: D. T. Silverman, "Urinary Bladder," in A. Harras (ed.), NIH Pub. 96-691 *Cancer Risks and Rates* (Bethesda, Md.: NCI, 1996), 197–99. Routine screening for bladder cancer is not done. Thus earlier detection or improved diagnostic techniques are unlikely explanations for recent increases in incidence rates. P. A. Schulte et al. (eds.), "Bladder Cancer Screening in High-Risk Groups," *Journal of Occupational Medicine* 32 (1990): 787–945.

257–258 *o*-toluidine: EPA, *1992 Toxics Release Inventory: Public Data Release*, EPA 745-R-001 (Washington, D.C.: EPA, 1994), 79.

258 quote from carcinogen report: USDHHS, *Seventh Annual Report on Carcinogens* (Research Triangle Park, N.C.: USDHHS, 1994), 389.

258 1996 study: E. M. Ward et al., "Monitoring of Aromatic Amine Exposure

in Workers at a Chemical Plant with a Known Bladder Cancer Excess," *JNCI* 88 (1996): 1046–52.

258 another recent investigation: R. Ouellet-Hellstromt and J. D. Rench, "Bladder Cancer Incidence in Arylamine Workers," *Journal of Occupational and Environmental Medicine* 38 (1996): 1239–47; J. D. Rench et al., *Cancer Incidence Study of Workers Handling Mono- and Di-arylamines Including Dichlorobenzidine, Ortho-toluidine, and Ortho-dianisidine* (Falls Church, Va.: SRA Technologies, 1995); "Study Finds Bladder Cancer Threat among Conn. Plant Workers," *Boston Globe*, 21 Sept. 1995, p. 42.

259 focus of cancer research: Francis Collins, Richard Klausner, and Kenneth Olden, statement on cancer, genetics, and the environment before the Senate Committee on Labor and Human Resources, 6 Mar. 1996 (USDHHS press release).

259 fewer than 10 percent of all malignancies involved inherited mutations: NCI, *Understanding Gene Testing*, NIH Pub. 96-3905 (Bethesda, Md.: NCI, 1995).

259 hereditary colon cancer: G. Marra and C. R. Boland, "Hereditary Nonpolyposis Colorectal Cancer: The Syndrome, the Genes, and Historical Perspectives," *JNCI* 87 (1995): 1114–25; N. Papadopoulos et al., "Mutation of a *mutL* Homolog in Hereditary Colon Cancer," *Science* 263 (1994): 1625–29.

259 definition of *sporadic*: Bert Vogelstein, "Heredity and Environment in a Common Human Cancer" (lecture at Harvard Univ. Medical School, 3 May 1995). In exploring the use of the term *sporadic* by cancer researchers, historian Robert Proctor observed, "The presumption is apparently that heredity is orderly, while environmental causation is chaotic, perhaps even indecipherable. . . . Genetics offers hope for new forms of therapy, but also seems to imply resignation with regard to the possibility of prevention." See R. N. Proctor, *Cancer Wars: How Politics Shapes What We Know and Don't Know About Cancer* (New York: Basic Books, 1995), 245.

259 breast cancer's hereditary connection: Five to 10 percent is the estimate most often cited. A recent prospective cohort study of more than 100,000 women placed this figure even lower—at about 2.5 percent. See G. A. Colditz, "Family History, Age, and Risk of Breast Cancer: Prospective Data from the Nurses' Health Study," *JAMA* 270 (1993): 338–43.

259–260 DNA repair and hereditary colon cancer: D. Holzman, "Mismatch Repair Genes Matched to Several New Roles in Cancer," *JNCI* 88 (1996): 950–51.

260 pink and blue brochure: "Cancer Prevention" (pamphlet) (Bethesda, Md.: USDHHS, n.d.).

260–261 genetics textbook: G. Edlin, *Human Genetics: A Modern Synthesis*, 2nd ed. (Boston: Jones & Bartlett, 1990). Quotes are from pp. 184–204.

262 quote by Martha Balshem: M. Balshem, *Cancer in the Community: Class and Medical Authority* (Washington, D.C.: Smithsonian Institution Press, 1993), 3.

262–263 lifestyle factors and cholera: C. E. Rosenberg, *The Cholera Years: The*

United States in 1832, 1849, and 1866 (Chicago: Univ. of Chicago Press, 1962), 1–60.

263 Some researchers argue that "delayed childbirth" among white women explains much of the elevated incidence of breast cancer in the northeastern states. See S. R. Sturgeon, "Geographic Variation in Mortality from Breast Cancer among White Women in the United States," *JNCI* 87 (1995): 1846–53.

263 majority of breast cancers unexplained by lifestyle: M. P. Madigan, "Proportion of Breast Cancer Cases in the United States Explained by Well-Established Risk Factors," *JNCI* 87 (1995): 1681–85.

263 quote by Robert Millikan: Personal communication.

263 total dietary fat and breast cancer: D. J. Hunter et al., "Cohort Studies of Fat Intake and the Risk of Breast Cancer—A Pooled Analysis," *NEJM* 334 (1996): 356–61; D. J. Hunter and W. C. Willett, "Diet, Body Size, and Breast Cancer," *Epidemiology Reviews* 15 (1993): 110–32; E. Giovannucci et al., "A Comparison of Prospective and Retrospective Assessments of Diet in the Study of Breast Cancer," *AJE* 137 (1993): 502–11. The role of dietary fat and creating breast cancer risk remains uncertain in part because the range of fat intake among the various groups of women studied has, so far, been relatively narrow.

263–264 as two leading researchers have observed, energy intake from fat has been declining as breast cancer has increased: Hunter and Willett, "Diet, Body."

264 Drs. Devra Lee Davis, Samuel Epstein, and Janette Sherman are among the researchers calling for a more ecological approach to diet. See S. S. Epstein, "Environmental and Occupational Pollutants Are Avoidable Causes of Breast Cancer," *International Journal of Health Services* 24 (1994): 145–50; and J. Sherman, *Chemical Exposure and Disease: Diagnostic and Investigative Techniques* (Princeton, N.J.: Princeton Scientific Publishing, 1994), 83.

264 Consumption of animal fat (or meat) is most strongly linked to colon and prostate cancers. See W. C. Willett, "Diet and Nutrition," in D. Schottenfeld and J. F. Fraumeni Jr. (eds.), *Cancer Epidemiology and Prevention*, 2nd ed. (Oxford, England: Oxford Univ. Press, 1996), 438–61.

264 reproduction, breast development, and the environment: N. Krieger, "Exposure, Susceptibility, and Breast Cancer Risk," *Breast Cancer Research and Treatment* 13 (1989): 205–23; S. G. Korenman, "Oestrogen Window Hypothesis of the Aetiology of Breast Cancer," *Lancet* 1980: 700–701.

264 organochlorine exposure and early puberty in rats: This topic is currently under exploration by Dr. Mary Wolff, who is interested in all factors—including childhood diet and level of physical activity—that contribute to the onset of puberty in girls. M. S. Wolff, "Organochlorines and Breast Cancers," presentation at the American Public Health Association, New York, 20 Nov. 1966. See L. M. Walters et al., "Purified Methoxychlor Stimulates the Reproductive Tract in Immature Female Mice," *Reproductive Toxicology* 7 (1993): 599–606; P. L. Whitten et al., "A Phytoestrogen Diet Induces the Premature Anovulatory Syndrome in Lactionally Exposed Female Rats," *Biology of Reproduction* 49 (1993): 1117–21; R. J. Gellert, "Uterotropic Activity of Polychlorinated Biphenyls and Induction

of Precocious Reproductive Aging in Neonatally Treated Female Rats," *Environmental Research* 16 (1978): 123–30.

264 grand arguments over cancer causes: See, for example, R. Doll and R. Peto, *The Causes of Cancer: Quantitative Estimates of Avoidable Risks of Cancer in the United States Today* (Oxford, England: Oxford Univ. Press, 1981); and rebuttal by S. S. Epstein and J. B. Swartz, "Fallacies of Lifestyle Cancer Theories," *Nature* 289 (1981): 127–30.

264–265 the percentages game: Described in R. N. Proctor, *Cancer Wars: How Politics Shapes What We Know and Don't Know about Cancer* (New York: Basic Books, 1995), 54–74. See also J. M. Kaldor and K. A. L'Abbé, "Interaction between Human Carcinogens," in H. Vainio et al. (eds.), *Complex Mixtures and Cancer Risk*, IARC Scientific Pub. 104 (Lyon, France: IARC, 1990), 35–43.

265 cancer control policies embrace lifestyle and downplay environment: The ACS does not discuss environmental factors in its recent report on cancer prevention. See ACS, *Cancer Risk Report: Prevention and Control, 1995* (Atlanta: ACS, 1995). See also K. R. McLeroy, "An Ecological Perspective on Health Promotion Programs," *Health Education Quarterly* 15 (1988), 351–77.

265–266 quote from Illinois cancer report: IDPH, *Cancer Incidence in Illinois by County, 1985–87*, Supplemental Report (Springfield, Ill.: IDPH, 1990), 7–8.

266 Rachel Carson on environmental human rights: Senate testimony hearings before the Subcommittee on Reorganization and International Organizations of the Committee on Government Operations, "Interagency Coordination in Environmental Hazards (Pesticides)," U.S. Senate, 88th Congress, 1st session, 4 June 1962.

266 Carson's belief: Carson, *Silent Spring*, (Boston, Mass.: Houghton Mifflin, 1962), 277–78.

268 we do not all bear equal risks: F. Perera, "Uncovering New Clues to Cancer Risk," *Scientific American*, May 1996, pp. 54–62; S. Venitt, "Mechanisms of Carcinogenesis and Individual Susceptibility to Cancer, *Clinical Chemistry* 40 (1994): 1421–25; G. W. Lucier, "Not Your Average Joe" (editorial), *EHP* 103 (1995): 10.

268–269 2 percent estimate: Harvard Center for Cancer Prevention, "Harvard Report on Cancer Prevention," *Cancer Causes and Control* 7, suppl. 1 (1996): 3–59; D. Trichopoulos et al., "What Causes Cancer?" *Scientific American*, Sept. 1996, pp. 80–87.

269 other estimates: Proctor, *Cancer Wars*.

269 10,940 is 2 percent of 547,000, the projected figure for total cancer deaths in 1995. See ACS, *Cancer Facts and Figures—1995*, rev. (Atlanta: ACS, 1995).

269 anonymity and homicide: The environmental analysts Paul Merrell and Carol Van Strum have argued that the concept of acceptable risk is tolerated only because of the anonymity of its intended victims. See P. Merrell and C. Van Strum, "Negligible Risk: Premeditated Murder?" *Journal of Pesticide Reform* 10 (1990): 20–22. Likewise, the molecular biologist and physician John Gofman has argued, "If you pollute when you DO NOT

KNOW if there is any safe dose (threshold), you are performing improper experimentation on people without their informed consent. . . . If you pollute when you DO KNOW that there is no safe dose with respect to causing extra cases of deadly cancers, then you are committing premeditated random murder" (J.W. Gofman, memorandum to the U.S. Nuclear Regulatory Commission, 21 May 1994).

269 quote from the ATSDR: ATSDR, *FY 1993 Agency Profile and Annual Report* (Atlanta: ATSDR, 1993), 15.

270 number of carcinogens in the environment: M. Eubanks, "Biomarkers: The Clues to Genetic Susceptibility," *EHP* 102 (1994): 50–56.

270 Rachel Carson's observation: Carson, *Silent Spring*, 248. See also M. J. Kane, "Promoting Political Rights to Protect the Environment," *Yale Journal of International Law* 18 (1993): 389–411.

270 precautionary principle: This principle was endorsed in 1987 by European environmental ministers in a meeting about the deterioration of the North Sea. (K. Geiser, "The Greening of Industry: Making the Transition to a Sustainable Economy," *Technology Review*, Aug.–Sept. 1991, pp. 65–72.) See also T. O'Riordan and J. Cameron (eds.), *Interpreting the Precautionary Principle* (London: Earthscan, 1994).

270 dead body approach: Devra Lee Davis, quoted in "Is There Cause for 'Environmental Optimism'?" *Environmental Science and Technology* 29 (1995): 366–69.

270 principle of reverse onus: This principle has been embraced by the International Joint Commission in their Eighth Biennial Report on Great Lakes Water Quality (Washington, D.C., and Ottawa, Ontario: International Joint Commission, 1996), 15–17. See also discussions of proof in T. Colborn et al., *Our Stolen Future: Are We Threatening Our Fertility, Intelligence, and Survival?—A Scientific Detective Story* (New York: Dutton, 1996); and G. K. Durnil, *The Making of a Conservative Environmentalist: With Reflection on Government, Industry, Scientists, the Media, Education, Economic Growth, and the Sunsetting of Toxic Chemicals* (Bloomington: Indiana Univ. Press, 1995).

270 principle of the least toxic alternative: My ideas on this topic are inspired in part by those of biologist Mary O'Brien. See M. H. O'Brien, "Alternatives to Risk Assessment: The Example of Dioxin," *New Solutions* (winter 1993): 39–42; and K. Geiser, "Protecting Reproductive Health and the Environment: Toxics Use Reduction," *EHP* 101, suppl. 2 (1993): 221–25.

270 quote by Dr. Mary O'Brien: personal communication.

271–272 abnormal changes in juvenile rats: M-H. Li and L. G. Hansen, "Enzyme Induction and Acute Endocrine Effects in Prepubertal Female Rats Receiving Environmental PCB/PCDF/PCDD Mixtures," *EHP* 104 (1996): 712–22.

INDEX

ABOUT THE AUTHOR

Sandra Steingraber, Ph.D., received her doctorate in biology from the University of Michigan. The author of *Post-Diagnosis*, a volume of poetry, and coauthor of a report on ecology and human rights in Africa, *The Spoils of Famine*, she has been called "a poet with a knife" (*Sojourner*). She has taught biology for several years at Columbia College, Chicago; held visiting fellowships at the University of Illinois, Radcliffe College, and Northeastern University; and was recently appointed to serve on the National Action Plan on Breast Cancer, administered by the U.S. Department of Health and Human Services. As an ecologist, she has conducted field work in northern Minnesota, East Africa, and Costa Rica.